Electronic Structure Crystallography and Functional Motifs of Materials

Electronic Structure Crystallography and Functional Motifs of Materials

Guo-Cong Guo
Xiao-Ming Jiang

WILEY-VCH

Authors

Prof. Guo-Cong Guo
FJIRSM, CAS
8 Gaoxin Avenue, Shangjie Town
Minhou County,
Fuzhou, 350608
China

Prof. Xiao-Ming Jiang
FJIRSM, CAS
8 Gaoxin Avenue, Shangjie Town
Minhou County,
Fuzhou, 350608
China

Library of Congress Card No.:
applied for

British Library Cataloguing-in-Publication Data
A catalogue record for this book is available from the British Library

Bibliographic information published by the Deutsche Nationalbibliothek
The Deutsche Nationalbibliothek lists this publication in the Deutsche Nationalbibliografie; detailed bibliographic data are available on the Internet at <http://dnb.d-nb.de>.

Print ISBN: 978-3-527-35220-3
ePDF ISBN: 978-3-527-84231-5
ePub ISBN: 978-3-527-84232-2
oBook ISBN: 978-3-527-84233-9

Typesetting Straive, Chennai, India
Printing and Binding CPI Group (UK) Ltd, Croydon, CR0 4YY

C9783527352203_090124

Contents

About the Authors

Prof. Guo-Cong Guo

Guo-Cong Guo is a professor at the Fujian Institute of Research on the Structure of Matter (FJIRSM), Chinese Academy of Sciences. He is director of the State Key Laboratory of Structural Chemistry, director of the Chinese Chemical Society, and vice president of the Chinese Crystallographic Society. He graduated from the Department of Chemistry, Xiamen University, in 1986 and received his Ph.D. from the Department of Chemistry at the Chinese University of Hong Kong in 1999 under the supervision of Professor Thomas Chung Wai Mak.

Prof. Xiao-Ming Jiang

Xiao-Ming Jiang is a professor at Fujian Institute of Research on the Structure of Matter (FJIRSM), Chinese Academy of Sciences. He received a Ph.D. in inorganic chemistry from FJIRSM in 2011. From 2011 to 2015, he was a postdoctoral researcher in physics at Nanjing University and the Technical University of Munich, Germany, focusing on experimental electronic structure characterization and structure–property relationships of functional materials and theory and methods in crystallography.

Foreword 1

Roentgen discovered X-rays in 1895 and started a new era of probing the inner world of matter. In 1912, Laue and the Braggs father and son team discovered that the diffraction of X-rays could be used to uncover the microstructure in a crystal, pioneering the X-ray crystallography. After more than 100 years of development, the current understanding of the material structure–property relationship at the atomic and molecular level has reached a high level. The property of a material, however, is derived from the electronic response process under the impact of an external field. Therefore, this understanding of the material structure–property relationship relies on the ability to explore its electronic structure. This book introduces both theoretical and experimental methods to explore material electronic structure in detail. It summarizes significant milestones in the field of structural chemistry and provides a theoretical foundation for the development of high-performance novel materials.

Material science is the foundation and forerunner of modern science and technology. The understanding of the material structure–property relationship directly determines the ability to research and develop novel materials. The material microstructure includes atomic arrangement structure, with crystal structure, local and defect structure, and electronic structure. The material electronic structure fundamentally determines the intrinsic material property. Currently, mature experimental technology is available to characterize the material atomic arrangement structure, but the experimental characterization of the material electronic structure remains in the exploratory stage. Although the electronic structure can be obtained from the first-principle calculation, the theoretical calculation is often based on multiple approximations and assumptions. Hence, the theoretical results usually deviate from reality, which has limited the use of these results to guide the design of high-performance materials. This book collects the basic theory, experimental technique, and refining model used to experimentally characterize the material electronic structure based on X-ray diffraction and other scattering methods, expanding the experimental research on material structure from the atomic to electronic level, and it represents a high level of study that remains to be explored by researchers around the world.

The in situ characterization of the crystal and electronic structure of a material at its initial and functional state under an external field, as well as the evolving process, can be achieved by combining the theoretical methods featured in electronic structure crystallography and in situ characterization techniques. Therefore, the structural origin of a functional material under light, temperature, force, and electricity can be studied, and the functional motif and structure–property relationship of a material can be revealed, based on which the theory of material functional motif can be established, thus providing the theoretical and technical foundation for the function-oriented material design and accelerating the material research and development process.

Electronic Structure Crystallography and Functional Motifs of Materials applies experimental techniques in crystallography to the characterization of electronic structure material and develops an understanding of the material structure–property relationship, representing the future direction and latest research frontier of structural chemistry.

July 2nd, 2023

Mao-Chun Hong
Member of the Chinese Academy of Sciences
Fujian Institute of Research on the Structure of Matter
Chinese Academy of Sciences

Foreword 2

Chemistry is a fundamental science subject that studies the composition, structure, and changes of matter at the molecular and atomic levels. The establishment of chemistry has a long history. The earliest chemistry practice began when humans learned to use fire. In modern chemistry, the understanding of chemical phenomena and laws has benefited from the advancement of theoretical methods and experimental techniques of structural chemistry, the most representative of which were the discovery of electrons and the establishment of nuclear atomic models. The development of quantum mechanics in the twentieth century laid the foundation for modern chemistry theories. The fundamentals of chemical bonds were revealed gradually as studies were conducted on the molecular structure of hydrogen. The valence bond theory, molecular orbital theory, and coordination field theory were created successively. Meanwhile, the chemical reaction theory was developed at the microscopic level.

The advancement of experimental techniques based on X-ray crystallography allowed researchers to "see" atoms in matter for the first time during the early twentieth century, which promoted the development of structural chemistry at the crystal level. In the subsequent hundreds of years, X-ray crystallography developed rapidly, and a significant amount of crystal structure data were accurately determined, which further deepened our understanding of fundamental chemical phenomena. These phenomena have included the formation of chemical bonds among adjacent atoms, chemical reactions being driven by the process of electron transfer, and the electron cloud distortion or transition of the functional material under an external field. Classic concepts and principles, such as atomic charge, chemical bond, covalency, iconicity, resonance structure, bond energy, and aromaticity, have been established and summarized to explain these phenomena. These concepts alone, however, cannot be clearly defined at the crystal structure level. Therefore, they still cannot accurately characterize some fundamental phenomena and laws in chemistry.

In fact, the atomic or molecular properties depend on their electron clouds, and some fundamental chemical phenomena can be explained based only on a fundamental understanding of electronic structure. Theoretical chemistry, especially the development of first-principle calculation methods, has advanced theoretical chemistry to the electronic structure level. The theoretical calculation of fundamental physical quantities, such as electron density, electron wave function, and density function, has provided effective methods to explain fundamental chemical phenomena and laws at the level of electronic structure. These chemical concepts and models did not have a solid experimental foundation until the establishment of electronic structure characterization technology. Therefore, electronic structure crystallography, with the aim of studying the material electronic structure based on crystallography experimental methods, is an important development in the field of structural chemistry. These methods extend a deeper understanding of some fundamental chemical phenomena and laws from an experimental perspective, promote the effective combination of theoretical and experimental chemistry, and provide new perspectives and approaches for solving major frontier scientific problems in this field.

The book proposes the concept of electronic structure crystallography as a promising new direction for structural chemistry and systematically discusses relevant theories and experimental techniques. *Electronic Structure Crystallography and Functional Motifs of Materials* is one of the most promising research fields in chemistry and material science. Relevant research results will have pioneering significance for the future advancement of the field of chemistry.

July 2nd, 2023

Jian-Nian Yao
Member of the Chinese Academy of Sciences
Beijing, China

Preface

Electronic structure is a central research topic in material science. According to the widely accepted density functional theory, the intrinsic properties and functions of a material are mainly determined by its electronic structure; thus, obtaining an accurate electronic structure is the key to studying the structure–property relationship. Electrons in materials are de Broglie (probability) waves in nature, satisfying the Pauli exclusion principle, with a density distribution in both the position space (referred to as electron density) and momentum space (referred to as momentum density). Electron density and momentum density are complementary and together determine the properties of a material. The electronic structure of a material usually refers to the electron and momentum densities, as well as the associated physical quantities that determine the collective behavior of electrons. Both the electron wavefunction and density matrix are fundamental physical quantities that describe the electronic structure of a material and are equivalent to pure states. Therefore, the electronic structure in this book contains information on three main aspects in real space and momentum space: electron density, density matrix, and wavefunction, where the terms electron density and charge density are used interchangeably in this book.

To date, the electronic structures of materials have been mainly obtained by first-principle calculations, and these electronic structures are usually a simplified and approximated representation of the true electronic structures, as the calculations involve many assumptions and approximations, ignore the working conditions of the materials, and rarely consider material-related experimental data. Compared to theoretical computations, the experimental electronic structures obtained by experimental means can be considered a more realistic representation of the actual electronic structures of materials.

X-ray, neutron, and electron scattering experiments are currently the main methods of probing the structure of matter at the (sub)atomic level, where scattering includes both coherent scattering (diffraction) and incoherent scattering (Compton scattering). Because electrons in materials can scatter X-rays and unpaired electrons can scatter neutrons, the scattering signal should contain information regarding the electronic structure of the material, suggesting that it is theoretically feasible to extract the electronic structure information of a material from the scattering signals, such as X-ray diffraction signals, through certain mathematical processes (e.g. electronic structure refinement). In fact, this has long attracted the attention of scientists. For example, the idea of using scattering experiments (mainly X-ray single-crystal diffraction) to determine the electron density distribution of materials was introduced by Nobel laureates Debye and Bragg and crystallographer Hirshfeld in 1915, 1922, and 1992, respectively.

> "It seems that the experimental study of scattered radiation, in particular from light atoms, should get more attention, since along this way it should be possible to determine the arrangement of the electrons in the atoms."
>
> – P. Debye (1915)

> "Measurement of the net charges was as yet impossible, but not beyond reach. It seems that crystal analysis must be pushed to a far greater degree of refinement before it can settle the point."
>
> – W. L. Bragg (1922)

> "An accurate set of nuclear coordinates and a detailed map of the electron density can be obtained, by X-ray diffraction, only jointly and simultaneously, never separately or independently."
>
> – F. L. Hirshfeld (1992)

Since the late twentieth century, with the rapid development of computer technology and automated equipment, an increasing amount of research has been conducted on the use of scattering techniques to determine the electronic structures of materials and become a research frontier of crystallography. Among the scattering techniques currently used to determine the electronic structures of materials, X-ray single-crystal diffraction is the most dominant method, followed by X-ray Compton scattering, polarized neutron diffraction, magnetic X-ray diffraction, and magnetic X-ray Compton scattering.

The mathematical processes used for experimental data to determine the microstructures (atomic or subatomic scales) of materials, whether crystal or electronic structures, are similar, as they all involve structure models with unknown parameters to fit the scattering experimental data, selecting the model that best fits the experimental data as the optimal structure model. Because the Fourier transform of electron density (also known as the structure factor, ignoring the phase for now) can be determined by the diffraction intensity, the structure factor bridges the structure model and the experimental data. Therefore, the best structure can be determined by minimizing the difference between the modeled value and the experimental value of the structure factor (or structure factor square) through least-squares iteration (or other minimization algorithms). Compared to crystal structures, experimental means for determining electronic structures differ significantly in terms of the structure model, refinement algorithm, data quality requirements, and structural analysis. ① Crystal structure determination involves independent-atom approximation and does not consider interatomic chemical bonding, whereas electronic structure models need to consider chemical bonding, as well as electron correlation effects, to some extent. ② Crystal structure can be viewed as a geometric structure of the arrangement of points with maximum electron densities. Given that the structure factor can be determined from diffraction experiments, the electron density and crystal structure can be fully determined by refining diffraction data. However, the refinement of diffraction data alone is not sufficient for wavefunctions and density matrices, on which it is necessary to impose constraints such as energy minimization, N-representability, and basis function orthogonality. Therefore, the refinement process must minimize the differences between the modeled and experimental values of the structure factor, as well as minimize the Lagrangian of the constraints. Alternatively, the electron density matrix can be obtained by joint refinement of the structure factor and the directional Compton-profile data, where the structure factor determines the diagonal elements of the density matrix in position space, while the Compton-profile data determine the off-diagonal elements of the density matrix in position space. ③ Crystal structure refinement typically requires only low-angle diffraction data (structure resolution $\lambda/2\sin\theta > \sim 0.83$ Å), whereas electronic structure refinement uses both low-angle and high-angle diffraction data ($\lambda/2\sin\theta > \sim 0.4$ Å) to accurately determine the temperature factor to obtain a reliable electronic structure, because the electronic structure parameters are strongly correlated with atomic temperature factors and the high-angle diffraction data are mainly contributed by the temperature factor with a small contribution of the valence electrons. It is of course possible to use neutron diffraction to determine the atomic temperature factor, but this method requires joint refinement of the X-ray diffraction data and neutron diffraction data to determine the electronic structure, which is more technically challenging than only refining the X-ray diffraction data. ④ Generally, small errors in the intensity of the X-ray diffraction data can have a significant impact on the final refined electronic structure; thus, the quality of diffraction data for electronic structure analysis (e.g. consistency factor R_{int} or R_{merge} should typically be less than 0.03) is higher than that for crystal structure refinement. ⑤ Experimentally determined crystal structures (e.g. unit cell parameters, atomic coordinates, and temperature factors) can be used directly for structural analysis and presentation, whereas experimental electronic structures (e.g. three-dimensional scalar data of electron densities) are not in data formats that chemists are familiar with. This makes it necessary to first obtain some intuitive characteristic quantities through topological analysis (e.g. Bader's quantum theory of atoms in molecules) before subsequent electronic structure analysis and structure–property relationship studies.

Of note, the electron wavefunction is generally not considered an experimentally observable physical quantity; therefore, it is not a rigorous statement to say that electron wavefunctions are determined exclusively by experimental methods. Rather, the experimental electron wavefunctions in this book were computed using refinement techniques with the experimental data and contain certain experimental information. Compared to an Nth-order reduced density matrix, a first-order reduced density matrix can be greatly simplified while still describing most of the material's physical and chemical properties of interest. Therefore, the density matrices mentioned in this book refer to first-order reduced density matrices unless otherwise specified.

Except for electronic structure in the real (or position) space, energy levels and band structures in the momentum space also govern the properties of materials. Therefore, the full understanding of the function origin of functional materials and energy spectroscopy methods such as angle-resolved photoelectron spectroscopy should also be covered besides scattering techniques. The electronic structure in momentum and energy spaces is particularly important for the understanding of the physical mechanisms of optical materials, which usually work in the excited/distorted states. Inspired by these ideas, the experimental extraction of the electronic structure of the ground and excited/distorted states in real and momentum/energy spaces can be termed electronic structure crystallography. This significantly differs from modern crystallography (referring to X-ray crystallography), which aims to determine the arrangement of molecules, atoms, and ions in solids.

The term quantum crystallography, introduced by Massa, Huang, and Karle in 1995, integrates quantum chemistry and diffraction/scattering experiments; it covers a much wide range of theoretical and experimental methods in quantum chemistry and crystallography. The majority of works in quantum crystallography are focused on the ground states of materials; however, the emergence and development of photocrystallography give more opportunities for exploring excited states with steady-state and time-resolved techniques under external optical fields, which is particularly important for the exploration of functional materials. Electronic structures of both ground and excited/distorted states in both real and momentum (or energy) space are the research focus in material science, which is one of the reasons why we propose electronic structure crystallography.

This book mainly elaborates on the theories and refinement techniques involved in using scattering methods (mainly X-ray single-crystal diffraction, while also including polarized neutron diffraction and Compton scattering) to determine the experimental electronic structure, including the experimental electron densities, electron wavefunctions, and electron density matrix of crystalline materials. This mainly includes the basic theories involved in scattering experiments, atomic thermal vibration analysis, theoretical models of electronic structure, refinement reconstruction, and topological analysis. In addition, among all electronic structure features/parts of a material, some may be more important than others in creating the desired properties and functions of the material. Such important electronic structure features/parts are introduced as functional electronic structures in the book, which is further generalized to functional motif to adapt to other types of microscopic structures such as crystal structure, magnetic (or spin) structure, and structures with modulated, statistical, and local defects. Functional motif of a material refers to the microstructural unit that plays a key role in the generation of material functions. The performances of a material depend on the functional motif and its arrangement in the material structure, and the understanding of the functional motif and its arrangement is crucial for understanding material properties and rationally designing new materials with specific functions.

Specifically, an overview of the history and some basic concepts of electronic structure crystallography are presented in Chapter 1. Chapter 2 elaborates on the first-principle computations of physical quantities related to experimental electronic structures (i.e. electron density functions), to build a bridge between crystallography and quantum chemistry, as many theoretical models of experimental electronic structures are based on quantum chemistry. Chapter 3 elaborates on the theories involved in the topological analysis of electronic structures, including the widely used quantum theory of atoms in molecules (QTAIM), the more general interacting quantum atoms (IQA) method, and the ω-restricted space partitioning method. Chapter 3 also presents a basic introduction to the computation methodology of other chemical interactions (e.g. source function, electron localization function, and reduced density gradient function). Chapter 4 elaborates on the experimental measurement techniques and refinement algorithms of electronic structures, with a focus on the treatment of atomic thermal vibrations. Chapters 5–7 present the pseudo-atom models of electronic structures (mainly focusing on the commonly used multipole model, classical X-ray atomic orbital model, the X-ray molecular orbital model, and the molecular orbitals with variable population numbers model), density matrix models, and wavefunction models (mainly focusing on the X-ray-constrained wavefunctions and extremely localized molecular orbitals models), with all these theoretical models serving as the basis for using scattering experimental data to reconstruct the electronic structures of materials. Chapter 8 describes functional electronic structure and functional motif of materials and illustrates the application of electronic structure crystallography in material research by presenting typical examples of studying nonlinear optical functional motifs with in situ electronic structure measurements.

Part of the work in this book was supported by the Youth Promotion Association of the Chinese Academy of Sciences Fund (project no. 2020303) and the National Natural Science Foundation of China (project no. 21827813, 22175172, and 21921001). We thank Mao-Chun Hong and Jian-Nian Yao for writing the forewords of this book. We hope that this book will assist graduate students and researchers in chemistry, physics, and material disciplines who are involved in the study of structures and properties of functional crystalline materials.

July 2nd, 2023 — From the authors at the FJIRSM,

Abbreviations

1-RDM	1-electron reduced density matrix
1-SRDM	spin-resolved 1-RDM
2-RDM	2-electron reduced density matrix
3D	three dimensional
ADP	anisotropic displacement parameters
AO	atomic orbital
BCA	syn-1,6:8-13-bicarbonyl[14]annulene
BCP	bond critical point
BO	bond order
BP	bond path
CBED	convergent-beam electron diffraction
CDW	charge density wave
CGDM	coarse-grained density matrices
CP	critical points
CPF	compton profile function
CSO	crystalline spin orbitals
DAC	diamond anvil cell
DCH	Dirac-Coulomb Hamiltonian
DFT	density functional theory
DKH	Douglas–Kroll–Hess
DMDD	dynamic model deformation density
EDF	electron population distribution function
EPMM	exact potential and multipole moment
ELF	electron localization function
ELI	electron localizability indicator
ELMO	extremely localized molecular orbital
EP	electrostatic potential
Full-CI	full configuration correlation
GASF	generalized atomic scattering factor
GTO	Gaussian-type orbitals
HAR	Hirshfeld atom refinement
HF	Hartree–Fock
IA	impulse approximation

IAM	independent atom model
IQA	interacting quantum atoms
IRI	interaction region indicator
KS	Kohn-Sham
LBO	LiB_3O_5
LCAO	linear combination of atomic orbitals
MEM	maximum entropy method
MM	multipole model
MO	molecular orbital
MOON	molecular orbitals with variable occupation numbers
MSO	molecular spin orbitals
NCP	nuclear critical point
NLO	nonlinear optical
NSO	natural spin orbitals
p.d.f.	probability density function
PCE	picture-change effect
PDF	probability distribution function
PAW	projected augmented waves
QTAIM	quantum theory of atoms in molecules
RDG	reduced density gradient
RDM	residual density map
RFF	reciprocal form factor
SMDD	static model deformation density
SOS	sum-over-state
UHF	spin-unrestricted HF
UKS	spin-unrestricted KS
UKS	unrestricted Kohn-Sham
VOM	valence orbital model
VSCC	valence shell charge concentration
WF	Wannier function
XAO	X-ray atomic orbital method
XC-ELMO	X-ray-constrained ELMO
XC-ELMO-VB	X-ray-constrained ELMO valence bond (method)
XCW	X-ray-constrained wavefunction
XMO	X-ray molecular orbital method
XRSC	X-ray-constrained spin-coupled (method)

Introduction

Electronic structure crystallography is an experimental crystallographic science that aims to study the electronic structures in real and momentum (or energy) space, combining experimental methods of X-ray/neutron/electron scattering and energy spectroscopy; it is an interdisciplinary discipline in the current frontier of crystallography. The intrinsic properties of a material are mainly determined by its electronic structure, which can be described in terms of electron density, electron wavefunction, or electron density matrix, where the Fourier transform (structure factor) of the electron density can be determined by scattering experiments. Therefore, a material's electron density can be obtained experimentally, and the electron density can be referred to as the experimental electron density. Moreover, by building a suitable theoretical model and using certain refinement techniques, the experimental electron wavefunction or experimental electron density matrix of a material can be further reconstructed to compute the material's physical and chemical properties. This book introduces the related theories and refinement techniques involved in using scattering methods (mainly X-ray single-crystal diffraction, as well as polarized neutron scattering and Compton scattering) to determine experimental electronic structures, including the experimental electron density, experimental electron wavefunction, and experimental electron density matrix of crystalline materials. Based on these theories and techniques, the authors proposed the ideas of functional electronic structures and functional motifs of materials, which refer to the microstructural units that play a key role in generating material properties and functions.

This book can be used as a textbook or reference book for graduate students and researchers in chemistry, physics, and material sciences for studying the structures and properties of functional crystalline materials.

1

Overview of Electronic Structure Crystallography

1.1 Introduction

Electrons in materials, which are by nature de Broglie (probability) waves, have distribution both in space (i.e. electron density distribution in position space, or electron density for short) and in energy (i.e. electron density distribution in momentum space, or momentum density for short). These two distributions are highly complementary and together determine the properties of the material. The concepts of orbitals (e.g. s, p, and d orbitals) and chemical bonds (e.g. σ and π bonds), which describe the spatial properties of electrons, and the concepts of energy levels and energy bands, which describe the energy properties of electrons, can explain various physical and chemical phenomena.

In principle, the electronic structure of a material and its properties can be determined by quantum-mechanics-based first-principles computations, which generally involve wavefunction and density functional theory (DFT) calculations for molecular and crystalline systems, respectively. The wavefunction approach uses various approximations to simplify the Schrödinger equation to obtain an equation that can be easily solved, such as the Hartree–Fock equation, and involves self-consistent field iteration to obtain the material's electron wavefunction, with the electron wavefunction taking the form of a single Slater determinant in the Hartree–Fock framework. The DFT approach was proposed by Hohenberg and Kohn [1] with the basic idea that all the ground-state properties of a physical system can be determined by its electron density alone. Both electron wavefunction and density matrice are fundamental physical quantities that describe the electronic structure, and a pure-state electron wavefunction can be equivalently presented as an electron density matrix. The density matrix obtained by integrating all of the coordinates of $N-1$ electrons in the complete N-electron density matrix can be denoted as a one-order reduced-density matrix (1-RDM) or a single-electron density matrix. Because all the properties of a single electron of interest can be obtained by the 1-RDM, any density matrix mentioned in this book actually refers to 1-RDM unless specified otherwise.

Although the electronic structures of materials are mainly obtained by first-principles computations, it is necessary to involve experimental measurements for determining electronic structures. First-principles computations typically introduce more approximations, leading to deviations between the theoretical

Electronic Structure Crystallography and Functional Motifs of Materials, First Edition.
Guo-Cong Guo and Xiao-Ming Jiang.

computations and experimental results. Moreover, theoretical computations do not accurately consider the effects of external fields, and usually do not yield the electronic structure of a material under the working conditions; however, these limitations can be overcome to some extent by experimental means. Because electrons scatter X-rays and unpaired electrons scatter neutrons, it is possible to use X-ray and neutron scattering experiments to probe the electronic state of a material. In addition, the electron density distribution in a crystal determines the electrostatic potential and magnetic field inside it, which in turn affects electron scattering, suggesting that electron diffraction can also be used to study the electronic structure of a material. Scattering occurs in the form of coherent scattering (diffraction) and incoherent scattering (Compton scattering), of which X-ray diffraction is the most commonly used method for the structural characterization of materials and is the main experimental research tool in modern crystallography (X-ray crystallography).

Modern crystallography is an experimental science that aims to determine the arrangement of atoms (or ions) in solids, mainly to study the formation, morphology, composition, and structure, as well as the physical and chemical properties of crystals and crystalloids. The main experimental methods used to characterize the crystal structures of materials are X-ray diffraction, neutron diffraction, and electron diffraction, among which X-ray diffraction is the most developed and widely used. Modern crystallographic methods typically provide structural information of materials only on the atomic scale.

Electronic structure crystallography can be defined as an experimental crystallographic science that aims to determine the electronic structures (electron density, electron wavefunction, or density matrix) of ground and excited/distorted states of solids in real and momentum (or energy) space with combined experimental techniques in crystallography and other closely relevant disciplines. Electronic structure crystallography combines the quantum theory of electronic structure and the experimental methodology of scattering and energy spectroscopy, and is an interdisciplinary discipline and research hotspot in the current frontier of crystallography for the following main reasons. ① Electronic structure crystallography links the experimental observations of a material to its basic physical quantities, and the obtained electron density and wavefunction, in principle, contain all the electronic structure information of the material, allowing it to definitively calculate all of the ground-state properties of a material. ② For some special material systems, such as those with numerous atoms or complex electronic structures, it is difficult to obtain accurate and reliable electronic structure information by first-principles computations. Instead, electronic structure crystallography can be used to study their experimental electronic structures, which is of great significance for investigating the structural nature of material functions and revealing the material structure-property relationships. ③ The experimental measurements of electronic structures can be used to verify the results of theoretical computations, which is beneficial for improving the theoretical methods and computational models and promoting the further development of first-principles computations.

Reconstructing the experimental electronic structure of a material from scattering data first requires the construction of a parametric theoretical model of the electronic structure, which can be an electron density model, a density matrix model, or a wavefunction model, such as a pseudo-atom multipole model, a single Slater determinant wavefunction, an extremely localized molecular orbital (ELMO) wavefunction, or a 1-RDM model, all of which have currently attracted considerable attention. The unknown parameters in these models are associated with the observed quantities in scattering experiments, such as the X-ray diffraction structure factor associated with the electron density (diagonal element of the position-space density matrix) and the X-ray Compton profile associated with the electron momentum density (off-diagonal element of the position-space density matrix or diagonal element of the momentum-space density matrix). Constructing equations based on these relations, least-squares refinements using diffraction structure factors and/or Compton profile data can be performed to obtain an electronic structure model that best fits the experimental observations. These optimal models can be referred to as the experimental electronic structures. Because electron density can be fully determined by the experimental structure factor, and similarly, the 1-RDM can be fully determined by the structure factor and Compton profile, the electron density and 1-RDM obtained by refinement of the experimental scattering data can be referred to as the experimental electron density and experimental density matrix, respectively. Unlike the electron density and 1-RDM, the electron wavefunction is generally not considered an experimentally observable physical quantity; therefore, it is not a rigorous statement to say that the electron wavefunction can be determined exclusively by experimental methods. Rather, the experimental electron wavefunctions in this book were derived using refinement techniques with experimental data and contain certain experimental information. Because the number of refinement parameters in the electron wavefunction model is large, with a limited number of experimental structure factors, constraints such as energy minimization must be introduced during the experimental refinement of the wavefunction.

Because the concepts of electron density, density matrix, and wavefunction are not very intuitive and difficult for scientists in chemistry and materials science to use directly, after reconstructing the experimental electronic structure of a material from experimental data, as described above, it is necessary to further perform topological analysis on an experimental electronic structure to extract its topological indicators related to canonical concepts, such as topological atomic properties (charge, volume, and multipole moment) and topological bond properties (electron density, Laplacian value, and energy density at the critical point of the bond). Quantum theory of atoms in molecules (QTAIM) is the most widely used theory for the topological analysis of electronic structures.

In this chapter, we describe some basic concepts, principles, and methods for electronic structure crystallography and give a brief history of the development of electronic structure crystallography. To establish a connection between crystallography and quantum chemistry, Chapter 2 covers the first-principles computation methods for physical quantities (electron density functions) relevant to experimental electronic structures. Chapter 3 covers the theories for the topological analysis of electronic structures, including the commonly used QTAIM, the more generalized interacting quantum atoms (IQA) method, and the ω-restricted space partitioning method. Chapter 3 also presents an introduction to the computational methods of intermolecular interaction energies, as well as other analysis methods for chemical interactions (e.g. source function, electron localization function [ELF], and reduced density gradient function). Chapter 4 covers the experimental measurement techniques and refinement algorithms of electronic structures, with a special focus on the treatment of atomic thermal vibrations. The pseudo-atom models of electronic structures (mainly including the commonly used multipole models, with some descriptions of the classical X-ray atomic orbital (XAO) model, the X-ray molecular orbital (XMO) model, and the molecular orbitals with variable population numbers model, density matrix models, and wavefunction models (mainly including the X-ray constrained wavefunction (XCW) and ELMO wavefunction models) are described in detail in Chapters 5–7. Electronic structure crystallography has wide applications in inorganic and organic materials. Chapter 8 describes the concepts of functional electronic structure and functional motif of materials and illustrates the application of electronic structure crystallography in material research by presenting typical example of studying nonlinear optical functional motifs with in-situ electronic structure measurements.

1.1.1 History of Electronic Structure Crystallography

Although the concept of electronic structure crystallography was not formally introduced until recently [2], relevant research dates to the beginning of X-ray crystallography and quantum mechanics in the late nineteenth and early twentieth centuries.

1.1.2 The Beginnings of X-ray Crystallography and Quantum Mechanics

Electron structure and X-ray crystallography have been closely linked since the beginning of the two disciplines. The year 1895 saw the discovery of X-rays by Roentgen, which started a new era of probing the inner world of matter. In 1912, Laue discovered the diffraction of X-rays after they penetrated a crystal, and in the same year, the Bragg father and son team discovered that X-ray diffraction photographs could be used to determine the positions of atoms in a crystal; thus, pioneering the X-ray crystallography. During the same period, important breakthroughs were made in the understanding of electrons and the study of quantum theory, with Thomson's discovery of the electron in 1897 and Bohr's proposal of the atomic structure model in 1913, which introduced the concept of quantization for the study of electron motion within the atom. Recognizing the scientific revolution that could be brought about by X-ray crystallography techniques, Debye predicted in 1915 that "It seems that the experimental study of scattered radiation, in particular from light atoms, should get more attention, since along this way it should be possible to determine the arrangement of the electrons in the atoms." [3] This prediction gave impetus to the establishment and development of a theoretical atomic model that could be tested by diffraction experiments. Due to the limitations of the atomic scattering factor in Bohr's atomic model, the computed diffraction intensity was shown to differ significantly from the experimental value; thus, it has been difficult to use for structure determination. It was not until the quantum model of electron density in atoms was developed [4] that the determination of the structure of matter became possible. Subsequent developments in crystallography and chemical bonding theory enhanced the connection between crystallography and quantum chemistry, as well as facilitated advances in information technology for extracting the electronic structures of materials from experimental data.

1.1.3 The Nascent Period of Experimental Electronic Structure Research

It was not until 1958 that research related to electronic structure crystallography made new progress. Richard Weiss and his collaborators took the lead in conducting pioneering work on the experimental charge density, momentum density, and spin density of single-crystal samples by obtaining the outer electronic configurations of metal atoms Cu, Ni, Co, Fe, and Cr, determining the atomic scattering factors using X-ray single-crystal diffraction measurements [5], determining the electron momentum density of diamond, graphite, and carbon black by Compton scattering experiments [6], and by calculating the spin density information for Co from the magnetic form factors measured by polarized-neutron diffraction experiments [7]. The main purpose of the research efforts made by Richard Weiss was to extract electron wavefunction information directly from X-ray scattering experiments. In a book [8], Richard Weiss suggested that the experimentally determined scattering intensity could be used to correct the Hartree–Fock wavefunction, which itself can be computed theoretically without considering the correlation between opposite spin electrons. Richard Weiss' intuition provided inspiration for other researchers, whereby integrating crystallographic experimental methods and quantum chemical theoretical approaches, as well as electron density information in the position space and Fourier transform space (momentum space), it is feasible to reconstruct the electron wavefunction from X-ray and neutron scattering experiments.

1.1.4 Developments of Pseudo-atom Models

Beginning in the late 1960s, technology related to electronic structure crystallography branched out into two directions, namely, reconstructing electronic structures from experimental data based on atomic orbitals versus molecular orbitals. The atomic orbital-based approach originated from the generalized atomic scattering factor (GASF), which was developed by Stewart and was a concept first proposed by Dawson [9]. The atomic scattering factor is also known as the atomic form factor. Stewart's basic idea was to first calculate the GASF by projecting the quantum-mechanically computed electron density of a molecule onto the atomic orbital basis functions, and then use this GASF as an alternative to the isolated-atom scattering factor for structural refinement [10, 11]. The isolated-atom scattering factor can be obtained by quantum mechanical computations on isolated atoms without considering the interactions between the atoms present in the actual crystal, and has been widely used in crystal structure refinement. By contrast, the GASF takes into account the interactions between atoms, and can be obtained by projecting molecular electron density. An atomic model obtained by projecting the electron density of the molecule is referred to as a pseudo-atom model. In contrast to the almost spherical distribution of the outer electron orbitals of isolated atoms, where no interatomic interactions are considered, the outer electron orbitals of a pseudo-atom are stretched to various extents in different directions due to chemical bonding and can be expressed as multipolar spherical harmonics centered on the atom of interest. Therefore, a pseudo-atom model is also known as an aspherical or multipole model. Considering the complexity of theoretical computations of a molecular orbital wavefunction and its atomic orbital projection, the Stewart model eventually involved a relatively simple set of atomic orbital parameters to fit the experimental values of the structure factor [12]. Hence, in the structure refinement process, the electron density of a crystal unit cell can be projected on the atomic orbital wavefunction, followed by refining the atomic orbital parameters to minimize the difference between the theoretical and experimental structure factors. Similar to Stewart's idea, Hansen and Coppens [13] designed a modified multipole model in which the electron density of each atom can be expressed as the sum of the core electron density, the valence electron density, and the non-spherical electron density, where the core electron density is set to a constant value, assuming that the core electron density is less affected by chemical bonds and consequently not involved in structural refinement. The valence shell can be adjusted by refining the number of valence electrons and the contraction factor. Non-spherical electron density has been mainly used to describe the bonding states of atomic orbitals and can be further expressed as the sum of a series of spherical harmonics, providing better modeling of the expansion of valence electron orbitals in three dimensions when atoms form chemical bonds. Thus, by refining the number of electrons occupying each orbital and the orbital contraction factor, detailed information on atoms and bonding orbitals can be obtained. Meanwhile, the introduction of a local atomic coordinate system also confers good portability to the pseudoatom model, i.e. the atomic orbital parameters of a crystal or molecule obtained from refinement can be directly applied to another crystal or molecule, as long as the target atoms in both the crystals and molecules have similar chemical environments.

Among all versions of multipole models, the Hansen and Coppens multipole model is currently the most widely used pseudo-atom model. This model has undergone multiple rounds of development and improvement over the years to adapt it to different situations. In the 1990s and at the beginning of the twenty-first century, research in this area mainly focused on improving the quality of the radial part of the valence electron density and the non-spherical electron density [14]. This has also led to the development of certain refinement methods for contraction factors [15], as contraction factors are difficult to determine using experimental diffraction intensity during the refinement process. Over the past decade or so, efforts focused on improving multipole models have centered around refining the electron density distribution and spin density distribution while correcting for the distortion of the core electron shell. Meanwhile, some databases consisting of multipole model parameters have been established, which contain portable parameters of electron density. These can be used in the construction of electron density models for large or complex molecules (especially biological macromolecules), for which accurate electron density analysis is difficult.

These multipole model improvements stem from the flexibility of the Hansen and Coppens model, specifically due to the fact that the model defines each atom in a local coordinate system as independent of the crystal orientation matrix, allowing for the transfer of multipole moment parameters from one atom to another, as long as the two atoms have sufficiently similar chemical environments. The development of multipole models was originally intended to make model parameters portable; however, it was not until many years later that this concept was introduced [16]. The importance of multipole model portability was fully demonstrated only when the experimentally determined data of atomic multipole moments were sufficiently large to construct a database for a number of similar molecules [17], especially to improve the refinement results of some samples with poor diffraction quality and construct electrostatic potentials and moments. This approach was later adopted for the theoretical computations of atomic multipole moments with the advantage of computing the electron density of large molecules, for which experimental electron density models were generally difficult to construct [18–20]. These methods are now relatively well established and have found some new applications, such as in model construction for the electron diffraction-based analysis of crystal structures [21].

The flexibility of multipole models offers the possibility of probing special electronic structure features, such as studying small distortions in the electron density of atomic core electron shells in molecules or covalent crystals. For this reason, an extended Hansen and Coppens model was proposed [22]. In this model, the non-spherical electron distribution includes not only the valence shell, but also the core electron shell. The contraction factor of the core electron shell can also be refined during the refinement process, to not only provide a chemically reasonable interpretation, but to also correct for the temperature factor. For example, the carbon atoms in diamonds show a small yet significant contraction of the core electron shell, which can be interpreted as an indication of chemical bonding with the involvement of core electrons [22].

As another very important improvement, multipole models can be simultaneously used to describe the electron and spin densities, due to the development of X-ray diffraction and elastic neutron scattering techniques, allowing for the study of electron and spin densities, respectively. The neutron scattering cross-section is the superposition of cross-sections of both atomic scattering and magnetic scattering. In addition, magnetic scattering is affected not only by the atomic spin magnetic moment, but also by the atomic orbital angular momentum; thus, polarized neutron diffraction can be used to determine the spin density. Data from both X-ray and neutron diffraction experiments can be jointly refined using the same model, wherein the electron density can be split into spin-up and spin-down components. The high number of parameters in the joint refinement model can worsen the parameter correlation problem due to the very limited data generated by polarized neutron diffraction and their linear dependence on X-ray diffraction parameters. Even though current X-ray diffractometers can provide diffraction data at very high angles, only a small fraction of the low-angle diffraction data contains information on the electron density of valence shells. As a result, this physical limit cannot be fundamentally broken, which limits the flexibility of multipole models. However, joint refinement of charge and spin densities is possible for some paramagnetic metallic compounds or magnetic inorganics when certain constraints are imposed, with further progress laying a foundation for the application of multipole models in spintronics [23–25].

Moreover, the flexibility of multipole models allows them to be used for studying certain substances under unconventional conditions such as laser irradiation, electric fields, high pressures, or temperatures, and even some powder samples [26]. In these cases, certain typical conditions beneficial for obtaining accurate electron densities cannot be met. For example, atomic vibrations dramatically intensify at high temperatures, seriously compromising the feasibility of obtaining electron densities from thermal motion by inverse convolution. Sometimes, it is the external field that allows the material to possess interesting properties, as demonstrated by some studies of crystals under high pressure [27–29]. Unlike high-temperature conditions, high pressure is not a factor that affects the accurate acquisition of electronic structures of crystals. However, the diamond anvil cell (DAC) commonly used in high-pressure studies can severely affect the quality of diffraction data, particularly by affecting the region of the reciprocal space where diffraction data can be measured, producing severe background signals. Another disadvantage is that pressure above a certain threshold cannot be uniformly transmitted to the crystal sample under investigation and can likely result in fracturing of the sample, thus, affecting the collection of diffraction data. Pressurization reduces atomic vibrations, especially for molecular crystals without strong supramolecular interactions. Studies [27, 28, 30] have shown that isostatic pressing under 5–10 GPa is equivalent to exposing a sample to a low temperature of 100 K in terms of decreasing the amplitude of the temperature factor. Low-temperature measurements under the purging of liquid nitrogen are a benchmark of experimental electronic structure studies. Though, to achieve the same effect that helium purging temperature has on the temperature factor, an isostatic pressure of more than 100 GPa is required, which is almost impossible for diffraction experiments in practice. Despite some difficulties and drawbacks, examples of the successful determination of electronic structures of high-pressure molecular crystals have been reported by means of multipole refinement [28].

In the 1970s, a number of other models were proposed in addition to the most famous multipole model proposed by Stewart and Coppens, all of which were based on the same general ideas but lacked the quantum mechanical basis of the Stewart model. Among these models, the Hirshfeld atom model is well-established [31]. This model uses the theoretical electron density of an isolated ground-state atom (spherical) as the basis for projection computations, similar to the atomic electron density used in traditional crystal structure refinement. In projection computations, the electron density of the whole molecule or crystal can be expressed as the sum of spherical atom electron densities and then decomposed onto each atom according to the projection weights to obtain the scattering factors of Hirshfeld atoms for subsequent structural refinement. The Hirshfeld atom refinement (HAR) model provides an operationally easy decomposition scheme for different types of electron density data, such as those originating from multipole model refinement or wavefunction computations. In electronic structure crystallography, Hirshfeld atoms are mainly used for three purposes: (i) to perform atomic charge population analysis on experimental or DFT electron densities [32], (ii) to construct molecular Hirshfeld surfaces [33], and (iii) to perform HAR [34].

In contrast to research on atomic orbital-based electronic structure reconstruction, research on molecular orbital-based electronic structure reconstruction faces more obstacles. The main idea of the second method is to obtain molecular orbital wavefunctions by direct refinement against diffraction or other scattering experimental data. Coppens and his collaborators made attempts in this regard [35], but found that the implementation of the process, even for medium-sized molecules, was very complicated, mainly because the computational process required refinement of the coefficients of the product of two atomic-orbital wavefunctions, which was not possible due to the relatively severe correlation between the wavefunction parameters.

1.1.5 Developments of Experimental Electron-density Matrix Models

Since 1969, the systematic method proposed by Clinton, Massa, and their collaborators has shown to be a landmark advance in electronic structure crystallography for extracting the 1-RDM from theoretically computed electron densities or structure factors [36–43]. This method constructs an iterative computation equation containing a density matrix and structure factor by introducing the electrostatic theorem, the virial theorem, and the Hellman–Feynman theorem, indicating the electron density that best conforms to the structure factor by solving the equation. Because the density matrix must obey the Pauli exclusion principle, it must be obtainable by integrating an N-electron antisymmetric wavefunction, a type of constraint referred to as N-representability conditions. A 1-RDM satisfying the N-representability conditions can be used to fully determine the experimental wavefunction. To make the obtained density matrix satisfy N-representability conditions, Clinton and Massa imposed certain constraints on the density matrix, such as using a single-Slater-determinant wavefunction and requiring the 1-RDM to be at least Hermitian, idempotent (i.e. the density matrix must be a projection operator), and normalized (the trace of the density matrix is the number of electrons N) [44], with the normalization condition ensuring an eigenvalue of the 1-RDM between 0 and 1.

The methods proposed by Clinton and Massa were later improved upon to some extent. For example, this strategy was extended to single-determinant wavefunctions for open-shell systems [45], in which Frishberg and Massa used high-quality theoretical X-ray structure factors to fit wavefunctions for a number of simple atomic and molecular systems, obtaining more accurate single-electron properties than the variational approach, despite the inability to obtain two-electron properties. In addition, Pecora obtained density matrices that not only satisfied the idempotency condition but also simultaneously provided a better fit of the experimental observables by introducing the steepest descent algorithm [46]. The most important advance in this line of research was made by Massa et al., who directly obtained the idempotent density matrix of beryllium from the analysis of X-ray diffraction data for the first time [47].

Howard and his collaborators constructed a variant of Clinton's scheme and used a simulated annealing algorithm to fit single-determinant wavefunctions for large systems, such as methylamine and formamide [48]. Snyder and Stevens used high-resolution single-crystal X-ray diffraction data to obtain an idempotent density matrix for azide anion in potassium azide [49]. Notably, the deformation density fitted using the density matrix showed a negative peak near the nucleus, a feature that was predicted for theoretical deformation density but not observed in the deformation density refined by the multipole model, suggesting that density matrix refinement could provide finer electronic structure features than multipole model refinement. According to X-ray diffraction structure factors, Tanaka proposed the XAO method [50, 51] to simulate heavy atomic orbital distortions due to crystal field effects. Based on this achievement, the XMO method was subsequently proposed[52].

The Clinton method, as well as similar methods described above, all impose idempotency constraints on the density matrices (i.e. single-Slater-determinant wavefunctions). To overcome this drawback, Hibbs, Waller, and their collaborators designed a novel method using experimental structure factors to optimize the occupation numbers of a predetermined set of (occupied or empty) molecular orbitals [53, 54]. This method, known as the molecular orbitals with variable occupation numbers (MOON) method, has the advantage of linearly increasing the number of refinement parameters with the scale of the refinement system, making it easy to perform property computations, even for very large systems.

In some more advanced methods, the idempotency condition can be replaced by more stringent N-representability conditions. In 1985, researchers noticed that the chemical bonding information hidden in the inelastic Compton scattering data could be used to reconstruct the 1-RDM [55, 56]. Subsequently, Weyrich [57], Gillet [58, 59], and their collaborators systematically proposed a method of reconstructing the 1-RDM through joint refinement of the diffraction and Compton scattering data, i.e. determining the diagonal elements of the density matrix by the structure factor of X-ray diffraction and determining the off-diagonal elements by the Compton scattering profile. For the spin density matrix, the diagonal matrix elements can be determined by polarized neutron diffraction and, in principle, by magnetic X-ray diffraction, while the off-diagonal matrix elements can be determined by magnetic Compton scattering experiments.

In 2007, Gillet designed a method to extend the Hansen–Coppens multipole model of charge density distribution to a single-particle density matrix [60]. Although tested only in simple diatomic systems (HF and CO), this approach allows the main features of chemical bonding to be extracted from complementary (diffraction and Compton scattering) experimental data. De Bruyne and Gillet used orthogonal atomic basis functions to describe the single-electron density matrix and introduced N-representability conditions into the least-squares refinement of the density matrix by using the semidefinite programming method. This approach was subjected to basic testing with dry ice crystals and was found to be in good agreement with the periodic quantum-mechanical computations [61].

Based on the fundamental research of the spin density of $YTiO_3$ in the coordinate space and momentum space by polarized neutron diffraction and magnetic X-ray Compton scattering [62, 63], a more advanced model, called the spin-resolved 1-RDM (1-SRDM) model, was proposed [64]. In this model, the density matrix was expanded by using atomic Gaussian basis functions, similar to the approach of De Bruyne and Gillet, in dealing with the closed-shell case [61]. However, unlike in the De Bruyne and Gillet approach, the exponent of the Gaussian function in 1-SRDM is not fixed and can be adjusted during the refinement process. Once the optimal exponent of the Gaussian function is obtained, the matrix elements of 1-SRDM can be obtained by minimizing the difference between the computed and experimental values (magnetic structure factor and magnetic Compton profile). Similarly, a quantum-mechanically rigorous 1-SRDM can be obtained by satisfying N-representability conditions. The study observed that the joint refinement indeed led to more accurate results than those obtained from polarized neutron diffraction data alone when the model was applied to an artificially modified magnetic crystal model based on a urea structure. The researchers also observed that the magnetic Compton profile not only affected the off-diagonal 1-SRDM elements but also improved the diagonal matrix elements, which was beneficial for obtaining a generally more accurate estimate of spin density. This technique was later used to determine the 1-SRDM of $YTiO_3$, and the refinement results could explain the magnetic properties of the Ti–O–Ti chemical bonds [65].

Cassam-Chenaï pointed out the importance of using a density matrix ensemble that satisfies N-representability conditions, considering the relatively large number of parameters to be refined by the density matrix approach and the fact that the amount of X-ray or neutron diffraction data is limited by the Ewald sphere, where a density matrix ensemble can be described by a pre-computed wavefunction, using experimental structure factors to refine a small number of coupling coefficients [66]. These coupling coefficients provide information on the electronic structure in terms of intra- and intermolecular electronic correlations as well as spin-orbit coupling. This method was later used to analyze polarized neutron diffraction data to obtain the magnetic moment density of $CoCl_4^{2-}$ polyanions in Cs_3CoCl_5 crystal [67].

Of note, for the density matrix model, satisfying N-representability conditions may still be insufficient. Indeed, according to the theories of Gilbert [68] and Coleman [69], a given electron density distribution can, in principle, be expressed by an infinite number of 1-RDMs that satisfy the N-representability conditions. Thus, fitting the electron density data alone, as in the case of all previous fitting strategies, is insufficient to ensure that the experimental wavefunction or density matrix obtained by refinement is physically realistic. To address this problem, Henderson and Zimmerman suggested that among all the single-Slater-determinant wavefunctions that fit experimental electron densities, the best one would minimize the Hartree–Fock energy [70]. Inspired by this idea, a new strategy for improving the Clinton equations was proposed and applied to the analysis of theoretical structure factor data of LiH. A new alternative scheme was subsequently proposed by

Levy and Goldstein [71], who argued that if a single-Slater-determinant wavefunction could not be uniquely determined by the experimental data, the wavefunction should allow for the minimization of electron kinetic energy. Following this idea, Parr and his collaborators [72–74] designed methods to extract Kohn–Sham orbitals from the electron density data obtained from theoretical wavefunctions or quantum Monte-Carlo computations. These strategies have also been used to generate new DFT functionals.

1.1.6 Developments of Experimental Electron Wavefunction Models

In recent years, inspired by the Levy-constrained search method [75], Jayatilaka proposed a technique that is easier to implement than the Henderson and Zimmerman idea, called the X-ray constrained wavefunction (XCW) method [76–83], where an exact wavefunction can reproduce the experimental electron density distribution while also minimizing the total energy of a given material. By using the Lagrange multiplier method and minimizing a new summation function, this method can obtain a single-Slater-determinant wavefunction that minimizes energy while also fitting the experimental structure factor within the experimental error. This summation function is the sum of the single-Slater-determinant energy and the agreement factor between calculated and experimental diffraction data. Lagrange multipliers can also be used to adjust the weights of experimental diffraction data in computation minimization, and the iterative computation process can be performed by gradually increasing the Lagrange multiplier until the theoretical and experimental structure factors are in best agreement. In practice, the XCW technique can be usually combined with the HAR technique [84], namely, to alternately perform HAR and XCW refinement computations until the structure converges. The HAR technique mainly refines crystal structure parameters (i.e. atomic positions and temperature factors), while the XCW technique mainly refines electronic structure parameters (e.g. coefficients of a single-Slater-determinant molecular orbital wavefunction). Some refinement examples of amino acids and peptides have demonstrated that this new XCW approach can yield crystal structures and charge densities that are more consistent with experimental diffraction data than traditional multipole model refinement. Of note, Jayatilaka's XCW and other similar methods use the term "constrained," referring to minimizing the wavefunction energy under the constraints of the experimental scattering data (mainly X-ray diffraction data).

Jayatilaka's XCW method is likely the most promising among all the constrained methods developed so far. In addition to obtaining the experimental electron density distributions of practical relevance, the XCW method has certain advantages, namely, it can successfully obtain the physical and chemical properties of molecular crystals, such as dipole moments, polarizabilities, and refractive indices [85–87]. By contrast, because multielectron interactions are not considered, the electron density distribution obtained by multipole model refinement cannot be used to accurately determine the polarizability and refractive index of a molecule or crystal. In principle, XCW does not satisfy the sum-over-state (SOS) requirements for polarizability computations (involving excited states); however, with reasonable approximation, a ground-state wavefunction can suffice to estimate the polarizability of a molecule.

Moreover, XCW considers the chemical environment of a crystal; thus, it can be regarded as a realistic ground-state wavefunction, with good agreement between the refractive indices of certain molecular crystals computed using XCW and experimental measurements. Later, Cole et al. further exploited the Jayatilaka method and applied it to the research of certain nonlinear optical crystals [86, 88]. These studies showed that when the nature of the two-electron effect contribution had to be considered, the XCW method outperformed the multipole model refinement technique in capturing information on solid-state effects from experimental data when properties depending on two-electron terms were concerned. In addition, the XCW method was successfully applied to reveal the functional motifs [89] of nonlinear optical crystals by Jiang and Guo et al. For example, the B–O motif $[B_3O_5]-$ was experimentally confirmed as the NLO functional motif of LiB_3O_5 (LBO) by comparing the in-situ electron density and wavefunction of LBO in the dark versus under laser irradiation at 360 and 1064 nm, where an NLO functional motif served as the structural unit with a large micro-polarizability and was the main contributor to the macroscopic NLO properties of the crystal. This work was the first to experimentally confirm the functional motif of an NLO material [90].

Macchi et al. investigated the potential of the Jayatilaka method for obtaining information regarding the crystal field and electron correlation effects [91, 92]. Because the Jayatilaka method can be regarded as a modification of the Hamiltonian quantities for both effects, the obtained experimental electron wavefunctions contain information on both the crystal field and electron correlation effects. Bučinský et al. developed unrestricted open-shell and relativistic XCW [93–96], further extending the application system of the XCW method. This unrestricted open-shell approach can obtain experimental spin densities, corresponding to those obtained by multipole models and 1-RDM. Still, the relativistic XCW method paves the way for studies on the experimental electron wavefunctions of compounds containing heavy atoms, as heavy atoms are strongly affected by relativistic effects.

However, the chemical interpretation of the electron density distribution obtained by the Jayatilaka method is not sufficiently intuitive, as the molecular orbitals are completely delocalized. Jayatilaka et al. attempted to overcome this drawback, e.g. by directly extracting the ELF [97], the electron localization index [98, 99], and the Roby index [100] from the obtained XCW. The direct introduction of traditional, chemically intuitive interpretations to XCW methods has been an important research direction in recent years, with most of the advances in this area based on the concept of ELMO [101–107], i.e. dividing a molecule into many fragments according to predetermined rules or chemical intuition, and describing each fragment using an ELMO localized on the fragment in question. Therefore, by introducing this fragmentation into the XCW framework of Jayatilaka, X-ray-constrained ELMOs (XC-ELMOs) strictly correspond to atoms, chemical bonds, lone pairs of electrons, and functional groups can be obtained [108–111]. Hence, it is possible to obtain X-ray-constrained molecular orbitals corresponding to the traditional chemical intuition of chemists. This endows the XCW technique with an ability originally available only in the multipole model, namely, regarding the total electron density of a molecule or crystal as the sum of the contributions of the individual fragments. One could even argue that the ELMO concept is as important to XCW as the pseudo-atom concept to the multipole model. In fact, XC-ELMOs act as transferable units as multipole model pseudo-atoms, i.e. the XC-ELMO parameters of a molecular fragment refined in one molecule or crystal can be directly used to refine the electronic structures of the same molecular fragments in other molecules or crystals. Theoretical ELMO libraries have been developed [112–114] and applied to the development of multiscale quantum chemistry embedding techniques [115, 116], the rapid detection of noncovalent interactions in large systems [117], and the refinement of crystal structures of peptides and small proteins in the framework of HAR [118].

ELMOs have also been used to develop the so-called X-ray-constrained ELMO valence bond method (XR-ELMO-VB) [27, 119], which can be considered the first prototype of the multideterminant XCW technique. Based on X-ray diffraction data, this strategy can be used to determine the weights of resonant molecular structures, i.e. the system wavefunctions can be written in the form of linear combinations of multiple ELMOs with different resonance structures. This technique has been used to study the charge density of syn-1,6:8-13-bicarbonyl[14]annulene (BCA) under isostatic pressure, a compound whose aromaticity has been shown to be partially suppressed with increasing pressure [28]. XR-ELMO-VB computations have confirmed the trend that, at ambient pressure the two resonance structures of BCA are basically equivalent, at high pressure, one of the two becomes clearly predominant [27].

To extract useful chemical information from X-ray diffraction data, the more recent X-ray constrained spin-coupled method (XRSC) has been introduced, resulting from the coupling of the Jayatilaka method with the spin-coupled technique of the valence bond theory [120–124]. This method can be considered a step forward compared to the previous XR-ELMO-VB strategy. In fact, in this novel XRSC method, it is possible to extract both the orbitals and resonance structure weights from the X-ray diffraction data without requiring any localization schemes or *a priori* wavefunctions. Several computations have shown that the spin-coupled orbital and resonance structure weights obtained by the XRSC method present a non-negligible difference compared to those obtained through conventional gas-phase spin-coupled computations. This has further demonstrated the intrinsic capability of the Jayatilaka method in accounting for the crystal field effects on electronic structures.

1.1.7 Developments in Electron Diffraction-Based Studies of Electronic Structures

In addition to X-ray and neutron scattering, the use of electron diffraction techniques to explore the experimental electronic structure of materials has long attracted the attention of researchers. The convergent-beam electron diffraction method (CBED) [125] converges electron beams with a certain cone angle into a small, unbent region with a uniform thickness on the sample. In addition to the usual diffraction spots, diffraction disks are produced, with an integrated diffraction intensity comparable to the theoretical values computed from the electron diffraction dynamics. Therefore, diffraction disks can be used to directly and accurately determine the low-index structure factor, and this advantage allows them to be used in reconstructing the electron density of certain simple inorganic substances, allowing CBED to outperform X-ray diffraction in terms of determining the crystal systems and space groups of materials [126]. Nakashima showed that the charge density distribution of metallic Al extracted by the quantitative CBED method was more reliable than theoretical or X-ray diffraction results, and matched well with the experimental results of the anisotropic elasticity coefficients of Al [127]. Zuo et al. studied the electron density information of Cu—Cu bonds in Cu_2O and interpreted the deformation density around Cu as an electron shift from Cu-3d to Cu-4s orbitals [128]. In addition, Palatinus et al. used precession electron diffraction tomography to determine the exact position of H atoms in nanocrystals [129]. These studies demonstrated the potential of the CBED method for studying the experimental electronic structures of materials.

1.2 Basic Descriptors of Electronic Structure

1.2.1 Electron Density

In quantum chemistry, electron density is a function of the wavefunction. For a molecular system containing M atoms and N electrons, the ground-state electron wavefunction ψ_{el} is considered a function of the electron spin and spatial coordinates ($\boldsymbol{t}_i = (\boldsymbol{s}_i, \boldsymbol{r}_i)$; $i = 1, 2, \cdots, N$). Under the Born–Oppenheimer approximation, when the spatial coordinates of nucleuses ($\boldsymbol{R}_k$; $k = 1, 2, \cdots, M$) are fixed and the influence of other electrons is not considered, the probability of finding an electron in the volume element $\mathrm{d}\boldsymbol{r}$ at any location $\boldsymbol{r}$, $\rho(\boldsymbol{r})\mathrm{d}\boldsymbol{r}$ can be written as

$$\rho(\boldsymbol{r})\mathrm{d}\boldsymbol{r} = N \int |\psi_{\mathrm{el}}|^2 \,\mathrm{d}\boldsymbol{t}'\mathrm{d}\boldsymbol{r}, \tag{1.1}$$

where $\mathrm{d}\boldsymbol{t}'$ denotes the integration over the spin and spatial coordinates of all but one electron, $\rho(\boldsymbol{r})$ is the density of electrons in the position space or electron density (electron density), and the total charge density is the sum of the charge densities of electrons and nuclei in the position space. When addressing electron density distribution, the term “charge density” can be used interchangeably with the term “electron density,” as is the case in this book.

In crystallography, the electron density is closely related to the structure factor. X-ray diffraction is a phenomenon of electron–photon interactions, and in principle, it cannot be understood well without knowledge of quantum mechanics. However, some reasonable simplifications are often made in practical applications. In kinematical theory, the amplitude of coherent elastic scattering (diffraction), i.e. the structure factor $F(\boldsymbol{H})$, is the Fourier transform of the unit cell electron density $\rho(\boldsymbol{r})$. Thus, $\rho(\boldsymbol{r})$ can be obtained by the inverse Fourier transformation of $F(\boldsymbol{H})$:

$$\rho(\boldsymbol{r}) = \int F(\boldsymbol{H}) \exp(-2\pi \boldsymbol{i}\boldsymbol{H} \cdot \boldsymbol{r})\mathrm{d}\boldsymbol{H}, \tag{1.2}$$

where $F(\boldsymbol{H})$ is the complex structure factor, which has been corrected for anomalous scattering. In addition, the scattering vector $\boldsymbol{H} = h\boldsymbol{a}_1^* + k\boldsymbol{a}_2^* + l\boldsymbol{a}_3^*$ corresponds to certain discrete lattice points in the reciprocal space, where h, k, l is an integer and $\boldsymbol{a}^*_{i,i=1,3}\left(\boldsymbol{a}_i\boldsymbol{a}^*_j = \delta_{ij}\right)$ is the reciprocal lattice basis vector.

Because $F(\boldsymbol{H})$ can be defined over the discrete $\boldsymbol{H}$, the integral Eq. (1.2) can be converted into summation form:

$$\rho(\boldsymbol{r}) = V^* \sum_{\boldsymbol{H}} F(\boldsymbol{H}) \exp(-2\pi \boldsymbol{i}\boldsymbol{H} \cdot \boldsymbol{r}), \tag{1.3}$$

where V^* is the unit cell volume of the reciprocal lattice, i.e. $1/V$, where V is the unit cell volume in real space.

Because electron density is a real function, the right-hand side of Eq. (1.3) must also be real, and $F(\boldsymbol{H})$ can now be written in complex form:

$$F(\boldsymbol{H}) = |F(\boldsymbol{H})| \exp i\,\varphi(\boldsymbol{H}) \equiv A(\boldsymbol{H}) + iB(\boldsymbol{H}), \tag{1.4}$$

$$A(\boldsymbol{H}) = |F(\boldsymbol{H})| \cos\varphi; B(\boldsymbol{H}) = |F(\boldsymbol{H})| \sin\varphi, \tag{1.5}$$

where $\varphi(\boldsymbol{H})$ is the phase of the structure factor. After substituting Eq. (1.4) into Eq. (1.3), the expression can be reorganized to remove the imaginary part. Moreover, the $F(\boldsymbol{H})$ and $F(-\boldsymbol{H})$ contributions can be summed up by using relationships $A(\boldsymbol{H}) = A(-\boldsymbol{H})$ and $B(\boldsymbol{H}) = -B(-\boldsymbol{H})$, as suggested in Eq. (1.5). Thus, the unit cell electron density can be written as

$$\rho(\boldsymbol{r}) = \frac{2}{V} \sum_{1/2} \{A(\boldsymbol{H}) \cos(2\pi \boldsymbol{H} \cdot \boldsymbol{r}) + B(\boldsymbol{H}) \sin(2\pi \boldsymbol{H} \cdot \boldsymbol{r})\}, \tag{1.6}$$

or

$$\rho(\boldsymbol{r}) = \frac{2}{V} \sum_{1/2} [|F(\boldsymbol{H})| \cos\{2\pi \boldsymbol{H} \cdot \boldsymbol{r} - \varphi(\boldsymbol{H})\}]. \tag{1.7}$$

Thus, each structure factor contributes a plane wave with wave vector $\boldsymbol{H}$ and phase angle φ to the total electron density.

1.2.2 Residual Density

The density difference $\Delta\rho(\boldsymbol{r})$ between the total electron density $\rho(\boldsymbol{r})$ and reference density $\rho_{\mathrm{ref}}(\boldsymbol{r})$ reflects the degree of adequacy of the reference density to describe the system. Thus, ΔF can be defined as the difference between the observed value of the structure factor $F_{\mathrm{obs}}(\boldsymbol{H})$ and the computed value $F_{\mathrm{calc}}(\boldsymbol{H})$, i.e.

$$\Delta F = F_{\mathrm{obs}}(\boldsymbol{H})/k - F_{\mathrm{calc}}(\boldsymbol{H}), \tag{1.8}$$

where k is a scale factor. Because $F_{\mathrm{obs}}(\boldsymbol{H})$ and $F_{\mathrm{calc}}(\boldsymbol{H})$ are complex numbers, ΔF is considered a vector in the complex plane. In addition, $\Delta\rho(r)$ can be obtained by the Fourier transformation of ΔF:

$$\Delta\rho(\boldsymbol{r}) = \rho_{\mathrm{obs}}(\boldsymbol{r}) - \rho_{\mathrm{calc}}(\boldsymbol{r}) = \frac{1}{V} \sum_{\boldsymbol{H}} \Delta F \exp(-2\pi i \boldsymbol{H} \cdot \boldsymbol{r}). \tag{1.9}$$

Similar to Eq. (1.9), the density difference $\Delta\rho(\boldsymbol{r})$ can be written as:

$$\Delta\rho(\boldsymbol{r}) = \frac{2}{V} \left\{ \sum_{1/2} (A_{\mathrm{obs}} - A_{\mathrm{calc}}) \cos 2\pi H \cdot r + \sum_{1/2} (B_{\mathrm{obs}} - B_{\mathrm{calc}}) \sin 2\pi \boldsymbol{H} \cdot \boldsymbol{r} \right\}. \tag{1.10}$$

If the structure model used to calculate F_{calc} is obtained by the least-squares refinement of the observed structural factor, and if the phase angle of F_{calc} is specified, the density map obtained from Eqs. (1.9) and (1.10) can be referred to as the residual density map (RDM). In structural analysis, residual density is a very useful tool from which the least-squares refinement process and the soundness of the structure model can be evaluated.

$F(\tau)$ can be defined as the Fourier transform of $f(t)$. In the discussion below, Parseval's theorem is given by:

$$\int_{-\infty}^{\infty} |f(t)|^2 \, dt = \int_{-\infty}^{\infty} |F(\tau)|^2 \, d\tau. \tag{1.11}$$

Because $\Delta\rho$ is the Fourier transform of ΔF, Eq. (1.11) indicates that minimizing $\int(\rho_{obs} - \rho_{calc})^2 \, d\boldsymbol{r}$ is equivalent to minimizing $\int(F_{obs} - F_{calc})^2 \, d\boldsymbol{H}$; thus, minimization of the structure factor will also lead to the minimization of residual density. When only some of the values of $f(t)$ are known, Eq. (1.11) still holds. In diffraction experiments, only structure factors no more than $\boldsymbol{H}_{max}$ will be experimentally available due to the limited structural resolution.

The Patterson function $D(\boldsymbol{u})$ can be defined as the autocorrelation of $\Delta\rho$:

$$\begin{aligned} D(\boldsymbol{u}) &\equiv \int \Delta\rho(\boldsymbol{r})\Delta\rho(\boldsymbol{r}-\boldsymbol{u})d\boldsymbol{r} = \Delta\rho(\boldsymbol{r}) * \Delta\rho(-\boldsymbol{r}) \\ &= \frac{1}{V}\sum_{\boldsymbol{H}} \Delta F(\boldsymbol{H}) \cdot \Delta F(-\boldsymbol{H}) \exp(-2\pi \boldsymbol{H} \cdot \boldsymbol{u}) \\ &= \frac{2}{V}\sum_{1/2} (\Delta F)^2 \cos 2\pi \boldsymbol{H} \cdot \boldsymbol{u} \end{aligned} \tag{1.12}$$

The first line of the above equation follows the relationship between autocorrelation and convolution (* denotes convolution), while the second line uses the convolution theorem of Fourier transform, i.e. the Fourier transform of the convolution of two functions is equal to the product of the Fourier transforms of the individual functions. Moreover, the third line does not contain the imaginary part, because $\Delta\rho$ is a real number. At $\boldsymbol{u} = 0$, Eq. (1.12) becomes:

$$D(0) = \frac{2}{V}\sum_{1/2} (\Delta F)^2 = \int \Delta\rho(\boldsymbol{r})^2 d\boldsymbol{r}. \tag{1.13}$$

In fact, minimization computation in the position space (or direct space) is not strictly equivalent to the reciprocal space, because during minimization, additional weight factors are added to each structure factor in the reciprocal space to participate in the computation, as is the case for electron density in the position space.

1.2.3 Deformation Density

The deformation density is defined as the difference between the total electron density and the electron density computed by the reference model. As is the case of computing the residual density according to Eq. (1.9), the deformation density can also be obtained through Fourier transformation, except F_{calc} is computed from the reference model and compared to the experimental electron density.

A commonly used reference density, also known as the promolecule density, is the superposition of spherically averaged atomic charge densities, which are each centered on the coordinates of the corresponding nucleus, where a promolecule represents a collection of independent atoms without the existence of interatomic interactions in a real crystal, only forming a hypothetical entity. However, electrostatic interactions will occur between the atoms in a promolecule, and a promolecule will be stable if interatomic interactions other than electrostatic interactions are absent. The deformation density computed by using the promolecule density as a reference density is the commonly used (standard) deformation density, i.e. the difference between the total density $\rho(\boldsymbol{r})$ and total density $\rho_{\text{pro}}(\boldsymbol{r})$ of the spherical ground-state atoms at the $\boldsymbol{R}_i$ position:

$$\Delta\rho(\boldsymbol{r}) = \rho(\boldsymbol{r}) - \rho_{pro}(\boldsymbol{r}) = \rho(\boldsymbol{r}) - \sum_{\text{all}} \rho(\boldsymbol{R}_i). \tag{1.14}$$

Deformation density can be used to determine whether a covalent bond has formed. The presence of charge aggregation (bond charge maximum) between two adjacent atoms is an indicator of two atoms forming a covalent bond, but not vice versa, i.e. the absence of a bond charge maximum between the two atoms does not necessarily indicate the absence of bond formation. This is because atoms in promolecules are neutral and for an element with both half-filled and fully filled orbitals, the number of electrons in a half-filled orbital (where there is only one electron) is mistakenly subtracted by more than 1. For example, when subtracting the electron density of a spherical O atom with electron configuration $(1s)^2$ $(2s)^2$ $(2p)^4$ from an O atom in an actual crystal, each valence orbital will be subjected to subtraction of 1.333 electrons, resulting in a lack of bonding charge when the extra 0.333 electrons exceed the bonding charge. This explains why the O—O bond in hydrogen peroxide has a lack of bond charge. Sometimes some other reference states, such as an oriented atomic state and a molecular fragment state, must be used to gain insight into the formation of chemical bonds.

Deformation density can be obtained by different experimental methods. The first method involves a combination of X-ray diffraction and neutron diffraction techniques. In this method, the exact atomic coordinates and temperature factors required to compute the reference-state density can be obtained by neutron diffraction, followed by difference Fourier transformation to derive the deformation density. However, this method involves both X-ray diffraction data and neutron diffraction data, and the experimental conditions are not the same for both sets of data, thus causing large systematic errors. The second method uses the same set of X-ray diffraction data. Because valence-electron scattering mainly occurs in the low-angle region (also known as the low-order region or low-resolution region) [130] and that high-angle (i.e. high-order or high-resolution) diffraction data are mainly contributed by core-electron scattering, it is possible to obtain exact atomic coordinates and temperature factors by refining the high-angle diffraction data, and then deriving the deformation density by performing difference Fourier transformation of the full-resolution structure factor:

$$\rho_{\text{deformation}}(\boldsymbol{r}) = \frac{1}{V}\sum_{\boldsymbol{H}}(F_{\text{obs}} - F_{\text{calc,high order}})\exp(-2\pi i\boldsymbol{H}\cdot\boldsymbol{r}). \tag{1.15}$$

The distinction between high- and low-angle scattering is compound-specific, and it is difficult to obtain the widely accepted criterion. It is generally recommended that the cut-off scattering angle satisfy the conditions of $\sin\theta/\lambda = 0.5$ Å^{-1} or $\lambda/2\sin\theta = 1.0$ Å.

In electron density analysis, dynamic model deformation density (DMDD) and static model deformation density (SMDD) are also commonly used, where the term dynamic means that the effect of atomic thermal vibrations is accounted for, while the term static means that the effect of atomic thermal vibrations is not considered.

DMDD can be defined as:

$$\Delta\rho_{\text{model}}(\boldsymbol{r}) = \rho_{\text{model}}(\boldsymbol{r}) - \rho_{\text{reference}}(\boldsymbol{r}), \tag{1.16}$$

where the total model density is given by:

$$\rho_{\text{model}}(\boldsymbol{r}) = \frac{1}{V}\sum_{\boldsymbol{H}} F_{\text{calc,model}}(\boldsymbol{H})\exp(-2\pi i\boldsymbol{H}\cdot\boldsymbol{r}), \tag{1.17}$$

where the model-calculated structure factor $F_{\text{calc,model}}$ takes into account the effect of atomic thermal vibrations.

SMDD generally refers to the deformation density of a multipole model and can be expressed as:

$$\Delta\rho_{\text{model}}(\boldsymbol{r}) = \sum_{i}^{\text{all atoms}} \left\{P_{i,c}\rho_{\text{core}}(r) + P_{i,v}\kappa^3\rho_{\text{valence}}(\kappa_i\boldsymbol{r}) + \sum_{l=0}^{l_{\max}} \kappa'^3_i R_{i,l}\left(\kappa'_i r\right)\sum_{m=0}^{l} P_{i,lm\pm}d_{lm\pm}(\boldsymbol{r}/r)\right\} - \rho_{\text{reference}}(\boldsymbol{r}) \tag{1.18}$$

Please refer to Chapter 5 for additional details regarding the multipole model. Because the structure model acts as a noise filter during the refinement process, the bond charge maximum will be higher in the SMDD map than in the DMDD map.

The true phase of a structure factor generally differs from one computed by an independent-atom model, and the difference will introduce errors into the computation of deformation density. The phase of structure factor is fixed at 0 or π in centrosymmetric structures, whereas in non-centrosymmetric structures, only the phases of certain diffraction spots formed by centrosymmetric projection are fixed, such as the $hk0$, $h0l$, and $0kl$ diffraction spots in the $P2_12_12_1$ space group.

1.2.4 Electron Wavefunction and Density Matrix

In the method of linear combination of atomic orbitals (LCAO), a molecular orbital χ_i can be described as a linear combination of atomic basis functions ϕ_μ:

$$\chi_i = \sum_{\mu} C_{i\mu}\phi_\mu. \tag{1.19}$$

To satisfy the Pauli exclusion principle, the many-electron wavefunction must be antisymmetric with respect to the permutation of electrons. This wavefunction can be written in the form of a Slater determinant, i.e. antisymmetric combinations of occupied molecular orbitals. In the Hartree–Fock method, a wavefunction contains only a single-Slater determinant, and each electron is assumed to be influenced by the average potential field of the other electrons.

For an n-electron system, the single-determinant wavefunction can be written by:

$$\psi = (n!)^{-1/2} \begin{vmatrix} \chi_1(1) & \cdots & \chi_n(1) \\ \vdots & & \vdots \\ \chi_1(n) & \cdots & \chi_n(n) \end{vmatrix}, \tag{1.20}$$

where the number i in parenthesis denotes the i-th electron (i = 1, 2 ...n), and the coefficient $(n!)^{-1/2}$ in front of the determinant is a normalization factor. The single-determinant wavefunction can often be abbreviated in the following form:

$$\psi = |\chi_1(1)\chi_2(2)\cdots\chi_n(n)\rangle. \tag{1.21}$$

According to the rule of determinant computation, Eq. (1.20) can be expanded:

$$\psi = \sum_i \sum_{j \geqslant i} \hat{P}_{ij}(\chi_1(1)\chi_2(2)\cdots\chi_n(n)), \tag{1.22}$$

where $\hat{P}_{ij}$ is an electron permutation operator that permutates $\chi_1, \chi_2, \cdots, \chi_n$, whose eigenvalues −1 and +1 denote an odd and even number of electron permutations, respectively. The permutation of the two rows (or columns) in the Slater determinant changes the sign of the wavefunction, where the determinant value is zero when any two rows (or columns) are identical. Thus, the Slater determinant wavefunction obeys the fermionic Pauli exclusion principle.

Because the n-electron wavefunction is a function of $3n$ spatial coordinates and n spin variables, the single-electron density $\rho(\boldsymbol{r})$ can be obtained by integrating over the spatial coordinates and spin variables of all but one of the electrons:

$$\rho(\boldsymbol{r}) = \int |\Psi|^2 \left(\boldsymbol{r}_1, \boldsymbol{r}_2, \ldots, \boldsymbol{r}_n, s_1, s_2, \ldots, s_n\right) \mathrm{d}\boldsymbol{r}_2 \cdots \mathrm{d}\boldsymbol{r}_n \mathrm{d}s_1 \cdots \mathrm{d}s_n. \tag{1.23}$$

Since electrons are indistinguishable and molecular orbitals are orthogonal, the overlap integral of two molecular orbitals is 0, which allows $\rho(\boldsymbol{r})$ to be simplified:

$$\rho(\boldsymbol{r}) = \sum_i n_i \chi_i^2. \tag{1.24}$$

If χ_i is a spin-orbital that contains both space and spin components, then $n_i = 1$. If χ_i is a spatial orbital, spin is not considered; thus, $n_i = 2$.

To obtain the electron density of atomic orbitals, Eq. (1.19) can be substituted into Eq. (1.24) to obtain:

$$\rho(\boldsymbol{r}) = \sum_\mu \sum_\nu P_{\mu\nu} \phi_\mu(\boldsymbol{r}) \phi_\nu(\boldsymbol{r}), \tag{1.25}$$

where $\boldsymbol{P}$ is the density matrix, and the matrix element $P_{\mu\nu}$ is the population of the orbital product $\phi_\mu(\boldsymbol{r})\phi_\nu(\boldsymbol{r})$, i.e.

$$P_{\mu\nu} = \sum_i n_i C_{i\mu} C_{i\nu}. \tag{1.26}$$

The wavefunction ψ contains all of information regarding the joint probability density of electrons. For example, the two-electron density can be obtained by integrating all of the other spatial and spin variables, except for the two-electron spatial coordinates, to describe the joint probability of simultaneously finding electron 1 at $\boldsymbol{r}_1$ and electron 2 at $\boldsymbol{r}_2$.

A single-Slater determinant obeys the Pauli exclusion principle by considering the correlation between parallel-spin electrons, but ignores the correlation between opposite-spin electrons. Computations based on a single-Slater determinant will show that the probability of simultaneously finding two opposite-spin electrons at the same position is not zero, which is practically impossible.

More advanced methods than the Hartree–Fock method include excited-state determinants in the wavefunctions. If the occupied orbitals are denoted by a, b, c while denoting the empty orbitals by r, s, t, a single excited configuration generated by the promotion of an electron from a to r can be expressed by:

$$|\psi_a^r\rangle = |\chi_1\chi_2 \cdots \chi_r\chi_b \cdots \chi_n\rangle. \tag{1.27}$$

A multi-configuration wavefunction can be expressed by:

$$|\psi\rangle = c_0|\psi_0\rangle + \sum_{ra} c_a^r |\psi_a^r\rangle + \sum_{\substack{a<b \\ r<s}} c_{ab}^{rs} |\psi_{ab}^{rs}\rangle + \cdots. \tag{1.28}$$

For a multi-Slater determinant wavefunction, molecular orbitals can be defined similarly to Eqs. (1.24) and (1.25), and these molecular orbitals are known as natural spin orbitals, where n_i is no longer necessarily an integer, but can be a fractional number between 0 and 1. Accordingly, Eq. (1.25) contains both one-center and two-center terms, with the same nucleus position for ϕ_μ and ϕ_ν in the one-center term and different nucleus positions for ϕ_μ and ϕ_ν in the two-center term. The two-center term represents the bond charge density, with observable values only when ϕ_μ and ϕ_ν are in the same spatial region, which can be neglected for distantly separated atoms.

1.3 Experimental Characterization of Electronic Structure

Scattering data generated by a variety of instruments or large-scale scientific facilities can be used for the refinement of experimental electronic structures. Typical examples of these datasets include X-ray diffraction data, polarized neutron diffraction data, and X-ray Compton scattering data, which are generated and collected by X-ray single-crystal diffractometers (in a laboratory setting), synchrotron X-ray single-crystal diffractometers, or neutron single-crystal diffractometers. Among these instruments, synchrotron and neutron sources are expensive and time-consuming. The most readily available instruments are small laboratory X-ray single-crystal diffractometers; however, X-ray single-crystal diffractometers used for experimental electronic structure measurement have more rigorous requirements for some technical specifications compared to instruments used for conventional crystal structure measurements, such as the technical specifications of the intensity and wavelength of X-ray light sources. Specifically, for in-situ electron structure measurements under an external field, the diffractometer host requires some additional external devices, such as laser and cryogenic systems.

Electron structure refinement requires near-perfect diffraction data, where changes by a few thousandths of diffraction intensity can have a significant impact on the shape of the refined electron cloud. To obtain high-precision diffraction data, it is necessary to strictly control the crystal quality, instrument accuracy (mainly depending on X-ray source quality, goniometer accuracy, and detector sensitivity), and data correction.

1.3.1 Experimental Electronic Structure Measurement with X-ray Single-crystal Diffractometer

1.3.1.1 X-ray Source

The fourth-generation synchrotron radiation source and rotating X-ray anodes are cutting-edge X-ray light sources used in large-scale scientific facilities and laboratories, respectively. Significant advances have been achieved in X-ray free-electron lasers at synchrotron radiation facilities; however, more in-depth research is needed to explore whether this technology can benefit experimental electron density studies. For laboratory X-ray sources, an important advancement is the use of microfocus X-ray spots, which allows for the generation of X-rays with higher optical density, but with a lower uniformity of spot intensity distribution compared to conventional (non-microfocus) sources.

Refinement of high-precision electronic structures requires X-ray single-crystal diffraction data with high structural resolution ($\lambda/2\sin\theta < 0.4$ Å). According to the Bragg law $d_{hkl} = \lambda/2\sin\theta$, structural resolution can be improved by using a short-wavelength ($\lambda = 0.71073$ Å) Mo-target X-ray source and collecting high-angle diffraction spots with sufficient intensity. Because X-ray diffraction intensity decays sharply with increasing diffraction angle, the high-angle diffraction spot intensity can be increased by increasing the exposure time, using a high-power X-ray generator and high-performance 2D detector, as well as using X-ray focusing and enhancing techniques.

Highly-precision diffraction data place high demands on the monochromaticity, uniformity, and high intensity of the X-ray source. Compared to synchrotron radiation, the X-rays from the Mo target of a laboratory X-ray single-crystal diffractometer is a mixture of $K_{\alpha1}$ (0.7093 Å), $K_{\alpha2}$ (0.71359 Å), and K_β (0.632288 Å) lines, which degrades the monochromaticity. When improving the monochromaticity, a monochromator can be used to filter out the K_β line and even the $K_{\alpha2}$ line, leaving only the most intense line –the $K_{\alpha1}$ line. However, this will greatly reduce the intensity of the X-ray to be used. Because a Mo-target X-ray source is a multiwavelength source, a diffraction spot will split into several at different locations, making it difficult to locate the diffraction spots and integrate the intensity, especially at high angles. However, this problem can be overcome by improving the integration algorithm.

Alternatively, a microfocus light source can be used to increase the X-ray intensity, but the intensity of the microfocus light source in the radial direction (i.e.

in the cross-section perpendicular to the direction of X-ray incidence) will follow a Gaussian distribution instead of a constant profile; thus, the uniformity will not be satisfactory. By nature, using a microfocus light source involves sacrificing monochromaticity and uniformity in favor of high intensity, which is necessary for collecting high-angle diffraction data from crystals, because such data are usually weak. High data intensity due to a microfocus light source can greatly reduce the diffraction data collection time and allow for the measurement of samples with relatively weak diffraction ability.

1.3.1.2 Goniometer

Based on the Bragg equation ($d_{hkl} = \lambda/2\sin\theta$), when using the Mo-target X-ray with $\lambda = 0.71073$ Å to achieve the structural resolution required for high-precision electronic structure refinement ($\lambda/2\sin\theta < 0.4 =$ Å), the diffraction angle 2θ should reach 125°, which is much higher than the diffraction angle of 50.7° required for conventional measurement of crystal structures (typically requiring a structural resolution of 0.83 Å). Therefore, the X-ray source, goniometer, detector, beamstop, and camera components must be mounted in a compact manner to ensure that the detector can be turned relatively close to this angle without component collisions.

The accuracy of the goniometer is also an important factor to consider, especially for microfocus X-ray sources. If the goniometer is not sufficiently accurate, the crystal will be more or less off-center during crystal measurement, resulting in a negative impact on the intensity and absorption correction of the diffraction data. Although this effect can be mitigated to some extent through complex absorption corrections, a high-precision goniometer is still desirable, as it can produce high-quality diffraction data. In general, the accuracy of goniometers decreases with use due to mechanical wear.

1.3.1.3 X-ray Detector

It is important to accurately measure the intensity of the Bragg peak, and unlike earlier point detectors, where only one diffraction spot could be measured at a time, a 2D digital detector can quickly collect diffraction photographs of the entire reciprocal space. A significant advantage of 2D detectors is that they can obtain high-redundancy data, which greatly improves statistical accuracy; however, this requires more corrections for absorption, attenuation, source instability, and dispersion.

Two important parameters of 2D detectors are the dynamic range of the readings and the reading time. The exposure time of a single photograph during diffraction data collection for electronic structure refinement is generally long, and it may be very long for generating some particularly strong diffraction spots, suggesting that the dynamic range of the detector's individual pixels should be wide. Otherwise, the intensity of the diffraction spots on these pixels may exceed their thresholds; thus, accurate measurement will be difficult.

1.3.1.4 Cryogenic Systems

Low temperatures are important for obtaining high-quality electron density distributions for two main reasons: (i) atomic vibrations reduce the scattering intensity, whereas low temperatures allow atoms to vibrate as close to the equilibrium position as possible, and (ii) with large thermal vibration factors, it is difficult to separate the electron density parameters from the thermal vibration parameters during refinement. However, exactly how low a temperature is needed for accurate electron density analysis is difficult to define, depending on the material under study. In general, the lower the temperature, the better the result. However, even when collecting data at 0 K, it is still impossible to completely avoid thermal vibrations of atoms because of the presence of zero-point vibrations. Modern cryogenic techniques in laboratories are sophisticated, including measurement at temperatures near the boiling point of liquid nitrogen, which generally generates stable and reliable data, and measurement at temperatures near the boiling point of liquid helium, which represents a more advanced technique.

Low temperatures play multiple roles in experimental electron density analysis. (i) Most importantly, low temperatures allow us to reduce the thermal motion of atoms, making the pseudo-atom approach a reasonable assumption in structural refinement, where the smaller the thermal vibrational factor, the smaller the correlation coefficients between the parameters, and the more reliable the refinement results. (ii) In addition, a smaller thermal vibrational factor makes the simple harmonic vibrational model of atoms more effective. Of note, although higher-order models of atomic vibrations, such as the Gram-Charlier expansion, can be used, this introduces too many parameters. Mallinson et al. showed that peak intensity distribution of residual map of non-harmonic vibrations was somewhat similar to that of deformation charges, making it difficult to identify true electron-density features [131]. Therefore, the physical significance of the Gram-Charlier parameter must be verified when approximating the refined temperature factor by high-order nonharmonics. Another advantage of low temperatures is the high accuracy of the diffraction intensity that can be obtained, especially for high-angle diffraction spots. Although the refinement of bond-charge density features generally does not require high-angle data (since high-angle diffraction spots are mainly a result of the scattering from core electrons), high-angle, high-intensity diffraction data are important for the refinement accuracy, especially when refining atomic positions and thermal vibration factors (except for H atoms). Moreover, low temperatures can significantly reduce the dispersion of high-angle diffraction spots; thus, allowing for high-precision intensity integration.

1.3.2 Key Aspects of Experimental Electronic Structure Measurement

1.3.2.1 Single-crystal Samples

Nearly defect-free single-crystal samples with high quality are crucial for obtaining high-quality diffraction data, as high-precision analysis of experimental electronic structure data is generally difficult when atomic disorder exists in the crystal structure. In practice, a single-crystal sample to be measured should not be too large, in most cases (e.g. mean size of less than 0.3 or 0.5 mm), otherwise, the sample may have strong X-ray absorption, degrading the accuracy of diffraction data. However, the sample should not be too small, such as mean size of less than

0.1 mm; otherwise, the diffraction intensity will be significantly reduced, and a longer exposure time will be required to obtain the diffraction data with a high signal-to-noise ratio. In general, to increase the intensity noise (S/N) ratio by a factor of n for the same sample using a single photograph, the exposure time of the photograph must be increased by a factor of n^2. Of note, an infinitely long exposure time will not lead to an indefinitely large S/N ratio, as there is an inherent limit to the S/N ratio achieved by the instrument. Meanwhile, an appropriate sample size will also depend on the diffracting power of the sample, the X-ray spot size, the light source intensity, and the acceptable duration of data collection.

1.3.2.2 Measurement Process

Diffraction data collection requires that the crystal sample be relatively stable and not shifted during goniometer rotation. Collecting a set of electron structure diffraction data by a laboratory light source typically takes a long time, ranging from 3–5 days to 1–2 weeks. Ensuring that the crystal position does not shift during such a long test time is not easy, especially under low-temperature gas purging, as it usually requires that the crystal carrier is in good condition and that the glass filament (or other materials such as metal rods) and glue used to adhere the crystal are strong.

Collecting a large number of diffraction photographs and achieving a data/parameter ratio above 10 are desirable; however, the data/parameter ratio will be much lower than 10 in many practical circumstances. Inter-parameter correlations during electronic structure refinement can be well reduced if a sufficient number of high-angle diffractions are available.

1.3.2.3 Data Correction

Raw diffraction data collected on a single-crystal diffractometer should be carefully corrected for various errors before subsequent electronic structure refinement. The main sources of diffraction intensity errors include the intensity instability of the X-ray source, the non-monochromaticity and non-uniformity of the X-ray source, the spherical aberration of the goniometer, the X-ray absorption by the crystal sample, and the irregularities of the crystal shape. Therefore, a suitable error model is required to correct the raw diffraction data.

The X-ray absorption coefficients are generally large for crystals containing heavier atoms, and the diffraction data usually need to be corrected for absorption before they can be used for electronic structure refinement. When using a microfocus light source, the X-ray intensity will have a distorted Gaussian distribution in the radial direction of the light, causing far greater X-ray absorption near the center of the crystal than at the crystal edge when the X-rays pass through the crystal; thus, the distorted Gaussian distribution should be taken into account when correcting for absorption.

Multiple scattering is less likely to occur in single-crystal diffraction experiments; however, because the volume of diffraction data required for electronic structure refinement is generally large, multiple scattering can significantly alter the intensity of a few diffraction spots. However, it should be mentioned that this effect cannot be eliminated through scale factor correction and absorption correction. The intensity and error of diffraction data should satisfy the requirements for general electronic structure refinement, which were indicated by a number of statistical metrics such as consistency factors R_{int} and R_{sigma} around 0.03 in the full resolution range, χ^2 around 1, $I/sig(I)$~log10(I) curves in a sinusoidal pattern [132], and the normalized scale factors within ±0.3 around 1.

1.3.2.4 Examination of the Quality of Electronic Structure Refinement

Electron structure refinement includes electron density refinement, density matrix refinement, and wave function refinement. Compared to the conventional crystal structure model, the number of refinement parameters in the electronic structure model is generally higher; thus, some statistical indicators have to be examined after the least-squares refinement to determine whether the refinement results are reliable. The statistical metrics for the multipole model refinement should normally have the following features: symmetric DRK plots with their central parts passing through the origin, mirror-symmetric plots of fractal dimensions of residual electron density, and errors within ±0.05 for the curve of F_{obs}/F_{calc} versus the structure resolution.

References

1 Hohenberg, P. and Kohn, W. (1964). Inhomogeneous electron gas. *Phys. Rev. B* 36: 864–871.

2 Jiang, X.-M., Deng, S., Whangbo, M.-H., and Guo, G.-C. (2022). Material research from the viewpoint of functional motifs. *Natl. Sci. Rev.* 9: nwac017.

3 Debye, P. (1915). Zerstreuung von Röntgenstrahlen. *Ann. Phys.* 46: 809–823.

4 Waller, I. and Hartree, D.R. (1929). On the intensity of total scattering of X-rays. *Proc. Royal. Soc. Lond. Ser. A* 124: 119–142.

5 Weiss, R.J. and De Marco, J.J. (1958). X-ray determination of the number of 3d electrons in Cu, Ni, Co, Fe, and Cr. *Rev. Mod. Phys.* 30: 59–62.

6 Weiss, R.J. and Phillips, W.C. (1968). X-ray retermination of the electron momentum density in diamond, graphite, and carbon black. *Phys. Rev.* 176: 900–904.

7 Weiss, R.J. (1963). Spin density in cobalt. *Phys. Rev. Lett.* 11: 264–265.

8 Weiss, R.J. and AzároffL, V. (1967). X-ray determination of electron distributions. *Phys. Today* 20: 103.

9 Dawson, B. (1967). A general structure factor formalism for interpreting accurate X-ray and neutron diffraction data. *Proc. R. Soc. Lond. Ser. A Math. Phys. Sci.* 298: 255–263.

10 Stewart, R.F. (1969). Generalized X-ray scattering factors. *J. Chem. Phys.* 51: 4569–4577.

11 Stewart, R.F. (1973). Electron population analysis with generalized X-ray-scattering factors—Higher multipoles. *J. Chem. Phys.* 58: 1668–1676.

12 Stewart, R.F. (1976). Electron population analysis with rigid pseudoatoms. *Acta Cryst. A: Cryst. Phys. Diffr. Theor. Gen. Crystallogr.* 32: 565–574.

13 Hansen, N.K. and Coppens, P. (1978). Testing aspherical atom refinements on small-molecule data sets. *Acta Cryst. A* 34: 909–921.

14 Volkov, A., Abramov, Y.A., and Coppens, P. (2001). Critical examination of the radial functions in the Hansen-Coppens multipole model through topological analysis of primary and refined theoretical densities. *Acta Cryst. A* 57: 395–405.

15 Volkov, A., Abramov, Y.A., and Coppens, P. (2001). Density-optimized radial exponents for X-ray charge-density refinement from ab initio crystal calculations. *Acta Cryst. A* 57: 272–282.

16 Pichon-Pesme, V., Lecomte, C., and Lachekar, H. (1995). On building a data bank of transferable experimental electron density parameters applicable to polypeptides. *J. Phys. Chem.* 99: 6242–6250.

17 Domagała, S., Fournier, B., Liebschner, D. et al. (2012). An improved experimental databank of transferable multipolar atom models–ELMAM2. construction details and applications. *Acta Cryst. A* 68: 337–351.

18 Volkov, A., Li, X., Koritsanszky, T., and Coppens, P. (2004). Ab initio quality electrostatic atomic and molecular properties including intermolecular energies from a transferable theoretical pseudoatom databank. *J. Phys. Chem. A* 108: 4283–4300.

19 Dittrich, B., Hubschle, C.B., Luger, P., and Spackman, M. (2006). Introduction and validation of an invariom database for amino-acid, peptide and protein molecules. *Acta Cryst. A* 62: 1325–1335.

20 Jarzembska, K.N. and Dominiak, P.M. (2011). New version of the theoretical databank of transferable aspherical pseudoatoms, UBDB2011 – towards nucleic acid modelling. *Acta Cryst. A* 68: 139–147.

21 Gruza, B., Chodkiewicz, M.L., Krzeszczakowska, J., and Dominiak, P.M. (2020). Refinement of organic crystal structures with multipolar electron scattering factors. *Acta Cryst. A* 76: 92–109.

22 Fischer A., Tiana D., Scherer W. et al. (2011) Experimental and theoretical charge density studies at subatomic resolution. *J. Phys. Chem. A*, 115: 13061–13071.

23 Deutsch, M., Claiser, N., Pillet, S. et al. (2012). Experimental determination of spin-dependent electron density by joint refinement of X-ray and polarized neutron diffraction data. *Acta Cryst. A* 68: 675–686.

24 Deutsch, M., Gillon, B., Claiser, N. et al. (2014). First spin-resolved electron distributions in crystals from combined polarized neutron and X-ray diffraction experiments. *IUCrJ* 1: 194–199.

25 Voufack, A.B., Claiser, N., Lecomte, C. et al. (2017). When combined X-ray and polarized neutron diffraction data challenge high-level calculations: spin-resolved electron density of an organic radical. *Acta Cryst. B* 73: 544–549.

26 Wahlberg, N., Bindzus, N., Bjerg, L. et al. (2015). Powder X-ray diffraction electron density of cubic boron nitride. *J. Phys. Chem. C* 119: 6164–6173.

27 Casati, N., Genoni, A., Meyer, B. et al. (2017). Exploring charge density analysis in crystals at high pressure: data collection, data analysis and advanced modelling. *Acta Cryst. B* 73: 584–597.

28 Casati, N., Kleppe, A., Jephcoat, A.P., and Macchi, P. (2016). Putting pressure on aromaticity along with in situ experimental electron density of a molecular crystal. *Nat. Commun.* 7: 10901.

29 Gajda, R., Stachowicz, M., Makal, A. et al. (2020). Experimental charge density of grossular under pressure—A feasibility study. *IUCrJ* 7: 383–392.

30 Eikeland, E.Z., Borup, M., Thomsen, M.K. et al. (2020). Single-crystal high-pressure X-ray diffraction study of host structure compression in clathrates of dianin's compound. *Cryst. Growth Des.* 20: 4092–4099.

31 Hirshfeld, F.L. (1977). Space partitioning of the charge density. *Isr. J. Chem.* 16: 198–201.

32 Bultinck, P., Van Alseneoy, C., Ayers, P.W., and Carbó-Dorca, R.J. (2007). Critical analysis and extension of the Hirshfeld atoms in molecules. *Chem. Phys.* 26: 144111.

33 Spackman, M.A. and Jayatilaka, D. (2009). Hirshfeld surface analysis. *CrystEngComm* 11: 19–32.

34 Jayatilaka, D. and Dittrich, B. (2008). X-ray structure refinement using aspherical atomic density functions obtained from quantum-mechanical calculations. *Acta Cryst. A* 64: 383–393.

35 Coppens, P., Willoughby, T.V., and Csonka, L.N. (1971). Electron population analysis of accurate diffraction data. I. Formalisms and restrictions. *Acta Cryst. A* 27: 248–256.

36 Clinton, W.L., Nakhleh, J., and Wunderlich, F. (1969). Direct determination of pure-state density matrices. I. Some simple introductory calculations. *Phys. Rev.* 177: 1–6.

37 Clinton, W.L., Galli, A.J., and Massa, L.J. (1969). Direct determination of pure-state density matrices. Ii. Construction of constrained idempotent one-body densities. *Phys. Rev.* 177: 7–13.

38 Clinton, W.L., Henderson, G.A., and Prestia, J.V. (1969). Direct determination of pure-state density matrices. Iii. Purely theoretical densities via an electrostatic-virial theorem. *Phys. Rev.* 177: 13–18.

39 Clinton, W.L. and Lamers, G.B. (1969). Direct determination of pure-state density matrices. Iv. Investigation of another constraint and another application of the p equations. *Phys. Rev.* 177: 19–27.

40 Clinton, W.L., Galli, A.J., Henderson, G.A. et al. (1969). Direct determination of purestate density matrices. V. Constrained eigenvalue problems. *Phys. Rev.* 177: 27–33.

41 Clinton, W.L. and Massa, L.J. (1972). The cusp condition: constraint on the electron density matrix. *Int. J. Quantum Chem.* 6: 519–523.

42 Clinton, W.L. and Massa, L.J. (1972). Determination of the electron density matrix from X-ray diffraction data. *Phys. Rev. Lett.* 29: 1363–1366.

43 Clinton, W.L., Frishberg, C.A., Massa, L.J., and Oldfield, P.A. (2009). Methods for obtaining an electron-density matrix from X-ray diffraction data. *Int. J. Quantum Chem.* 7: 505–514.

44 McWeeney, R. (1960). Some recent advances in density matrix theory. *Rev. Mod. Phys.* 32: 335–369.

45 Frishberg, C. and Massa, L.J. (1981). Idempotent density matrices for correlated systems from X-ray diffraction structure factors. *Phys. Rev. B* 24: 7018–7024.

46 Pecora, L.M. (1986). Determination of the quantum density matrix from experiment: an application to positron annihilation. *Phys. Rev. B* 33: 5987–5993.

47 Massa, L., Goldberg, M., Frishberg, C. et al. (1985). Wave functions derived by quantum modeling of the electron density from coherent x-ray diffraction: beryllium metal. *Phys. Rev. Lett.* 55: 622–625.

48 Howard, S.T., Huke, J.P., Mallinson, P.R., and Frampton, C.S. (1994). Density matrix refinement for molecular crystals. *Phys. Rev. B* 49: 7124–7136.

49 Snyder, J.A. and Stevens, E.D. (1999). A wavefunction and energy of the azide ion in potassium azide obtained by a quantum-mechanically constrained fit to X-ray diffraction data. *Chem. Phys. Lett.* 313: 293–298.

50 Tanaka, K. (1988). X-ray analysis of wavefunctions by the least-squares method incorporating orthonormality. i. general formalism. *Acta Cryst. A* 44: 1002–1008.

51 Tanaka, K., Makita, R., Funahashi, S. et al. (2008). X-ray atomic orbital analysis. i. quantum-mechanical and crystallographic framework of the method. *Acta Cryst. A* 64: 437–449.

52 Tanaka, K. (2018). X-ray molecular orbital analysis. I. Quantum mechanical and crystallographic framework. *Acta Cryst. A* 74: 345–356.

53 Hibbs, D.E., Howard, S.T., Huke, J.P., and Waller, M.P. (2005). A new orbital-based model for the analysis of experimental molecular charge densities: an application to (z)-n-methyl-c-phenylnitrone. *Phys. Chem. Chem. Phys.* 7: 1772–1778.

54 Waller, M.P., Howard, S.T., Platts, J.A. et al. (2006). Novel properties from experimental charge densities: an application to the zwitterionic neurotransmitter taurine. *Chem. Eur. J.* 12: 7603–7614.

55 Cooper, M. (1985). Compton scattering and electron momentum determination. *Rep. Prog. Phys.* 48: 415–481.

56 Loupias, G. and Chomilier, J. (1986). Electron momentum density and compton profiles: an accurate check of overlap models. *Z. Phys. D Atoms Mol. Clusters* 2: 297–308.

57 Schmider, H., Smith, V.H., and Weyrich, W. (1992). Reconstruction of the one-particle density matrix from expectation values in position and momentum space. *J. Chem. Phys.* 96: 8986–8994.

58 Gillet, J.M., Cortona, P., and Becker, P.J. (2001). Joint refinement of a local wave-function model from Compton and Bragg scattering data. *Phys. Rev. B* 63: 235115.

59 Gillet, J.M. and Becker, P.J. (2004). Position and momentum densities. Complementarity at work: refining a quantum model from different data sets. *J. Phys. Chem. Solids* 65: 2017–2023.

60 Gillet, J.M. (2007). Determination of a one-electron reduced density matrix using a coupled pseudo-atom model and a set of complementary scattering data. *Acta Cryst. A* 63: 234–238.

61 De Bruyne, B. and Gillet, J.-M. (2020). Inferring the one-electron reduced density matrix of molecular crystals from experimental data sets through semidefinite programming. *Acta Cryst. A* 76: 1–6.

62 Kibalin, I.A., Yan, Z., Voufack, A.B. et al. (2017). Spin density in $YTiO_3$: I. Joint refinement of polarized neutron diffraction and magnetic x-ray diffraction data leading to insights into orbital ordering. *Phys. Rev. B* 96: 054426.

63 Yan, Z., Kibalin, I.A., Claiser, N. et al. (2017). Spin density in $YTiO_3$: II. Momentum-space representation of electron spin density supported by position-space results. *Phys. Rev. B* 96: 054427.

64 Gueddida, S., Yan, Z., and Gillet, J.-M. (2018). Development of a joint refinement model for the spin-resolved oneelectron reduced density matrix using different data sets. *Acta Cryst. A* 74: 131–142.

65 Gueddida, S., Yan, Z., Kibalin, I. et al. (2018). Joint refinement model for the spin resolved one-electron reduced density matrix of $YTiO_3$ using magnetic structure factors and magnetic Compton profiles data. *J. Chem. Phys.* 148: 164106.

66 Cassam-Chenaï, P. (1995). Ensemble representable densities for atoms and molecules. i. general theory. *Int. J. Quantum Chem.* 54: 201–210.

67 Cassam-Chenaï, P., Wolff, S., Chandler, G., and Figgis, B. (1996). Ensemble-representable densities for atoms and molecules. ii. application to $CoCl_4^{2-}$. *Int. J. Quantum Chem.* 60: 667–680.

68 Gilbert, T.L. (1975). Hohenberg-Kohn theorem for nonlocal external potentials. *Phys. Rev. B* 12: 2111–2120.

69 Coleman, A.J. (1963). Structure of fermion density matrices. *Rev. Mod. Phys.* 35: 668–686.

70 Henderson, G.A. and Zimmerman, R.K. (1976). One-electron properties as variational parameters. *J. Chem. Phys.* 65: 619–622.

71 Levy, M. and Goldstein, J.A. (1987). Electron density-functional theory and x-ray structure factors. *Phys. Rev. B* 35: 7887–7890.

72 Zhao, Q. and Parr, R.G. (1992). Quantities Ts[n] and Tc[n] in density functional theory. *Phys. Rev. A* 46: 2337–2343.

73 Zhao, Q. and Parr, R.G. (1993). Constrained-search method to determine electronic wave functions from electronic densities. *J. Chem. Phys.* 98: 543–548.

74 Zhao, Q., Morrison, R.C., and Parr, R.G. (1994). From electron densities to kohn-sham kinetic energies, orbital energies, exchange-correlation potentials, and exchange correlation energies. *Phys. Rev. A* 50: 2138–2142.

75 Levy, M. (1979). Universal variational functionals of electron densities, first-order density matrices, and natural spin-orbitals and solution of the v-representability problem. *Proc. Natl. Acad. Sci. U.S.A.* 76: 6062–6065.

76 Jayatilaka, D. (1998). Wave function for beryllium from X-ray diffraction data. *Phys. Rev. Lett.* 80: 798–801.

77 Jayatilaka, D. and Grimwood, D.J. (2001). Wavefunctions derived from experiment. i. motivation and theory. *Acta Cryst. A* 57: 76–86.

78 Grimwood, D.J. and Jayatilaka, D. (2001). Wavefunctions derived from experiment. ii. a wavefunction for oxalic acid dihydrate. *Acta Cryst. A* 57: 87–100.

79 Bytheway, I., Grimwood, D., and Jayatilaka, D. (2002). Wavefunctions derived from experiment. iii. topological analysis of crystal fragments. *Acta Cryst. A* 58: 232–243.

80 Bytheway, I., Grimwood, D.J., Figgis, B.N. et al. (2002). Wavefunctions derived from experiment. iv. investigation of the crystal environment of ammonia. *Acta Cryst. A* 58: 244–251.

81 Grimwood, D.J., Bytheway, I., and Jayatilaka, D. (2003). Wavefunctions derived from experiment. v. investigation of electron densities, electrostatic potentials, and electron localization functions for noncentrosymmetric crystals. *J. Comput. Chem.* 24: 470–483.

82 Hudák, M., Jayatilaka, D., Perašínová, L. et al. (2010). X-ray constrained unrestricted hartree-fock and douglas-kroll-hess wavefunctions. *Acta Cryst. A* 66: 78–92.

83 Jayatilaka, D. (2012). Using wavefunctions to get more information out of diffraction experiments. In: *Modern Charge-Density Analysis* (ed. C. Gatti and P. Macchi), 213–257. Berlin: Springer.

84 Woińska, M., Jayatilaka, D., Dittrich, B. et al. (2017). Validation of X-ray wavefunction refinement. *ChemPhysChem* 18: 3334–3351.

85 Whitten, A.E., Jayatilaka, D., and Spackman, M. (2006). Effective molecular polarizabilities and crystal refractive indices estimated from X-ray diffraction data. *J. Chem. Phys.* 125: 174505.

86 Hickstein, D.D., Cole, J.M., Turner, M.J., and Jayatilaka, D. (2013). Modeling electron density distributions from X-ray diffraction to derive optical properties: constrained wavefunction versus multipole refinement. *J. Chem. Phys.* 139: 064108.

87 Jayatilaka, D., Munshi, P., Turner, M.J. et al. (2009). Refractive indices for molecular crystals from the response of X-ray constrained Hartree–Fock wavefunctions. *Phys. Chem. Chem. Phys.* 11: 7209–7218.

88 Cole, J.M. and Hickstein, D.D. (2013). Molecular origins of nonlinear optical activity in zinc tris(thiourea)sulfate revealed by high-resolution x-ray diffraction data and ab initio calculations. *Phys. Rev. B* 88: 184105.

89 Guo, G.-C., Yao, Y.-G., Wu, K.-C. et al. (2001). Studies on the structure-sensitive functional materials. *Prog. Chem.* 13: 151–155.

90 Jiang, X.-M., Lin, S.-J., He, C. et al. (2021). Uncovering functional motif of nonlinear optical material by in situ electron density and wavefunction studies under laser irradiation. *Angew. Chem. Int. Ed.* 60: 11799–11803.

91 Ernst, M., Genoni, A., and Macchi, P. (2020). Analysis of crystal field effects and interactions using X-ray restrained ELMOs. *J. Mol. Struct.* 1209: 127975.

92 Genoni, A., Dos Santos, L.H.R., Meyer, B., and Macchi, P. (2017). Can X-ray constrained Hartree-Fock wavefunctions retrieve electron correlation? *IUCrJ* 4: 136–146.

93 Hudák, M., Jayatilaka, D., Perašínová, L. et al. (2009). X-ray constrained unrestricted Hartree–Fock and Douglas–Kroll–Hess wavefunctions. *Acta Cryst. A* 66: 78–92.

94 Bucinsky, L., Biskupič, S., and Jayatilaka, D. (2012). Study of the picture change error at the 2nd order Douglas Kroll Hess level of theory. Electron and spin density and structure factors of the Bis[bis(methoxycarbimido) aminato] copper (II) complex. *Chem. Phys.* 395: 44–53.

95 Bucinsky, L., Jayatilaka, D., and Grabowsky, S. (2016). Importance of relativistic effects and electron correlation in structure factors and electron density of diphenyl mercury and triphenyl bismuth. *J. Phys. Chem. A* 120: 6650–6669.

96 Bučinský, L., Jayatilaka, D., and Grabowsky, S. (2019). Relativistic quantum crystallography of diphenyl and dicyano mercury. theoretical structure factors and hirshfeld atom refinement. *Acta Cryst. A* 75: 705–717.

97 Jayatilaka, D. and Grimwood, D. (2004). Electron localization functions obtained from x-ray constrained hartree-fock wavefunctions for molecular crystals of ammonia, urea and alloxan. *Acta Cryst. A* 60: 111–119.

98 Grabowsky, S., Jayatilaka, D., Mebs, S., and Luger, P. (2010). The electron localizability indicator from x-ray diffraction data—a first application to a series of epoxide derivatives. *Chem. Eur. J.* 16: 12818–12821.

99 Grabowsky, S., Weber, M., Jayatilaka, D. et al. (2011). Reactivity differences between α, β-unsaturated carbonyls and hydrazones investigated by experimental and theoretical electron density and electron localizability analyses. *J. Phys. Chem. A* 115: 12715–12732.

100 Grabowsky, S., Luger, P., Buschmann, J. et al. (2012). The significance of ionic bonding in sulfur dioxide: bond orders from X-ray diffraction data. *Angew. Chem. Int.* 51: 6776–6779.

101 Stoll, H., Wagenblast, G., and Preu, H. (1980). On the use of local basis sets for localized molecular orbitals. *Theor. Chem. Accounts* 57: 169–178.

102 Fornili, A., Sironi, M., and Raimondi, M. (2003). Determination of extremely localized molecular orbitals and their application to quantum mechanics/molecular mechanics methods and to the study of intramolecular hydrogen bonding. *J. Mol. Struct. THEOCHEM* 632: 157–172.

103 Genoni, A. and Sironi, M. (2004). A novel approach to relax extremely localized molecular orbitals: the extremely localized molecular orbital valence bond method. *Theor. Chem. Acc.* 112: 254–262.

104 Genoni, A., Fornili, A., and Sironi, M. (2005). Optimal virtual orbitals to relax wavefunctions built up with transferred extremely localized molecular orbitals. *J. Comput. Chem.* 26: 827–835.

105 Genoni, A., Ghitti, M., Pieraccini, S., and Sironi, M. (2005). A novel extremely localized molecular orbitals based technique for the one-electron density matrix computation. *Chem. Phys. Lett.* 415: 256–260.

106 Sironi, M., Genoni, A., Civera, M. et al. (2007). Extremely localized molecular orbitals: theory and applications. *Theor. Chem. Acc.* 117: 685–698.

107 Sironi, M., Ghitti, M., Genoni, A. et al. (2009). DENPOL: a new program to determine electron densities of polypeptides using extremely localized molecular orbitals. *J. Mol. Struct. THEOCHEM* 898: 8–16.

108 Genoni, A. (2013). Molecular orbitals strictly localized on small molecular fragments from X-ray diffraction data. *J. Phys. Chem. Lett.* 4: 1093–1099.

109 Genoni, A. (2013). X-ray constrained extremely localized molecular orbitals: theory and critical assessment of the new technique. *J. Chem. Theory Comput.* 9: 3004–3019.

110 Dos Santos, L.H.R., Genoni, A., and Macchi, P. (2014). Unconstrained and X-ray constrained extremely localized molecular orbitals: analysis of the reconstructed electron density. *Acta Cryst. A* 70: 532–551.

111 Genoni, A. and Meyer, B. (2016). X-Ray constrained wave functions: fundamentals and effects of the molecular orbitals localization. *Adv. Quantum Chem.* 73: 333–362.

112 Meyer, B., Guillot, B., Ruiz-Lopez, M.F., and Genoni, A. (2016). Libraries of extremely localized molecular orbitals. 1. model molecules approximation and molecular orbitals transferability. *J. Chem. Theory Comput.* 12: 1052–1067.

113 Meyer, B., Guillot, B., Ruiz-Lopez, M.F. et al. (2016). Libraries of extremely localized molecular orbitals. 2. comparison with the pseudoatoms transferability. *J. Chem. Theory Comput.* 12: 1068–1081.

114 Meyer, B. and Genoni, A. (2018). Libraries of extremely localized molecular orbitals. 3. construction and preliminary assessment of the new databanks. *J. Phys. Chem. A* 122: 8965–8981.

115 Macetti, G. and Genoni, A. (2019). Quantum mechanics/extremely localized molecular orbital method: a fully quantum mechanical embedding approach for macromolecules. *J. Phys. Chem. A* 123: 9420–9428.

116 Macetti, G., Wieduwilt, E.K., Assfeld, X., and Genoni, A. (2020). Localized molecular orbital-based embedding scheme for correlated methods. *J. Chem. Theory Comput.* 16: 3578–3596.

117 Arias-Olivares, D., Wieduwilt, E.K., Contreras-Garcia, J., and Genoni, A. (2019). NCI-ELMO: a new method to quickly and accurately detect noncovalent interactions in biosystems. *J. Chem. Theory Comput.* 15: 6456–6470.

118 Malaspina, L.A., Wieduwilt, E.K., Bergmann, J. et al. (2019). Fast and accurate quantum crystallography: from small to large, from light to heavy. *J. Phys. Chem. Lett.* 10: 6973–6982.

119 Genoni, A. (2017). A first-prototype multi-determinant X-ray constrained wavefunction approach: The X-ray constrained extremely localized molecular orbital–valence bond method. *Acta Cryst. A* 73: 312–316.

120 Genoni, A., Franchini, D., Pieraccini, S., and Sironi, M. (2018). X-ray constrained spin-coupled wavefunction: a new tool to extract chemical information from X-ray diffraction data. *Chem. A Eur. J.* 24: 15507–15511.

121 Genoni, A., Macetti, G., Franchini, D. et al. (2019). X-ray constrained spin-coupled technique: theoretical details and further assessment of the method. *Acta Cryst. A* 75: 778–797.

122 Cooper, D.L., Gerratt, J., and Raimondi, M. (1991). Applications of spin-coupled valence bond theory. *Chem. Rev.* 91: 929–964.

123 Cooper, D.L., Gerratt, J., and Raimondi, M. (1986). The electronic structure of the benzene molecule. *Nature* 323: 699–701.

124 Cooper, D.L., Gerrat, J., Raimondi, M. et al. (1993). Expansion of the spin-coupled wavefunction in Slater determinants. *Theor. Chim. Acta* 85: 261–270.

125 Spence, J.C.H. (1993). On the accurate measurement of structure-factor amplitudes and phases by electron diffraction. *Acta Crystallogr., Sect. A: Found. Crystallogr.* 49: 231–260.

126 Tanaka, M. (1994). Convergent-beam electron diffraction. *Acta Cryst. A* 50: 261–286.

127 Nakashima, P.N.H., Smith, A.E., Etheridge, J., and Muddle, B.C. (2011). The bonding electron density in aluminum. *Science* 331: 1583–1586.

128 Zuo, J.M., Kim, M., O'Keeffe, M., and Spence, J.C.H. (1999). Direct observation of d-orbital holes and Cu–Cu bonding in Cu_2O. *Nature* 401: 49–52.

129 Palatinus, L., Brázda, P., Boullay, P. et al. (2017). Hydrogen positions in single nanocrystals revealed by electron diffraction. *Science* 355: 166–169.

130 Zheng, J.-C., Zhu, Y., Wu, L., and Davenport, J.W. (2005). On the sensitivity of electron and X-ray scattering factors to valence charge distributions. *J. Appl. Cryst.* 38: 648–656.

131 Mallinson, P.R., Koritsanszky, T., Elkaim, E. et al. (1988). The Gram-Charlier and multipole expansions in accurate X-ray diffraction studies: can they be distinguished? *Acta Cryst. A* 44: 336–343.

132 Diederichs, K. (2010). Quantifying instrument errors in macromolecular X-ray data sets. *Acta Cryst. D* 66: 733–740.

2

First-Principles Calculations of the Electron Density Functions

2.1 Introduction

The methods for first-principle calculations of crystalline electronic structures (a theory that does not depend on empirical parameters) have been well developed. As one of the fundamental equations that describe the nonrelativistic motion of microscopic particles, the Schrödinger equation can be applied to material systems to obtain the Schrödinger equation for multi-particle systems containing electrons and nuclei. Because the material properties in question have little to do with nuclei, the multi-electron Schrödinger equation can be obtained by separating the electrons and nuclei under the Born-Oppenheimer approximation. In addition, because the multi-electron Schrödinger equation is complex and difficult to solve mathematically, it can be simplified to a one-electron Schrödinger equation (Hartree equation) by introducing the assumption of no interactions between the electrons. The system wave function in the Hartree equation is the product of the wave functions of each electron. Because electrons are fermions, their wave functions need to satisfy the exchange antisymmetry condition, which is a necessary requirement of the Pauli exclusion principle. The famous Hartree–Fock (HF) equation can be obtained by writing the wave function as a single Slater determinant and introducing the exchange antisymmetry condition, where the HF equation has one more exchange interaction term than the Hartree equation. The Eigen energy of the HF equation signifies one-electron energy (Koopmans' theorem), and the HF equation has difficulty achieving symmetry due to a contradiction between the antisymmetry requirement and the minimum energy principle. Thus, the HF equation only considers part of the electron correlation effect. To consider a complete electron correlation effect, many post-HF methods have been developed, such as the configuration interaction, many-body perturbation, and coupled cluster. The system wave function can be rigorously calculated by the aforementioned quantum chemical ab initio calculations; however, the corresponding computational effort increases dramatically with the size of the system.

The HF equation occupies an important place in electronic structure crystallography for two main reasons: (i) the HF equation is simpler in form and requires less computation demand than post-HF methods, facilitating the computation of large

Electronic Structure Crystallography and Functional Motifs of Materials, First Edition.
Guo-Cong Guo and Xiao-Ming Jiang.

systems; and (ii) although the HF method is inferior to the post-HF methods in terms of computational accuracy, combined with correction performed using data from scattering experiments, HF method may obtain results that are closer to the experimental results, reflecting the basic idea of electronic structure crystallography.

According to Hohenberg-Kohn's density functional theory (DFT), all properties of the system are uniquely determined by electron density. Based on this theory, the Kohn–Sham (KS) equation can be obtained by variation of the energy function with respect to the density function. The KS equation can rigorously describe the electron density of the system, where the full complexity of multi-particle interaction is contained in the exchange-correlation energy, and the eigen energy of the KS equation does not have the meaning of one-electron energy, which differs from that of the HF equation.

Because electron interactions occur in pairs, the Hamiltonian of the system contains only one-electron (or one-body) and two-electron (or double-body) operators, with no other many-body effects. Thus, all one- and two-electron properties can be calculated from the first- and second-order (reduced) density matrices, where one-electron properties can be uniquely determined by the electron density. Electron density is an experimentally observable physical quantity and a main object in studying electronic structure crystallography. This chapter aims to establish a bridge between crystallography and quantum chemistry, focusing on density matrices and methods for first-principles calculations of related functions [1], such as density functions of electron density, spin density, and momentum density, which can be determined by crystallographic scattering experiments. A comprehensive introduction to the methods of first-principles calculations of electronic structures can be found elsewhere [2].

2.2 Basic Framework and Assumptions of the First-Principles Calculations

In this section, we introduce the crystal density matrix and ab initio calculation methods for related functions such as the functions of electron, spin, and momentum densities. A variety of ab initio calculation methods exist, which can be divided into four levels depending on the problem under study and the utilized assumption approximation. The first level describes all aspects of the system in the most rigorous manner, including the consideration of relativistic effects, spin-orbit coupling, external field effects, nuclear motion, and thermal effects. If these aspects are not considered, this leads to the second level, with a nonrelativistic description of the electronic ground state in an equilibrium configuration for fixed nuclei. Further ignoring the effect of electron motion correlation leads to the third level, where the electronic ground state can be described by using single determinants or KS wave functions. On this basis, if more approximations, such as the choice of KS exchange-correlation potentials and representative basis sets, are used, then the fourth level will be reached, where approximate single determinants or KS wave functions are employed to describe the electronic ground state.

Our main goal is to deal with the electronic ground state problem at the second level, to obtain the exact ground state density matrix and density function by solving the following Schrödinger equation in the absence of an external field:

$$\hat{H}_{\text{el}}\Psi_0 = E_0\Psi_0. \tag{2.1}$$

Because the crystal of interest is a system containing an infinite number of atoms, we had to use the mean-field approximation to solve the problem at the third level. Of course, to obtain a reasonable single-determinant solution, it is often necessary to use certain computational approximations; thus, we actually solved the problem at the fourth level.

Assuming an electron system with M nuclei, charge Z_A, and coordinates $\boldsymbol{R}_A$, the nonrelativistic electrostatic Hamiltonian under the electrostatic field of these nuclei could be written as

$$\hat{H}_{\text{el}} = \sum_{n=1}^{N}\frac{-\nabla_n^2}{2} + \sum_{n=1}^{N}\sum_{A=1}^{M}\frac{-Z_A}{r_{nA}} + \frac{1}{2}\sum_{n,m=1}^{N}{}'\frac{1}{r_{nm}} + \frac{1}{2}\sum_{A,B=1}^{M}{}'\frac{Z_AZ_B}{r_{AB}}. \tag{2.2}$$

Atomic units were used here, with summation notation with a prime $'$ indicating that the diagonal terms of $n = m$, $A = B$ were removed in the summation. $\hat{H}_{\text{el}}$ was related to the nuclear coordinates $\{\boldsymbol{R}\}$, the atomic number $\{Z\}$, and charge Z_A, and so were E_0 and ψ_0. In addition, ψ_0 denoted the antisymmetric wave function of the N-electron space-spin coordinates $\boldsymbol{x}_n = \boldsymbol{r}_n, \sigma_n$ $(n = 1, \cdots, N)$. In the nuclear coordinates $\{\boldsymbol{R}^{\text{eq}}\}$ at equilibrium, it had a minimal energy of $E_0[\{\boldsymbol{R}\}, \{Z\}]$, and the corresponding eigenfunction could be written as:

$$\psi_0^{\text{eq}}(\ldots, \boldsymbol{x}_n, \ldots) \equiv \psi_0(\ldots, \boldsymbol{x}_n, \ldots; [\{\boldsymbol{R}^{\text{eq}}\}, \{Z\}]). \tag{2.3}$$

We only dealt with charge-neutral periodic systems and considered a lattice vector $\boldsymbol{T}_{\text{m}}$, which was formed by a linear combination of individual primitive cell basis vectors a_i in the original space:

$$\boldsymbol{T}_{\boldsymbol{m}} = \sum_{i=1}^{D} m_i\boldsymbol{a}_i, \tag{2.4}$$

where m is an integer, D is the periodic dimension (e.g. $D = 3$ for a three-dimensional or 3D crystal), and $\boldsymbol{a}_i$ defines the unit cell of the crystal. It was assumed that the coordinates and charges of all nuclei in the crystal could be obtained by translation of an irreducible finite set $\{\boldsymbol{R}_{A,0}; Z_{A,0}\}$, namely,

$$\begin{aligned} \boldsymbol{R}_{A,\boldsymbol{m}} &= \boldsymbol{R}_{A,0} + \boldsymbol{T}_{\boldsymbol{m}} \\ Z_{A,\text{m}} &= Z_{A,0}. \end{aligned} \tag{2.5}$$

To solve the Schrödinger Eq. (2.1), the Born-von Karman boundary condition had to be introduced, where ψ_0 was assumed to be a periodic function of the superlattice vector $\overline{\boldsymbol{W}}_{\boldsymbol{m}}$ defined by Eq. (2.4) (i.e. the system state could be restored after a cycle of $\overline{\boldsymbol{W}}_{\boldsymbol{m}}$). However, a D-dimensional superlattice basis vector $\overline{\boldsymbol{A}}_i = w_i\boldsymbol{a}_i$ was used, where if $\boldsymbol{x}_n = (\boldsymbol{r}_n, \sigma_n), \boldsymbol{x}_n' = (\boldsymbol{r}_n + \overline{\boldsymbol{W}}_{\boldsymbol{m}}, \sigma_n)$, then $\Psi_0(\ldots, \mathbf{x}_n, \ldots) = \Psi_0\left(\ldots, \mathbf{x}_n', \ldots\right)$ for any n, m. The integer w_i defined the effective number of electrons in the system, i.e.

$N_{\mathrm{eff}} = WN_0$ where $W = \prod_i w_i$ and $N_0 = \sum_A Z_{A,0}$, with N_0 number of electrons per unit cell. If w_i is large enough, the energy-related physical quantities will be almost unaffected; however, we must be aware of the influence of the Born-Kaman condition on the density matrix and function.

2.3 Density Matrix and Density Function

2.3.1 Basic Definition

The generalized density function or the exact position-spin density matrix of a single particle can be defined as:

$$\gamma(\boldsymbol{x};\boldsymbol{x}') = N \int \Psi_0(\boldsymbol{x},\boldsymbol{x}_2,\ldots,\boldsymbol{x}_N) \times (\Psi_0(\boldsymbol{x}',\boldsymbol{x}_2,\ldots,\boldsymbol{x}_N))^* \mathrm{d}\boldsymbol{x}_2 \ldots \mathrm{d}\boldsymbol{x}_N. \tag{2.6}$$

In principle, it is also possible to define higher-order density matrices; however, they are not of interest and, hence, not discussed here. Furthermore, $\gamma(\boldsymbol{x};\boldsymbol{x}')$ has a wealth of information as it contains six spatial coordinates and two spin coordinates. Based on this density matrix, the ground state expectation of any one-electron operator $\hat{F} = \sum_n \hat{f}(\boldsymbol{x}_n)$ can be calculated as:

$$\langle \hat{F} \rangle_0 = \int \{\hat{f}(\boldsymbol{x})\gamma(\boldsymbol{x};\boldsymbol{x}')\}_{(\boldsymbol{x}'=\boldsymbol{x})} \mathrm{d}\boldsymbol{x}. \tag{2.7}$$

By separating the spatial coordinates $\boldsymbol{r}$ and spin σ coordinates of $\boldsymbol{x}$ and considering the dual coordinates of the component with spin in the z direction, i.e. $(\sigma, \sigma' = \alpha, \beta)$, the density matrix of the spin projection (or referred to as the spin density matrix), $\boldsymbol{P}^\sigma$ can be defined by:

$$\begin{aligned} \boldsymbol{P}^\sigma(\boldsymbol{r};\boldsymbol{r}') &= \gamma(\boldsymbol{r},\sigma;\boldsymbol{r}',\sigma) \\ \boldsymbol{P}(\boldsymbol{r};\boldsymbol{r}') &= \boldsymbol{P}^\alpha(\boldsymbol{r};\boldsymbol{r}') + \boldsymbol{P}^\beta(\boldsymbol{r};\boldsymbol{r}') \\ \boldsymbol{Q}(\boldsymbol{r};\boldsymbol{r}') &= \mathbf{P}^\alpha(\boldsymbol{r};\boldsymbol{r}') - \boldsymbol{P}^\beta(\boldsymbol{r};\boldsymbol{r}'), \end{aligned} \tag{2.8}$$

where $\boldsymbol{P}(\boldsymbol{r};\boldsymbol{r}')$ is known as the position density matrix and can be obtained by integrating $\gamma(\boldsymbol{x};\boldsymbol{x}')$ over the two spinors, and $\boldsymbol{Q}(\boldsymbol{r};\boldsymbol{r}')$ is referred to as the net spin density matrix (the part of α exceeding β). The momentum density matrix is a 6D Fourier transform of $\boldsymbol{P}(\boldsymbol{r};\boldsymbol{r}')$, according to:

$$\overline{\boldsymbol{P}}(\boldsymbol{p};\boldsymbol{p}') = \int \boldsymbol{P}(\boldsymbol{r};\boldsymbol{r}')\exp(-i\boldsymbol{r}\cdot\boldsymbol{p})\exp(i\boldsymbol{r}'\cdot\boldsymbol{p}')\mathrm{d}\boldsymbol{r}\mathrm{d}\boldsymbol{r}'. \tag{2.9}$$

Density functions are the diagonal elements of the density matrix:

$$\begin{aligned} \rho(\boldsymbol{r}) &= \boldsymbol{P}(\boldsymbol{r};\boldsymbol{r}) \\ \rho^\sigma(\boldsymbol{r}) &= \boldsymbol{P}^\sigma(\boldsymbol{r};\boldsymbol{r}) \\ \zeta(\boldsymbol{r}) &= \boldsymbol{Q}(\boldsymbol{r};\boldsymbol{r}) = \rho^\alpha(\boldsymbol{r}) - \rho^\beta(\boldsymbol{r}) \\ \pi(\boldsymbol{p}) &= \overline{\boldsymbol{P}}(\boldsymbol{p};\boldsymbol{p}), \end{aligned} \tag{2.10}$$

where $\rho(\boldsymbol{r})$, $\rho^{\sigma}(\boldsymbol{r})$, $\zeta(\boldsymbol{r})$, $\pi(\boldsymbol{p})$ are the electron density, spin density, net spin density, and electron momentum density, respectively. In quantum mechanics, these quantities are objectively observable because they are all ground state expectations of the one-electron operator $\hat{F}^{y} = \sum_{n}\delta(y - y_n)$.

In a periodic system, when changing all $\boldsymbol{r}_n$ by a $\boldsymbol{T}_m$, the corresponding ground state wave function must change by only one phase factor. Thus, according to Eq. (2.6), we can obtain

$$\begin{aligned}\gamma(\boldsymbol{r},\sigma;\boldsymbol{r}',\sigma') &= \gamma(\boldsymbol{r}+\boldsymbol{T}_m,\sigma;\boldsymbol{r}'+\boldsymbol{T}_m,\sigma')\\ \boldsymbol{P}(\boldsymbol{r};\boldsymbol{r}') &= \boldsymbol{P}(\boldsymbol{r}+\boldsymbol{T}_m;\boldsymbol{r}'+\boldsymbol{T}_m).\end{aligned} \tag{2.11}$$

All information regarding the density matrix can be restricted to a unit cell for consideration. In the Fourier transform Eq. (2.9), the integral over $\boldsymbol{r}$ can be confined to a unit cell. According to this convention, $\pi(\boldsymbol{p})$ will be independent of the crystal size and satisfy $\int \pi(\boldsymbol{p})\mathrm{d}\boldsymbol{p} = N_0$, where N_0 is the number of electrons per unit cell. The Born-Kaman condition will also lead to more general properties of the density matrix:

$$\begin{aligned}\gamma(\boldsymbol{r},\sigma;\boldsymbol{r}',\sigma') &= \gamma(\boldsymbol{r},\sigma;\boldsymbol{r}'+\overline{\boldsymbol{W}}_m,\sigma')\\ \boldsymbol{P}(\boldsymbol{r};\boldsymbol{r}') &= \boldsymbol{P}(\boldsymbol{r};\boldsymbol{r}'+\overline{\boldsymbol{W}}_m)\end{aligned} \tag{2.12}$$

2.3.2 Electron Density

Electron density $\rho(\boldsymbol{r})$ is a very important physical quantity that characterizes the properties of a multi-electron system. As a simple function of spatial coordinates, it is real-valued, nonnegative, and has a finite value anywhere except at the nuclear position; thus, it can faithfully describe the chemical composition and geometry of the system. In fact, the nuclear position $\boldsymbol{R}_A$ and atomic number Z_A can also be obtained from Kato's pole condition for electron density, according to $Z_A = -\overline{\rho'(\boldsymbol{R}_A)}/(2\rho(\boldsymbol{R}_A))$, where $\overline{\rho'(\boldsymbol{R}_A)}$ is the average slope of the electron density around the pole. Therefore, electron density contains all Hamiltonian information in the Schrödinger equation for the system and is the basis of DFT and the KS equation. Bader's quantum theory of atoms in molecules (QTAIM) provides a good tool for topological analysis of the electron density, using the rich chemical properties of the system that can be extracted based on physical quantities such as electron density and density gradient.

The nondegenerate ground state wave function and corresponding electron density have the exact symmetry as the system Hamiltonian. The electron density of a crystal will be invariant under all symmetry operations of the space group, namely,

$$\{V/T_m + s_V\}\rho(\boldsymbol{r}) \equiv \rho(V^{-1}\boldsymbol{r} - T_m - s_V) = \rho(\boldsymbol{r}). \tag{2.13}$$

The Seitz notation can be used here to denote the symmetry operation, where V is the matrix representation of proper or improper rotation and $\boldsymbol{s}_V$ is the corresponding fractional translation, with $\boldsymbol{s}_V = 0$ for the conformal space group. All rotations will form a point group of the order h. According to (2.13), all information regarding electron density can be contained in the $1/h$ irreducible wedge-shaped part of the unit cell, and the integral of the electron density over this part will be equal to N_0/h.

The 3D Fourier transform of the electron density will yield the structure factor as follows:

$$F(\kappa) = \int \rho(\boldsymbol{r}) \exp(i\kappa \cdot \boldsymbol{r}) \mathrm{d}\boldsymbol{r}. \tag{2.14}$$

The reciprocal space basis vector $\boldsymbol{B}_j$ of the same dimension can be defined from the D-dimensional direct space basis vector $\boldsymbol{a}_i$, satisfying $\boldsymbol{a}_i \cdot \boldsymbol{B}_j = 2\pi\delta_{ij}(i, j = 1, \dots, D)$, and any point in the reciprocal space can be expressed as the sum of a periodic part $\kappa^{\parallel} = \sum_{i=1}^{D} \kappa_i \boldsymbol{B}_i$ and a nonperiodic part $\kappa^{\perp}$. Due to the symmetry of electron density, the structure factor will have the following properties.

1. Due to the translational invariance of electron density, $F(k)$ will be nonzero only on the reciprocal lattice vector $\boldsymbol{G}_{hkl} \equiv h\boldsymbol{B}_1 + k\boldsymbol{B}_2 + l\boldsymbol{B}_3$ (h, k, l are integers); thus, the structure factor can be defined as a set of infinite discrete data:

$$F_{hkl} = \int_{\text{unit cell}} \rho(\boldsymbol{r}) \exp(i\boldsymbol{G}_{hkl} \cdot \boldsymbol{r}) \mathrm{d}\boldsymbol{r}. \tag{2.15}$$

   According to the above equation, when h, k, $l = 0$, then $F_{000} = N_0$, where N_0 is the number of electrons in the conventional unit cell and typically used for the normalization of the structure factor.
2. Due to the rotational invariance of the electron density, $F(\boldsymbol{G}) = \exp(-i\boldsymbol{G} \cdot \boldsymbol{s}_V)F(V\boldsymbol{G})$, where the rotational operation of the structure factor adds a phase factor.
3. Given the symmetry of a space group, certain structure factors will be equal to zero, known as the so-called systematic absence.

2.3.3 Momentum Density

Electron momentum density $\pi(\boldsymbol{p})$ provides important additional information for studying the properties of electron density, where the $\pi(\boldsymbol{p})$ near the origin will be mainly contributed by slow valence electrons, and $\pi(\boldsymbol{p})$ at locations with large $|\boldsymbol{p}|$ will mainly reflect the properties of core electrons. The properties of the momentum density include being real-valued, positive, normalized to the number of electrons in the unit cell N_0, and rotationally invariant, as $\pi(\boldsymbol{p}) = \pi(V\boldsymbol{p})$ holds for all operations of the crystal point group.

The expected value of the total kinetic energy of ground state electrons in the single cell of the system can be obtained directly from the electron momentum density, i.e.

$$\langle \hat{T} \rangle_0 = \frac{1}{2} \int \pi(\boldsymbol{p}) p^2 \mathrm{d}\boldsymbol{p}. \tag{2.16}$$

According to the virial theorem, this will be applicable to electrostatic Hamiltonian in Eq. (2.2), and the total energy of the unit cell in the equilibrium configuration can be calculated from the momentum density:

$$\frac{1}{2} \int \pi(\boldsymbol{p})_{(\{\boldsymbol{R}\}=\{\boldsymbol{R}^{\text{eq}}\})} p^2 \mathrm{d}\boldsymbol{p} = -E_0^{\text{eq}}. \tag{2.17}$$

Equation (2.17) can be derived from the result of energy minimization for all coordinates and will not be applicable to the case of configuration optimization with constant volume. Similar to the structure factor obtained from the electron density by Fourier transform, the reciprocal form factor (RFF) $B(\boldsymbol{r})$ and Compton profile function (CPF) $J(\boldsymbol{q})$ can be defined from the momentum density:

$$\begin{aligned} B(\boldsymbol{r}) &= \int \pi(\boldsymbol{p}) \exp(-i\boldsymbol{p}\cdot\boldsymbol{r})\mathrm{d}\boldsymbol{p} = \frac{1}{W}\int P(\boldsymbol{r}';\boldsymbol{r}+\boldsymbol{r}')\mathrm{d}\boldsymbol{r}' \\ J(\boldsymbol{q}) &= \int \pi(\boldsymbol{p})\delta\left(\frac{\boldsymbol{p}\cdot\boldsymbol{q}}{|\boldsymbol{q}|} - |\boldsymbol{q}|\right)\mathrm{d}\boldsymbol{p}. \end{aligned} \tag{2.18}$$

The normalization factor W is the number of unit cells and $B(0) = N_0$, where CPF is the 2D integral of the momentum density $\pi(\boldsymbol{p})$ in a plane perpendicular to the $\boldsymbol{q}$ direction. This was of interest mainly because this quantity is correlated with the experimental Compton profile. Let us consider the directional Compton scattering profile:

$$\begin{aligned} J_{hkl}(q) &= J(q\boldsymbol{e}_{hkl}) \\ \boldsymbol{e}_{hkl} &= \frac{h\boldsymbol{a}_1 + k\boldsymbol{a}_2 + l\boldsymbol{a}_3}{|h\boldsymbol{a}_1 + k\boldsymbol{a}_2 + l\boldsymbol{a}_3|} \end{aligned} \tag{2.19}$$

Under the impulse approximation, this function will be proportional to the momentum loss distribution of the scattered photons along the $\boldsymbol{e}_{hkl}$ direction in the Compton scattering experiment. The directional RFF $[B_{hkl}(r) = B(r\boldsymbol{e}_{hkl})]$ and corresponding directional Compton profile can be related by 1D Fourier transform, namely:

$$B_{hkl}(r) = \int J_{hkl}(q) \exp(-irq)\mathrm{d}q. \tag{2.20}$$

Due to resolution limitations, the experimental directional Compton profile can be considered as the convolution of the ideal Compton profile with the experimental resolution function ($w(q)$), which is usually a Gaussian function:

$$J_{hkl}^{\exp}(q) \approx \int J_{hkl}(q')w(q-q')\mathrm{d}q'. \tag{2.21}$$

Therefore, according to the convolution theorem, where the Fourier transform of the convolution of two functions will be equal to the product of the individual Fourier transforms of the two functions, $B_{hkl}(r)$ is the Fourier transform of $J_{hkl}^{\exp}(q)$ divided by $\overline{w}(r)$, which is the Fourier transform of $w(q)$. In this way, the RFF can be obtained by Compton scattering experiments. Usually, the fine structure of the RFF will not be significantly affected by experimental errors, and the zero point can be measured with high accuracy.

Under the Born-von Karman condition, the calculated RFF will satisfy the translational periodicity of the cyclic superlattice, i.e.

$$B(\boldsymbol{r}) = B(\boldsymbol{r} + \overline{\boldsymbol{W}}_{\boldsymbol{m}}). \tag{2.22}$$

The RFF contains important electronic structure information, such as the oscillatory properties in direct space, the coordinates of the nodal planes, and the maxima and minima, which will be closely related to the chemical properties of the system.

2.4 Hartree–Fock (HF) and Kohn–Sham (KS) Methods

2.4.1 Basic Theoretical Framework

The Schrödinger Eq. (2.1) can be often solved using HF and KS schemes, both of which can describe periodic systems and obtain information about density matrices and density functions. HF and KS have many similar properties. In the case of spin-unrestricted HF and KS (UHF and UKS), both can obtain N one-electron functions, namely, the molecular spin orbitals (MSO) of the nonperiodic system and crystalline spin orbitals (CSO) of the periodic system with $\psi_j^{\mathrm{X}}(\mathbf{x}) = \phi_j^{\mathrm{X},\sigma}(\boldsymbol{r})\omega(\sigma)$, where σ is either α or β spin, satisfying the following equation:

$$\begin{aligned}\hat{h}^{\mathrm{X},\sigma}\phi_j^{\mathrm{X},\sigma}(\boldsymbol{r}) &= \left[-\frac{\nabla^2}{2} + \sum_A \frac{-Z_A}{|\boldsymbol{R}_A - \boldsymbol{r}|} + \int \frac{\rho^{\mathrm{X}}(\boldsymbol{r}')}{|\boldsymbol{r} - \boldsymbol{r}'|}\mathrm{d}\boldsymbol{r}' + \widetilde{V}^{\mathrm{X},\sigma}\right]\phi_j^{\mathrm{X},\sigma}(\boldsymbol{r}) \\ &= \varepsilon_j^{\mathrm{X},\sigma}\phi_j^{\mathrm{X},\sigma}(\boldsymbol{r}),\end{aligned} \tag{2.23}$$

where the effective Hamiltonian $\hat{h}^{\mathrm{X},\sigma}$ acts on individual MSOs and contains a kinetic energy term, a nuclear attraction term, a Hartree operator, which describes the Coulomb repulsion between electrons, and a corrective potential operator $\widetilde{V}^{\mathrm{X},\sigma}$. The corrective potential operators for HF and KS are different.

The single Slater determinant N-electron wave function can be defined by assigning N electrons on the MSOs, which respectively correspond to the lowest energy eigenvalues $\varepsilon_j^{\mathrm{X},\sigma}$ of the Hamiltonian in (2.23), generating the form of an antisymmetric product:

$$\Psi_0^{\mathrm{X}} = N^{-1/2}\sum_P (-1)^{sp}\hat{P}\left[\psi_1^{\mathrm{X}}(\mathbf{x}_1)\times\cdots\times\psi_N^{\mathrm{X}}(\mathbf{x}_N)\right] \equiv |\cdots j\cdots>, \tag{2.24}$$

where $\hat{P}$ is the permutation operator of order N acting on the electron coordinates, and sp is the corresponding parity.

In the following, we consider a spinless system, in which all occupied MSOs occur in pairs, with the same eigenvalue and same spatial part ϕ_n^{X}, and only differing in α and β spins, labeled as n^α and n^β ($n = 1, \cdots, N/2$), respectively. In this case, the spin labels in the Hamiltonian could be removed. According to the definitions in Eqs. (2.6) and (2.8), the position density matrix and the corresponding electron density could be simplified as:

$$\begin{aligned}P^{\mathrm{X}}(\boldsymbol{r};\boldsymbol{r}') &= 2\sum_{n=1}^{N/2}\phi_n^{\mathrm{X}}(\boldsymbol{r})\left(\phi_n^{\mathrm{X}}(\boldsymbol{r}')\right)^* \\ \rho^{\mathrm{X}}(\boldsymbol{r}) = P^{\mathrm{X}}(\boldsymbol{r};\boldsymbol{r}) &= 2\sum_{n=1}^{N/2}\left|\phi_n^{\mathrm{X}}(\boldsymbol{r})\right|^2.\end{aligned} \tag{2.25}$$

In the framework of HF theory, the exchange operator $\hat{V}_{\mathrm{exch}}$ acting on the function $\chi(\boldsymbol{r})$ can be used for $\widetilde{V}^{\mathrm{HF}}$ and defined as:

$$\hat{V}_{\mathrm{exch}}\chi(\boldsymbol{r}) = -\frac{1}{2}\int\frac{P^{\mathrm{HF}}(\boldsymbol{r};\boldsymbol{r}')\chi(\boldsymbol{r}')}{|\boldsymbol{r}-\boldsymbol{r}'|}\mathrm{d}\boldsymbol{r}'. \tag{2.26}$$

The corresponding ground-state energy E_0^{HF} can be written as:

$$
\begin{aligned}
E_0^{\mathrm{HF}} &\equiv \left\langle \Psi_0^{\mathrm{HF}} \middle| \hat{H}_{\mathrm{el}} \middle| \Psi_0^{\mathrm{HF}} \right\rangle \\
&= -\int \left[\frac{\nabla^2}{2} P^{\mathrm{HF}}(\boldsymbol{r};\boldsymbol{r}') \right]_{(\boldsymbol{r}'=\boldsymbol{r})} \mathrm{d}\boldsymbol{r} - \sum_A Z_A \int \frac{\rho^{\mathrm{HF}}(\boldsymbol{r})}{|\boldsymbol{R}_A - \boldsymbol{r}|} \mathrm{d}\boldsymbol{r} + \\
&\quad + \frac{1}{2} \int \frac{\rho^{\mathrm{HF}}(\boldsymbol{r})\rho^{\mathrm{HF}}(\boldsymbol{r}')}{|\boldsymbol{r} - \boldsymbol{r}'|} \mathrm{d}\boldsymbol{r}\mathrm{d}\boldsymbol{r}' - \frac{1}{4} \int \frac{|P^{\mathrm{HF}}(\boldsymbol{r};\boldsymbol{r}')|^2}{|\boldsymbol{r} - \boldsymbol{r}'|} \mathrm{d}\boldsymbol{r}\mathrm{d}\boldsymbol{r}' \\
&\quad + \frac{1}{2} \sum_{A,B=1}^{M}{}' \frac{Z_A Z_B}{r_{AB}} \geq E_0.
\end{aligned} \tag{2.27}
$$

In the KS equation in the DFT framework, for the electron density $\rho(\boldsymbol{r})$ of any N-electron system, we introduce two generalized functionals, namely, the exchange-correlation interaction $\varepsilon_{\mathrm{xc}}(\boldsymbol{r};\ [\rho])$ and exchange-correlation potential $V_{\mathrm{xc}}(\boldsymbol{r};\ [\rho])$, the latter of which can be obtained by taking the derivative of the former's functional, i.e. $V_{\mathrm{xc}} = \delta\varepsilon_{\mathrm{xc}}/\delta\rho$. If $\widetilde{V}^{\mathrm{KS}}$ in (2.23) is replaced by $V_{\mathrm{xc}}(\boldsymbol{r};\ [\rho]^{\mathrm{KS}})$, the electron density $\rho^{\mathrm{KS}}(\boldsymbol{r})$ obtained in (2.25) will be the exact ground state electron density $\rho(\boldsymbol{r})$. Based on the occupied KS orbitals, the exact ground-state energy can be calculated as follows, using the functional $\varepsilon_{\mathrm{xc}}(\boldsymbol{r};\ [\rho])$:

$$
\begin{aligned}
E_0^{\mathrm{KS}} &= -\int \left[\frac{\nabla^2}{2} P^{\mathrm{KS}}(\boldsymbol{r};\boldsymbol{r}') \right]_{(\boldsymbol{r}'=\boldsymbol{r})} \mathrm{d}\boldsymbol{r} - \sum_A Z_A \int \frac{\rho(\boldsymbol{r})}{|\boldsymbol{R}_A - \boldsymbol{r}|} \mathrm{d}\boldsymbol{r} + \\
&\quad + \frac{1}{2} \int \frac{\rho(\boldsymbol{r})\rho(\boldsymbol{r}')}{|\boldsymbol{r} - \boldsymbol{r}'|} \mathrm{d}\boldsymbol{r}\mathrm{d}\boldsymbol{r}' + \int \rho(\boldsymbol{r})\varepsilon_{\mathrm{xc}}(\boldsymbol{r};[\rho])\mathrm{d}\boldsymbol{r} + \frac{1}{2} \sum_{A,B=1}^{M}{}' \frac{Z_A Z_B}{r_{AB}} = E_0.
\end{aligned} \tag{2.28}
$$

In both HF and KS methods, Eq. (2.23) must be solved by self-consistent calculations because the Hartree and the modified potential are defined by occupied orbitals. However, there are two important differences between HF and KS, as follows:

1. In the HF case, the modified potential (the nonlocal operator in Eq. (2.26)) has a rigorously defined expression. By contrast, there are no rigorous definitions of the local exchange-correlation potentials $V_{\mathrm{xc}}(\boldsymbol{r};\ [\rho])$ and $\varepsilon_{\mathrm{xc}}(\boldsymbol{r};\ [\rho])$ in KS; however, some approximate expressions can be used for these functionals.
2. The single determinant wave function Ψ_0^X defined in (2.24) has different meanings for the HF and KS cases, and Ψ_0^{HF} can be regarded as a zero-order approximation of the true ground state wave function in the post-HF methods, where instantaneous electron correlation interactions are introduced. In principle, Ψ_0^{KS} only mathematically describe the behavior of electrons without interactions in the effective potential field, with the electron density the same as in the actual situation, and where Ψ_0^{KS} can be used instead of Ψ_0. For example, $P^{\mathrm{KS}}(\boldsymbol{r};\boldsymbol{r}') \approx P(\boldsymbol{r};\boldsymbol{r}')$ can be used to describe the non-diagonal terms of the position density matrix.

2.4.2 Periodic Solutions of Hartree–Fock (HF) and Kohn–Sham (KS) Equations

Because the one-electron Hamiltonian $\hat{h}^X$ commutes with all symmetry operations of the space group, particularly the eigenfunction crystal orbitals of the pure translation subgroups, these crystal orbitals can be classified according to the irreducible representation of the groups. These crystal orbitals can be expressed as Bloch functions, with each containing a reciprocal space vector κ and satisfying the following property:

$$\phi_n^X(\boldsymbol{r} + \boldsymbol{T_m}; \kappa) = \phi_n^X(\boldsymbol{r}; \kappa)\exp(i\kappa \cdot \boldsymbol{T_m}). \tag{2.29}$$

Obviously, if two Bloch functions differ by only a reciprocal lattice vector G in κ, they will belong to the same irreducible representation. Among all equivalents of κ, a set of reciprocal spaces closest to the origin can be selected, with the smallest space known as the first Brillouin zone. Furthermore, the crystal orbitals must also satisfy the Born-von Kaman condition, i.e. $\left[\phi_n^X(\mathbf{r} + \overline{\boldsymbol{W}}_{\boldsymbol{m}}; \kappa) = \phi_n^X(\boldsymbol{r}; \kappa)\right]$, implying $\exp(i\kappa \cdot \overline{\boldsymbol{W}}_{\boldsymbol{m}}) = 1$; hence, κ must belong to the Monkhorst lattice:

$$\boldsymbol{\kappa_h} = \sum_{i=1}^{D}(h_i + s_i)\frac{\boldsymbol{B}_i}{w_i}, \tag{2.30}$$

where h_i is an integer, $s_i = 0$ or $1/2$. The exact solution can be obtained by choosing a finite w_i in line with the convention that the superlattice vector defines the same Bravais lattice as the original lattice and the origin shift should be determined depending on s_i. Thus, the $\boldsymbol{\kappa_h}$ vector will form a contracted reciprocal lattice with respect to the original lattice, where the number of $\boldsymbol{\kappa_h}$ in the first Brillouin zone is $W = \prod_i w_i$. Due to rotational symmetry, if two $\boldsymbol{\kappa_h}$ can be related by a point group operation, the corresponding eigenvalues will be the same, and the eigenfunctions will differ by only a rotational phase. These properties allow us to consider $\boldsymbol{\kappa_h}$ only the irreducible Brillouin zone when solving the Schrödinger equation.

Since the effective number of electrons in the system is $N = WN_0$, the number of occupied orbitals will be $WN_0/2$, i.e. with, on average, $N_0/2$ electrons at each $\boldsymbol{\kappa_h}$. The Fermi energy (E_F) can be defined by sorting energy eigenvalues in ascending order, with the $WN_0/2$ energy eigenvalues strictly satisfying the condition $\varepsilon_n^X(\kappa_\mathbf{h}) \leq E_\mathrm{F}$. Accordingly, insulators, including semiconductors, can be distinguished from metals.

For insulators, the $N_0/2$ energy eigenvalues will be strictly below E_F at each $\boldsymbol{\kappa_h}$, and these orbitals will form $N_0/2$ fully occupied energy bands. This can be easily extended to spin-polarized systems, where the number of occupied energy bands varies for different spins.

The occupied crystal orbitals can be transformed to equivalent Wannier functions (WFs) by the unitary transformation matrix U:

$$\{\ldots, (n\kappa_{\boldsymbol{h}}), \ldots\} \overset{U}{\longleftrightarrow} \{\ldots, [\ell\boldsymbol{T_m}], \ldots\}[\ell\boldsymbol{T_m}] \equiv w_\ell(\mathbf{r} - \boldsymbol{T_m});$$
$$\ell = 1, \quad N_0/2, \quad 0 \leq m_i < w_i. \tag{2.31}$$

WFs$[\ell \boldsymbol{T_m}]$ can be specified as a real local function of the ordinary $\boldsymbol{T_m}$ unit cell, and we only need to consider W nonequivalent unit cells, namely those contained in the superlattice Wigner–Seitz primitive cell. With a total of $N_0/2$ WFs in this reference primitive cell, the others can be represented equivalently by translation. Because they can be obtained by the unitary transformation of the orthogonalized crystal orbitals through the transformation matrix U, the WFs themselves will constitute an orthogonal set. By choosing a suitable U matrix, these functions can be made to conform to the best localization characteristics while maximally reflecting the rotational symmetry of the system. The spinless ground state wave function ψ_0^X of an insulating crystal can be written in the form of the antisymmetric product of the spin WFs:

$$\psi_0^X \overset{\text{(i)}}{=} | \dots [\ell \boldsymbol{T_m}]^{\alpha} [\ell \boldsymbol{T_m}]^{\beta} \dots >. \tag{2.32}$$

The symbol $\overset{\text{(i)}}{=}$ indicates that it is valid only for spinless insulators. For metals, the number of eigenvalues below E_F will vary for different $\boldsymbol{\kappa_h}$, where some bands are partially occupied, and the interface separating occupied and unoccupied orbitals in the reciprocal space is called the Fermi surface of the metal.

2.4.3 Calculation of Crystal Density Matrix and Density Function

The density matrices of HF and KS can be expressed as follows, where $\overline{f}(\boldsymbol{p})$ is the Fourier transform of $f(\mathbf{r})$:

$$\begin{aligned}
\boldsymbol{P}^{X,\sigma}(\mathbf{r};\mathbf{r}') &= \sum_{\boldsymbol{h}} \sum_{n}{}' \phi_n^{X,\sigma}(\boldsymbol{r};\kappa_{\boldsymbol{h}}) \left[\phi_n^{X,\sigma}(\boldsymbol{r}';\kappa_{\boldsymbol{h}})\right]^* \\
\boldsymbol{P}^{X}(\boldsymbol{r};\boldsymbol{r}') &= \boldsymbol{P}^{X,\alpha}(\boldsymbol{r};\boldsymbol{r}') + \boldsymbol{P}^{X,\beta}(\boldsymbol{r};\boldsymbol{r}') \\
\overline{\boldsymbol{P}}^{X,\sigma}(\mathbf{p};\mathbf{p}') &= \sum_{\boldsymbol{h}} \sum_{n}{}' \overline{\phi}_n^{X,\sigma}(\mathbf{p};\kappa_{\boldsymbol{h}}) \left[\overline{\phi}_n^{X,\sigma}(\mathbf{p}';\kappa_{\boldsymbol{h}})\right]^* \\
\overline{\boldsymbol{P}}^{X}(\mathbf{p};\mathbf{p}') &= \overline{\boldsymbol{P}}^{X,\alpha}(\mathbf{p};\mathbf{p}') + \overline{\boldsymbol{P}}^{X,\beta}(\mathbf{p};\mathbf{p}'),
\end{aligned} \tag{2.33}$$

where summation notation with a prime $'$ means that it only works for the occupied crystal spin orbits ($\varepsilon_n^{X,\sigma}(\kappa_{\mathbf{h}}) < E_F$), and $\overline{\phi}_n^{X,\sigma}$ is the Fourier transform of the crystal orbitals. The density matrix of a spinless system can be simplified as follows:

$$\begin{aligned}
\boldsymbol{P}^{X}(\boldsymbol{r};\boldsymbol{r}') &= 2\sum_{\boldsymbol{h}} \sum_{n}{}' \phi_n^{X}(\boldsymbol{r};\kappa_{\boldsymbol{h}}) \left[\phi_n^{X}(\boldsymbol{r}';\kappa_{\mathbf{h}})\right]^* \overset{\text{(i)}}{=} 2\sum_{\ell,\boldsymbol{m}} w_\ell^X(\boldsymbol{r}-\boldsymbol{T_m}) w_\ell^X(\boldsymbol{r}'-\boldsymbol{T_m}) \\
\overline{\boldsymbol{P}}^{X}(\mathbf{p};\mathbf{p}') &= \frac{2}{W}\sum_{\boldsymbol{h}} \sum_{n}{}' \overline{\phi}_n^{X}(\mathbf{p};\kappa_{\boldsymbol{h}}) \left[\overline{\phi}_n^{X}(\mathbf{p}';\kappa_{\boldsymbol{h}})\right]^* \overset{\text{(i)}}{=} 2\sum \overline{w}_\ell^X(\mathbf{p}) \overline{w}_\ell^X(\mathbf{p}').
\end{aligned} \tag{2.34}$$

As mentioned earlier, the KS electron density and the corresponding structure factor will be those of a strictly true system, with the HF electron density containing systematic errors (for example, overestimation of the covalent bond charge density) compared to the true situation. Substitution of Eq. (2.34) into Eq. (2.18) will yield the RFF expressed by WFs:

$$\begin{aligned} B^X(\boldsymbol{r}) &\overset{\text{(i)}}{=} \frac{2}{W} \sum_{\ell,\mathbf{m}} \int w_\ell^X(\boldsymbol{r}' - \boldsymbol{T}_m) w_\ell^X(\boldsymbol{r}' + \boldsymbol{r} - \boldsymbol{T}_m) \mathrm{d}\boldsymbol{r}' \\ &= 2 \sum_\ell \int w_\ell^X(\boldsymbol{r}') w_\ell^X(\boldsymbol{r}' + \boldsymbol{r}) \mathrm{d}\boldsymbol{r}'. \end{aligned} \tag{2.35}$$

If $\boldsymbol{r} = \boldsymbol{T}_n \neq 0$, then the above integral will be zero because the overlap integral of two WFs belonging to different unit cells will be zero, indicating that the $B^X(\boldsymbol{r})$ of an insulator will be zero at all non-zero lattice points. In Compton scattering experiments, this condition may not be satisfied because of experimental errors or insufficient accuracy of the single-determinant representation of the ground state wave function.

2.4.4 Pseudopotentials

It is generally believed that only valence electrons are truly involved in the formation of compound molecules and crystals, while core electrons are barely affected. Thus, the wave function can be described approximately as the product of N_v valence electron wave functions $\psi_v(\ldots, \boldsymbol{x}_n, \ldots)$ and N_c core electron wave functions $\psi_c(\ldots, \boldsymbol{x}_m, \ldots)$. The latter can be described as the antisymmetric product of core electron wave functions Ψ_c^A, with A denoting the atoms forming the compound. Therefore, ψ_c^A can be obtained by solving for the isolated atoms and can be simply used by shifting appropriately according to the atomic coordinates in the compound. Using the method for separating core and valence electrons, the problem can be simplified for the determination of ψ_v^X only, resulting in several advantages.

1. The computational effort of HF and KS increases exponentially (N^3 in general) with the number of electrons in the system and can be significantly reduced by simplifying the treatment of core electrons.
2. The all-electron wave functions show abrupt changes near the nucleus; hence, higher cutoff energies are required to describe their plane waves.
3. In the vicinity of the nucleus, where the electron speed approaches the order of the speed of light, relativistic effects will be pronounced, especially for heavy atoms; thus, the use of nonrelativistic Hamiltonian to describe the electron density in the nuclear region will introduce large errors. In fact, the exchange-correlation potential $V_{\mathrm{xc}}(\boldsymbol{r}; [\rho]^{\mathrm{KS}})$ in the DFT framework has already included this effect. The separation of valence and core electrons allows for a simple relativistic treatment of the two using different techniques.

One of the simplest and most popular approaches in HF and KS frameworks involves replacing the nuclear Coulomb potential $A(-Z_A/|\boldsymbol{R}_A - \boldsymbol{r}|)$ in Eq. (2.23) by a nonlocal operator $\hat{V}_{\mathrm{ps}}^{X,A}$, called the pseudopotential of atom A, which can be solved according to the modified equation to obtain pseudo-orbitals and pseudo-eigenvalues. In this approach, the rest of the treatment remains unchanged, with only the lowest N_v pseudo-orbitals occupied. In general, pseudopotentials can be generated from solutions to the Schrödinger equation for isolated atoms with their core electrons containing relativistic corrections, often using the ground state atoms as a reference. The basic steps of generating pseudopotentials include replacing the true atomic valence orbitals with a set of pseudo-orbitals that are the same as the true valence orbitals in the region, where r is larger than a suitably chosen matching radius r_c (valence shell), but smoother than the true valence orbitals in the core region ($r < r_c$). The simplest treatment involves pseudo-orbitals with no nodes and satisfying the norm-conserving condition, i.e. the charge in the core region is consistent with that of the reference atom. By fixing the eigen-energy to the true valence eigenvalue and solving the KS equation backward, the effective potential can be obtained, called the norm-conserving pseudopotential. If the norm-conserving condition is not imposed on pseudo-orbitals, the obtained potential will be the supersoft pseudopotential.

There are a variety of pseudopotentials, which can be divided into local, semi-local, and nonlocal pseudopotentials according to the degree of localization. They can also be divided into large-core and small-core pseudopotentials according to the electron shells of the treated core and into other different types of pseudopotentials according to the parameter optimization rules. Despite the many advantages of pseudopotentials, there are serious drawbacks in using pseudopotentials to calculate density functions. This is because pseudo-valence electron orbitals are not orthogonal to the true core orbitals; hence, the sum of the obtained pseudo-valence electron density and the core electron density will not be the true electron density, as is the case of the momentum density calculated by using pseudopotentials. Therefore, it is necessary to reconstruct the electron density of the all-electron wave function using certain special techniques, such as the projector augmented waves (PAW) method. Using the PAW method, the all-electron wave function can be obtained by making the projection of the pseudo-orbital in the core the same as the all-electron orbital on the atomic ground state orbital and then augmenting the difference between the pseudo-orbital and the all-electron orbital on the pseudo-orbital.

2.4.5 Basis Set

As mentioned earlier, both HF and KS schemes solve the single-particle Schrödinger equation under a self-consistent effective potential field, consisting of the Hartree and exchange potential in HF and Hartree and exchange-correlation potential in KS. The KS problem is simple because the effective potential depends only on the electron density, whereas in HF, it depends on the density matrix $\boldsymbol{P}(\boldsymbol{r}, \boldsymbol{r}')$ and can be solved by self-consistent iterations as follows. First, a suitable initial effective potential can be estimated to calculate the single-particle orbitals, then the effective potential can be recalculated based on the orbitals, and so forth until convergence. In practical calculations, the orbitals are expanded using a suitable basis set. In periodic systems, it is convenient to use Bloch functions $f_\mu(\boldsymbol{r}; \boldsymbol{\kappa})$ as a basis set, reducing the problem of determining the crystal orbital $\phi_n^X(\boldsymbol{r}; \boldsymbol{\kappa})$ to a problem of solving the secular equation involving only a certain $\boldsymbol{\kappa}$ point. The choice of basis set is very important, and plane waves or atomic basis functions can be used as basis sets in most calculations.

Plane waves are traditionally chosen in solid-state physics to reflect the nonlocality of valence and conduction band electrons in crystals. Plane waves can form an infinite number of complete basis sets, according to:

$$f_G(\boldsymbol{r};\boldsymbol{\kappa}) = \frac{1}{\sqrt{\Omega}} \exp[i(\boldsymbol{\kappa} + \boldsymbol{G}) \cdot \boldsymbol{r}], \tag{2.36}$$

where Ω is the volume of the Born-von Karman box and $\boldsymbol{G}$ is the reciprocal lattice vector. The Bloch condition of Eq. (2.29) can be satisfied because of $\boldsymbol{G} \cdot \boldsymbol{T}_m = 2n\pi$, where n is an integer. A finite plane wave basis set can be obtained by considering only plane waves with the kinetic energy $(\boldsymbol{\kappa} + \boldsymbol{G})^2/2$ below an energy threshold E_c.

Plane waves have the following advantages:

1. They enable convenient numerical computation, lead to orthogonal results, and allow for the use of fast Fourier transform to speed up the calculation process.
2. Plane waves can form an unbiased basis set and do not depend on the types or coordinates of atoms in the compound; thus, they do not have force errors caused by basis set incompleteness or energy errors due to basis set overlap.
3. Convergence can be easily evaluated by varying the cutoff energy as a parameter. However, the plane wave method also suffers from some serious deficiencies, with the most obvious one being the inability to treat the electronic states of core electrons. The standard approach involves introducing pseudopotentials; however, a good approach for calculating the density function is to employ the PAW technique, which can calculate the true all-electron density using a small plane-wave basis set.

Atomic orbitals $\chi_\mu(\boldsymbol{r})$ are often used to construct basis sets in quantitative calculations, as they reflect the atomic nature of matter. In periodic systems, the Bloch sum of atomic orbitals can be used, according to:

$$f_\mu(\boldsymbol{r};\boldsymbol{\kappa}) = \frac{1}{\sqrt{W}} \sum_{\boldsymbol{T}} \exp[i\boldsymbol{\kappa} \cdot \boldsymbol{T}] \chi_\mu(\boldsymbol{r} - \boldsymbol{T}). \tag{2.37}$$

References

1 Pisani, C., Dovesi, R., Erba, A., and Giannozzi, P. (2012). Electron densities and related properties from the ab-initio simulation of crystalline solids. In: *Modern Charge-Density Analysis* (ed. C. Gatti and P. Macchi), 79–132. Berlin: Springer.

2 Levine, I.N. (2014). *Quantum Chemistry*, 7e. New York: Pearson Education.

3

Topological Indices and Properties of Electronic Structures

3.1 Introduction

We have observed certain facts where different compounds with the same functional groups or structural units have similar physicochemical properties and functions. This indicates that these functional groups and functional motifs have certain relatively independent electronic structural features in different structures and compounds, and they are not easily influenced by the surrounding chemical environment, endowing structures with these electronic structural features with corresponding properties. The portability and additivity of such functional motifs in different structures can be considered the basis for the design of materials based on functional motifs. Therefore, it is necessary to divide the electron density of a molecule or an asymmetric unit into the superposition of atomic and chemical bond contributions according to certain rules, which is very important for understanding the origin of molecule and compound properties.

Many classic concepts in chemistry, such as atomic charge, chemical bond, covalency, ionicity, resonance structure, bond energy, and aromaticity, are difficult to obtain from multi-electron wave functions and have a clear definition. As the accuracy of the available wave functions has increased, some of the vague chemical concepts have slowly fallen out of use. However, because we can carry out sophisticated quantitative analysis of electron density, some vague chemical concepts have started to be defined in a new and rigorous manner. Major advances include the density functional theory of Hohenberg and Kohn, as well as the QTAIM of Bader and his collaborators [1, 2], paving the way for the rigorous definition of chemical bonding-related concepts from electron density analysis. Accordingly, the topological analysis of electron density has gained wider applications since the electron density can be obtained by scattering experiments.

Topological analysis techniques based on scalar functions in real space can partition the 3D real space corresponding to electron density into different independent regions; thus, enhancing our deep understanding of chemical intuition. However, not all scalars can be obtained directly from the electron density, such as the local electron density function. Without a rigorous universal energy or electron repulsion-related density functional as a basis, many practical concepts

Electronic Structure Crystallography and Functional Motifs of Materials, First Edition.
Guo-Cong Guo and Xiao-Ming Jiang.

will inevitably rest on physical quantities that are much more complex than electron density.

Most of the concepts presented in this chapter rely on quantum mechanical objects that cannot be derived solely from electron density, such as operations based on the density matrix. Nevertheless, this does not mean that these methods are decoupled from experiments, as it is possible to make these complex quantities dependent only on the electron density through suitable approximations or to analyze them directly using experimental electron wave functions. In fact, as discussed in Chapter 6, the density matrix can also be determined experimentally. The so-called topological analysis of the electronic structure generally includes some scalar and vector-derived function properties obtained from the electron density itself or from the one- or two-electron density matrices.

This chapter first describes certain basic elements of the QTAIM theory [3], which is the basis for the practical topological analysis of the refined electronic structure. Then, the source function, electron localization function (ELF), and reduced density gradient (RDG) for chemical interaction analysis are introduced, followed by a description of the two schemes of 3D space partition: [4] ① the coarse-grained partition of space into various atomic regions, and ② the partition of space into an infinite number of small variable-sized units that possess certain invariant quantum mechanical properties.

The intermolecular interaction energy, an important property of electron density, will be presented at the end of this chapter [5]. In molecular compounds, the forces between structural units, molecules or ions are mainly derived from the classical electrostatic force between them. The classical electrostatic force is the main part of molecular crystal binding energy, and the anisotropic electrostatic force plays a critical role in the formation of the final crystal structure. Thus, advances in experimental electron density measurements have provided new avenues for the qualitative and quantitative study of molecular forces.

3.2 Analysis of Topological Atoms in Molecules

3.2.1 Topological Description of the Electron Density

The electron density distribution in a molecule is mainly influenced by the attraction of the nucleus, which occupies the position of the maximum electron density. Therefore, the space around a nucleus can be partitioned to this atom according to certain boundary conditions, and the electron density on the atomic boundary reflects a balance of the attraction of the surrounding nuclei.

In an electron density map, unique points can be observed where the electron density gradient $\nabla\rho$ (the first derivative of the electron density) is zero. These unique points are known as critical points (CP), namely:

$$\nabla\rho = \boldsymbol{i}\frac{\mathrm{d}\rho}{\mathrm{d}x} + \boldsymbol{j}\frac{\mathrm{d}\rho}{\mathrm{d}y} + \boldsymbol{k}\frac{\mathrm{d}\rho}{\mathrm{d}z}, \tag{3.1}$$

where $\nabla\rho$ is zero at CP and at infinity, and is generally not zero elsewhere.

The electron density $\rho(\boldsymbol{r})$ is a scalar function of the 3D space $\boldsymbol{r}$ and its gradient $\nabla\boldsymbol{\rho}$ is a vector that represents the direction of the greatest rate of increase for the value of $\rho(\boldsymbol{r})$ in a 3D space when it changes at different positions. The magnitude of the electron density gradient, $|\nabla\boldsymbol{\rho}|$, represents the rate of change of $\rho(\boldsymbol{r})$ in this direction, and the position of the nucleus is a type of critical point called the nuclear critical point (NCP).

The critical point of the electron density may be its maximum point, minimum point, or saddle point, and in each case $\nabla\boldsymbol{\rho}$ is zero. To further distinguish the three cases, we must calculate the second derivative of the electron density $\nabla\nabla\boldsymbol{\rho}$, which is a second-order tensor with nine components and can be written in the form of a 3×3 matrix, known as the Hessian matrix. At a critical point $\boldsymbol{r}_\mathrm{c}$, the Hessian matrix can be defined as

$$\boldsymbol{H}(\boldsymbol{r}_\mathrm{c}) = \begin{pmatrix} \dfrac{\partial^2\rho}{\partial x^2} & \dfrac{\partial^2\rho}{\partial x\partial y} & \dfrac{\partial^2\rho}{\partial x\partial y} \\ \dfrac{\partial^2\rho}{\partial y\partial x} & \dfrac{\partial^2\rho}{\partial y^2} & \dfrac{\partial^2\rho}{\partial y\partial z} \\ \dfrac{\partial^2\rho}{\partial z\partial x} & \dfrac{\partial^2\rho}{\partial z\partial y} & \dfrac{\partial^2\rho}{\partial z^2} \end{pmatrix}_{\boldsymbol{r}=\boldsymbol{r}_\mathrm{c}}. \tag{3.2}$$

Because the Hessian matrix has real-valued components and is symmetric, it can be diagonalized according to matrix theory. The diagonalization of the Hessian matrix will be equivalent to the rotation of the coordinate system $\boldsymbol{r}(x, y, z)$ through a rotation matrix $\boldsymbol{U}$ to a new coordinate system $\boldsymbol{r}' = \boldsymbol{r}\boldsymbol{U}$, with the three coordinate axes of $\boldsymbol{r}'(x', y', z')$ being the principal axes of curvature at $\boldsymbol{r}_\mathrm{c}$. $\boldsymbol{U}$ satisfies the eigenvalue equations $\boldsymbol{H}\boldsymbol{u}_i = \lambda_i\boldsymbol{u}_i$ $(i = 1, 2, 3)$, where $\boldsymbol{u}_i$ contains each of the three eigenvectors of $\boldsymbol{U}$, and λ_i is the corresponding eigenvalue. $\boldsymbol{H}$ can be transformed into a diagonal matrix form $\boldsymbol{\Lambda}$ by orthogonal transformation, i.e.

$$\boldsymbol{U}^{-1}\boldsymbol{H}\boldsymbol{U} = \boldsymbol{\Lambda} = \begin{pmatrix} \dfrac{\partial^2\rho}{\partial x'^2} & 0 & 0 \\ 0 & \dfrac{\partial^2\rho}{\partial y'^2} & 0 \\ 0 & 0 & \dfrac{\partial^2\rho}{\partial z'^2} \end{pmatrix}_{\boldsymbol{r}'=\boldsymbol{r}_\mathrm{c}} = \begin{pmatrix} \lambda_1 & 0 & 0 \\ 0 & \lambda_2 & 0 \\ 0 & 0 & \lambda_3 \end{pmatrix}, \tag{3.3}$$

where λ_1, λ_2, and λ_3 are the curvatures of the electron density on the three principal curvature axes x', y', and z' at $\boldsymbol{r}_\mathrm{c}$, respectively.

The Hessian matrix has a property where its trace (i.e. the sum of the elements on the main diagonal) remains unchanged after orthogonal transformation. The trace of the Hessian matrix can be defined as the Laplacian of the electron density, which is the sum of the three principal curvatures when $x = x'$, $y = y'$, $z = z'$, i.e.

$$\nabla^2\rho(\boldsymbol{r}) = \nabla \cdot \nabla\rho(\boldsymbol{r}) = \frac{\partial^2\rho(\boldsymbol{r})}{\partial x^2} + \frac{\partial^2\rho(\boldsymbol{r})}{\partial y^2} + \frac{\partial^2\rho(\boldsymbol{r})}{\partial z^2} = \lambda_1 + \lambda_2 + \lambda_3. \tag{3.4}$$

Critical points can be classified according to the ranks (ω) and signs (σ) of the three principal curvatures λ_1, λ_2, and λ_3 at the critical points. The classification criterion is usually expressed in the form of (ω, σ), where ω is the number of nonzero principal curvatures (for example, $\omega = 2$ if $\lambda_1 \neq 0$, $\lambda_2 \neq 0$, $\lambda_3 = 0$), and σ is the sum of the numbers of the positive signs (each assigned a value of 1) and negative signs (each assigned a value of −1) of the principal curvatures (for example, $\sigma = 1 + 1 - 1 = 1$ if $\lambda_1 > 0$, $\lambda_2 > 0$, $\lambda_3 < 0$). A critical point with $\omega < 3$ will be physically unstable and the perturbation from a small displacement of the nucleus will suffice to make this critical point disappear; thus, $\omega < 3$ rarely appears in the electron density map of an equilibrium system. Here, we only consider the case of $\omega = 3$. Depending on σ, the stable critical points of the electron density can be classified into four types, as shown below.

(ω, σ)	Principal curvature feature	Features of the electron density ρ	Corresponding structure type	Critical point symbol
(3, −3)	Three negative principal curvatures	A local maximum	Nuclear critical point	NCP
(3, −1)	Two negative principal curvatures	A maximum in the plane defined by the two principal curvature axes and a minimum in the direction of the third principal axis perpendicular to this plane	Bond critical point (the third principal axis is parallel to the bond direction)	BCP
(3, +1)	Two positive principal curvatures	A minimum in the plane defined by the two principal curvature axes and a maximum in the direction of the third principal axis perpendicular to this plane	Ring critical point, where the third principal axis is perpendicular to the torus	RCP
(3, +3)	Three positive principal curvatures	Local minimum point	Cage critical point	CCP

Different types of critical points may exist in the same molecule or crystal structure, and the numbers of these critical points (n) can constitute a feature set $\{n_{NCP}, n_{BCP}, n_{RCP}, n_{CCP}\}$ that satisfies different exact topological relations for isolated molecules and infinitely extended structures, where $n_{NCP} - n_{BCP} + n_{RCP} - n_{CCP}$ is 1 for isolated molecules (Poincaré-Hopf relationship) and 0 for infinitely extended crystal structures (Morse equation).

When performing topological analysis of the electron density, it is often necessary to find various critical points in the structure by a computational search, and the above topological relation can serve as a basis for determining the completeness of the critical points searched. If the relation does not hold, this indicates that certain critical points are missing in the search process or that some of the found critical points are problematic. If the relation holds, it does not necessarily indicate that the searched critical point feature set is complete because the relation will still hold if one more critical point is found or missing for each of n_{NCP} and n_{BCP} simultaneously.

Such is the case for n_{RCP} and n_{CCP}. However, in the practical analysis of electronic structures, the above topological relation can serve as useful evidence that a feature set is self-consistent and complete.

When several atoms form a ring, an RCP will often occur at the center of the ring; however, when several rings form an enclosed space, a CCP will appear. A bond path is a line that is formed by the points of maximum electron density between the nuclei of two atoms with chemical bond interactions. A molecular topological graph is a set of critical points and bond paths. It should be noted that the molecular topological graph here differs from other graphs drawn with atoms and chemical bonds (straight lines connecting bonded atoms). These two types of graphs may look similar but differ fundamentally, with the former based on the topological features of electron density in a molecule, and the latter is just a molecular model that shows geometric features.

3.2.2 Gradient Vector Field and Topological Atoms

In the electron density map, the locations with the maximum density correspond to nuclei. The entire electron density map can be divided into separate regions, with the nuclei as centers and regions Ω belonging to specific atoms, called topological atoms or atomic basins. At any position vector $\mathbf{r}$ on the atomic boundary $S(\Omega)$, the charge density gradient will satisfy the following condition:

$$\nabla \rho(\boldsymbol{r}) \cdot \boldsymbol{n}(\boldsymbol{r}) = 0, \tag{3.5}$$

where $\boldsymbol{r}$ is the position vector and $\boldsymbol{n}(\boldsymbol{r})$ is the unit normal vector of the surface $S(\Omega)$.

According to the above definition, atomic boundaries can be considered as the zero-flux surfaces of the electron density gradient vector field, i.e. without the intersection of any density gradient vectors. One end of each gradient field line belongs to an atom, accumulating to the nucleus of this atom as if the gradient field lines were attracted to the nucleus; hence, the nucleus is also known as a nuclear attractor, while the region covered by all gradient field lines connected to a nucleus is called the basin of an atom. The atoms in a molecule can be defined as a set of nuclei and corresponding basins, where each atomic basin is surrounded by one or more zero-flux surfaces. The zero-flux surfaces may extend to infinity; however, as they move away from the nuclei, the electron density near them will gradually decrease. In general, a small electron density can be taken as the cut-off value to avoid having a topological atom with an infinite volume.

In the electron density map of certain metal and semiconductor materials, maximum electron density will occur not only at nuclei but in other locations as well, known as nonnuclear attractors. Considering the topological analysis of electron density, nonnuclear attractors will be indistinguishable from nuclear attractors. Like nuclear attractors, nonnuclear attractors can also be associated with a basin, covered by electron density gradient field lines and surrounded by zero-flux surfaces. As a result, nonnuclear attractors are also called pseudo-atoms, which can form chemical bonds with other atoms or pseudo-atoms, i.e. they share the same zero-flux surfaces, bond critical points (BCP), and bond paths. Nonnuclear attractors and their basins are important for characterizing metallic bonds and have garnered attention from theorists.

Gradient field lines can start at infinity and end at BCPs. In addition, the zero-flux surface is a characteristic indicator of atomic bonding and can only appear between two bonded atoms.

3.2.3 Bond Path and Molecular Topological Graph

In addition to zero-flux surfaces, a topological feature also exists between atoms, as the points of maximum electron density between atoms can be connected to form a curve segment called a bond path. The bond path serves as a general indicator of all chemical bonding, such as covalent bonds, ionic bonds, hydrogen bonds, π–π stacking, and van der Waals interactions. The point on the bond path where the electron density is the smallest along the bond path is the BCP, and at this point, the bond path will intersect the zero-flux surface shared by the bonding atoms.

The graph formed by connecting the bonded atoms through the bond paths is called the electron density topological graph of a molecule, or simply the molecular topological graph. Unlike the molecular geometry diagram formed by connecting bonded atoms by straight lines, the molecular topographical graph can uniquely identify the electron density topographical features of a molecule, where the positions of the BCPs on the bond path reflect the changes in electron density around the bonded atoms. Under external fields such as light and electricity, the electron density of a molecule will be prone to changes, while the position of the nucleus will be stable; as a result, the molecular topological graph will change accordingly, while the molecular geometry diagram will be almost the same.

In a molecule, the difference in electron density at each location will lead to different electrostatic potential energy densities at that location, where the higher the electron density, the more negative the potential energy density. In the potential energy density diagram of a molecule, the most negative points of potential energy density between bonded atoms can be connected to a curve segment called a virial path. The graphs formed by connecting the bonded atoms through virial paths are known as virial graphs. Like the molecular topological graph, the virial graph can also uniquely identify the electron density topological features of a molecule.

Topological atoms in a molecule are those that can form chemical bonds and have nuclei connected by a single bond path and a virial path, with two nuclei sharing a critical point and zero-flux surface between them. An advantage of dividing the electron density of a molecule into topological atoms is that the properties of the molecule can be divided into the superposition of contributions of the atoms.

3.2.4 Laplacian

The Laplacian is the second derivative of electron density with respect to the spatial coordinates, and its sign determines whether the electron density at a certain position in the molecule is in an electron-depleted or electron-rich state relative to the electron density in its domain. When $\nabla^2\rho(\boldsymbol{r}) > 0$, the electron density at $\boldsymbol{r}$ will be smaller than the surrounding average electron density, where the local region at $\boldsymbol{r}$ will be in an electron-depleted state. Conversely, when $\nabla^2\rho(\mathbf{r}) < 0$, the electron density at $\boldsymbol{r}$ will be larger than the surrounding average electron density, and the local region at $\boldsymbol{r}$ will be in an electron-rich state. According to the Lewis theory of acids and bases, substances (molecules, ions, or atomic groups) that can accept/donate electron pairs are called acids/bases. The electron-rich region will behave as a Lewis base (electron donor), while the electron-depleted region will behave as a Lewis acid (electron acceptor).

The Laplacian can reveal the shell structure of electron density on an isolated atom, with a valence shell charge concentration (VSCC) on the outer shell of the atom. The spherical symmetry of the VSCC will break when a chemical bond forms. The chemical reaction corresponds to the combination of the VSCC electron cloud of the Lewis base and the electron-depleted region of the Lewis acid, and the bond charge will be concentrated between the two atoms when a covalent bond forms. In addition to bond charges, lone pairs of electrons will also correspond to nonbonded charge-rich regions. Therefore, as indicated, the Laplacian of electron density will correspond to localized electrons.

3.2.5 Topological Properties of Chemical Bonds

The zero-flux surface $\nabla\rho(\boldsymbol{r}) = 0$ is defined by the gradient trace starting at infinity and ending at the BCP. A BCP exists between two bonded atoms, where the two atoms are connected by a bond path and share the interatomic zero-flux surface. In addition to the traces that end at the BCP and define the atomic interface, a pair of traces exist from the BCP that point to the two bonded atoms separately and define the bond path. The chemical bond properties can be defined and classified by the electron density and energy density at the BCP.

3.2.5.1 Electron Density at Bond Critical Points

Bond order (BO) can be used to measure the strength of a chemical bond. BO is a function of electron density (ρ_b) at the BCP and can be defined as $\text{BO} = \exp[\text{A}(\rho_b - \text{B})]$, where A and B are constants related to bonded atoms. In general, $\rho_b > 0.2$ a.u. for electron-sharing bonds (e.g. covalent bonds) and $\rho_b < 0.1$ a.u. for closed-shell interactions (e.g. ionic bonds, van der Waals forces, hydrogen bonds, double hydrogen bonds, and H—H bonds). It has been shown that ρ_b is related to bond length and binding energy in certain types of chemical bonds.

3.2.5.2 Bond Radius and Bond Path Length

If atoms A and B are bonded and with a BCP, the length from A to the BCP is called the bond radius $r_b(\text{A})$ of atom A. If the bond path coincides with the A—B bond axis, the sum of the bond radii of atoms A and B will be equal to the bond length. However, in many cases, the bond path is bent, or chemical strain is present; thus, the bond path length will exceed the bond length.

3.2.5.3 Laplacian of Electron Density at the Bond Critical Points

The Laplacian $\nabla^2\rho_b$ at the BCP is the sum of the curvatures of the electron density in the three principal directions at the critical point, with curvatures λ_1 and λ_2 in the two directions perpendicular to the bond path that are negative (usually $|\lambda_1| > |\lambda_2|$ is specified), and a third curvature λ_3 along the bond path direction that is positive, i.e. $\lambda_3 > 0$. The negative curvatures (λ_1 and λ_2) depict the degree of electron density accumulation along the bond path, while the positive curvature (λ_3) depicts the degree of electron density depletion at the interface between the two atoms and the degree of accumulation in each atomic basin. Covalent bonds are dominated by the two negative curvatures; thus, $\nabla^2\rho_b < 0$. For example, $\nabla^2\rho_b = -1.1$ for a typical C–H bond. By contrast, for closed-shell bonding such as ionic bonds, hydrogen bonds, or van der Waals interactions, atomic bonding interactions are mainly manifested as electron density depletion in the region of contact between the two atoms; thus, $\nabla^2\rho_b > 0$. For example, $\nabla^2\rho_b = +0.03$ a.u. for $\mathrm{N-(H\cdots O)=C}$ hydrogen bonds. In strong polar bonds (e.g. C—X, X=O, N, F), similar to electron-sharing bonds, there will be significant accumulation of electron density between the two atoms; however, the Laplacian of such bonds can be positive or negative.

3.2.5.4 Ellipticity

Ellipticity (ε) depicts the degree of preferential accumulation of electron density in the plane containing the bond path and can be defined as:

$$\varepsilon = \frac{\lambda_1}{\lambda_2} - 1 \qquad (|\lambda_1| \geq |\lambda_2|), \tag{3.6}$$

If $\lambda_1 = \lambda_2$, then $\varepsilon = 0$, in which case the bond will be axisymmetric, such as the C—C single bond in ethane and the C≡C triple bond in acetylene. Thus, ellipticity measures the degree of π-bonding behavior of a chemical bond and reaches a maximum for double bonds. From double to triple bonds, the changing trend of ellipticity with BO reverses, where the ellipticity decreases with increasing BO as the bonds regain axisymmetry (two π-bonds in the vertical plane and an axisymmetric σ-bond) at BO=3. The ellipticity can be calculated as 0.23 for the aromatic bond in benzene and 0.45 for the C=C double bond in ethylene.

3.2.5.5 Energy Density of Bond Critical Points

The calculation of energy density requires the full information of the one-electron density matrix, not only the electron density of the diagonal elements. Energy density includes potential energy, kinetic energy, and total energy, which can be used to characterize the mechanical properties of chemical bonds. The potential energy density $V(\boldsymbol{r})$, also known as the virial field, denotes the average effective potential field felt by an electron at $\boldsymbol{r}$ in a multi-particle system. The virial field at any point in space is always negative, and the integral over the whole space is the total potential energy of the molecule. The localized formulation of the virial theorem describes the relationship between the virial field, kinetic energy density, and Laplacian, according to:

$$\left(\frac{\hbar^2}{4m}\right)\nabla^2\rho(\boldsymbol{r}) = 2G(\boldsymbol{r}) + \mathcal{V}(\boldsymbol{r}), \tag{3.7}$$

where

$$G(\boldsymbol{r}) = \frac{\hbar^2}{2m}N \int d\tau' \, \nabla\Psi^* \cdot \nabla\Psi, \tag{3.8}$$

where $G(\boldsymbol{r})$ is the gradient kinetic energy density and Ψ is the antisymmetric multi-electron wave function. Since $G(\boldsymbol{r}) > 0$ and $\mathcal{V}(\boldsymbol{r}) < 0$ always hold, by applying the local virial theorem to the BCP, a decrease in the local potential energy will lead to $\nabla^2\rho_b < 0$ and, conversely, an increase in the local kinetic energy will lead to $\nabla^2\rho_b > 0$. Abramov provided an approximate expression for calculating $G(\boldsymbol{r})$ (unit: a.u.) directly from the electron density $\rho(\boldsymbol{r})$: [6]

$$G(\boldsymbol{r}) = \frac{3}{10}(3\pi^2)^{2/3}\rho(\boldsymbol{r})^{5/3} + \frac{1}{6}\nabla^2\rho(\boldsymbol{r}). \tag{3.9}$$

To compare the kinetic energy density and potential energy density on the same scale, instead of using 2 : 1 from the virial ratio of potential energy to kinetic energy, Cremer and Kraka defined the total electron energy density (H_b) of the electron at the BCP as [7]

$$H_b = G_b + \mathcal{V}_b. \tag{3.10}$$

The integral of H_b over the whole space is the total electron energy. For shared electron bonding, H_b will be negative, with its magnitude reflecting the covalency of the interaction.

Based on the electron density ρ_b at the BCP, the Laplacian $\nabla^2\rho_b$, the ratio of the maximum negative electron density curvature perpendicular to the bond path direction of the electron density curvature along the bond path direction $|\lambda_1|/\lambda_3$, the total electron energy density H_b, and the ratio of kinetic energy density to the electron density G_b/ρ_b, overall chemical bond interactions can be divided into closed-shell atomic interactions (e.g. ionic bonds), shared interactions (e.g. covalent bonds), and intermediated interactions, with the following respective characteristics [8].

For closed-shell atomic interactions: $\rho_b < 0.3\text{e Å}^{-1}$; $\nabla^2\rho_b > 0$, $|\lambda_1|/\lambda_3 < 0.25$; $H_b > 0$ and $G_b/\rho_b > 1$;
For shared interactions: $\rho_b > 1.0\text{e Å}^{-1}$; $\nabla^2\rho_b < 0$, $|\lambda_1|/\lambda_3 > 1$; $H_b < 0$ and $G_b/\rho_b < 1$;
For intermediated interactions: $0.3 < \rho_b < 1.0\text{e Å}^{-1}$; $\nabla^2\rho_b > 0$, $|\lambda_1|/\lambda_3 > 0.2$; $H_b < 0$ and $G_b/\rho_b > 1$.

In fact, the diversity of chemical bond types in materials or compounds and a smooth transition between different bond types make it difficult to classify them in a unified manner based on the topological characteristics of chemical bonds, which have been discussed elsewhere [9].

3.2.5.6 Delocalization Index and Bond Order

The number of electron pairs shared between two atoms is often referred to as the BO. QTAIM provides a method for calculating the number of shared electron pairs, which can be obtained by integrating the exchange densities of the two atomic basins. In principle, this property can also be classified as a topological atomic property. The exchange amplitude between the electrons in atomic basin A and the electrons in the basin B is called the delocalization index of the two atoms, $\delta(A, B)$, which for a closed-shell system can be defined as:

$$\delta(A, B) = 2|F^{\alpha}(A, B)| + 2|F^{\beta}(A, B)|, \tag{3.11}$$

where the Fermi correlation $F^{\sigma}(A, B)$ can be defined as:

$$\begin{aligned} F^{\sigma}(A,B) &= -\sum_i \sum_j \int_A d\boldsymbol{r}_1 \int_B d\boldsymbol{r}_2 \left\{\phi_i{}^*(\boldsymbol{r}_1)\phi_j(\boldsymbol{r}_1)\phi_j{}^*(\boldsymbol{r}_2)\phi_i(\boldsymbol{r}_2)\right\} \\ &= -\sum_i \sum_j S_{ij}(A)S_{ji}(B) \end{aligned} , \qquad (3.12)$$

where $S_{ij}(\Omega) = S_{ji}(\Omega)$ is the overlap integral of the two spin-orbitals in the region Ω, and σ denotes α or β spin.

The integration of Eq. (3.12) performed only in the atomic basin A will yield the total Fermi correlation for electrons in the region A:

$$F^{\sigma}(A,A) = \int_A \mathrm{d}\boldsymbol{r}_1 \int_A \mathrm{d}\boldsymbol{r}_2 \rho^{\sigma}(\boldsymbol{r}_1) h^{\sigma}(\boldsymbol{r}_1, \boldsymbol{r}_2). \qquad (3.13)$$

The limiting value is the spin-σ charge $-N^{\sigma}(A)$ of the negative atom A, where such a large number of σ electrons will be completely localized in this atom because all other spin-σ electrons will be excluded from A. Hence, if this limit is reached, all electrons in A will not exchange with electrons outside A. Thus, the localization index $\lambda(A)$ can be defined as:

$$\lambda(A,A) = |F^{\alpha}(A,A)| + |F^{\beta}(A,A)|. \qquad (3.14)$$

In ionic systems, the limit of total localization can reach or even exceed 95%. It has been found that $|F^{\alpha}(A, A)| < N^{\sigma}(A)$, showing that the electrons in region A will often exchange to some extent with electrons outside the boundary of atom A, i.e. they become delocalized.

Because the Fermi correlation considers all electrons, half of the localization index and delocalization index is the total number of electrons in the molecule. This provides a method of calculating the number of electrons localized in atomic basins and delocalized between atomic basins:

$$N(A) = \lambda(A) + \frac{1}{2}\sum_{B \neq A} \delta(A,B). \qquad (3.15)$$

It is worth mentioning that the delocalization index can be calculated whether it is between bonded atoms or nonbonded atoms. For bonded atoms, the delocalization index can be used to measure the number of electron pairs (BO) shared between two atoms. Because ρ_b is closely related to the BO, Matta et al. established a calibration relationship between the two [10]:

$$\delta(A,B) = \exp[A(\rho_b - B)]. \qquad (3.16)$$

Equation (3.16) can be used to calibrate the experimental ρ_b based on the theoretically calculated delocalization index. The fitted equation can be used to estimate the electron-sharing information implied in the full-density matrix, and this information cannot be derived from the ρ_b obtained by conventional X-ray diffraction experiments.

3.2.6 Topological Atomic Properties

The average value of the property operator $O(\Omega)$ acting on an atomic basin Ω can be expressed by

$$O(\Omega) = \langle \hat{O} \rangle_{\Omega} = \frac{N}{2} \int_{\Omega} d\boldsymbol{r} \int d\tau' [\Psi^* \hat{O}\, \Psi + (\hat{O}\Psi)^* \Psi]. \tag{3.17}$$

where $\hat{O}$ is a one-electron operator or the sum of several one-electron operators.

3.2.6.1 Atomic Charges

The total number of electrons in an atom of a molecule can be calculated by setting $\hat{O} = 1$ in Eq. (3.17):

$$N(\Omega) = \int_{\Omega} \rho(\boldsymbol{r})\, d\boldsymbol{r}. \tag{3.18}$$

The number of electrons can also be obtained by summing the expected number of electrons in the α and β spin subsystems according to the following:

$$N(\Omega) = \sum_{i} \left[\langle \psi_i(\boldsymbol{r}) \mid \psi_i(\boldsymbol{r}) \rangle_{\Omega}^{\alpha} + \langle \psi_i(\boldsymbol{r}) \mid \psi_i(\boldsymbol{r}) \rangle_{\Omega}^{\beta} \right], \tag{3.19}$$

where

$$\langle \psi_i(\boldsymbol{r}) \mid \psi_i(\boldsymbol{r}) \rangle_{\Omega}^{\sigma} = \int_{\Omega} \psi_i^{\sigma *}(\boldsymbol{r}) \psi_i^{\sigma}(\boldsymbol{r})\, d\boldsymbol{r} \equiv S_{ii}^{\sigma}(\Omega), \tag{3.20}$$

where σ refers to the α or β spin, and $S_{ii}^{\sigma}(\Omega)$ is the ith diagonal element of the atomic overlap matrix.

The atomic charge $q(\Omega)$ can be obtained by subtracting the nuclear charge Z_{Ω} from the number of electrons in the atom:

$$q(\Omega) = Z_{\Omega} - N(\Omega). \tag{3.21}$$

3.2.6.2 Atomic Volume

The atomic volume is the volume of the region enclosed by the zero-flux surface around the atoms in the molecule and a very low isodensity surface outside the molecule (if the atomic basin extends to infinity). In principle, a molecule extends to infinity; however, in practice, an isodensity surface of $\rho(\boldsymbol{r}) = 0.001$ a. u. is often used as the boundary of an atom, and this isodensity surface well contains the experimental van der Waals volume of the gas molecule and more than 99% of the electron population in a molecule.

3.2.6.3 Atomic Kinetic Energy

The kinetic energy operator contains at least two forms, which can be used to calculate two types of mean atomic kinetic energy. One is the Schrodinger kinetic energy:

$$K(\Omega) = -\frac{\hbar^2}{4m} N \int_{\Omega} d\boldsymbol{r} \int d\tau' [\Psi \nabla^2 \Psi^* + \Psi^* \nabla^2 \Psi], \tag{3.22}$$

while the other contains the gradient kinetic energy:

$$G(\Omega) = \frac{\hbar^2}{2m} N \int_\Omega d\mathbf{r} \int d\tau' \nabla_i \Psi^* \cdot \nabla_i \Psi, \tag{3.23}$$

where τ' contains both spatial and spin coordinates.

For a whole system or properly defined system, Eqs. (3.22) and (3.23) yield the same value of kinetic energy, i.e. $K(\Omega) = G(\Omega) = T(\Omega)$. However, $K(\Omega)$ and $G(\Omega)$ for a particular atom in the molecule may be different, with a small but nonzero error between them, due to the accuracy of numerical integration, and the Laplacian can be used as a criterion for the accuracy of atomic integration.

3.2.6.4 Laplacian

The Laplacian has the dimensions of number of electrons × (length)$^{-5}$. Due to the boundary condition of the zero-flux surface, the atomic Laplacian will be the integral of the electron density Laplacian over an atomic basin and will generally be zero, as shown by the following derivation:

$$\begin{aligned} L(\Omega) &= K(\Omega) - G(\Omega) \\ &= -\frac{\hbar^2}{4m} \int_\Omega d\mathbf{r}[\nabla^2 \rho(\mathbf{r})] \\ &= -\frac{\hbar^2}{4m} \int_\Omega dS(\Omega, \mathbf{r}) \nabla \rho(\mathbf{r}) \cdot \mathbf{n}(\mathbf{r}) = 0 \end{aligned}, \tag{3.24}$$

where the above equation applies only to a whole system or system enclosed by a zero-flux surface.

The degree to which the Laplacian approaches zero is often regarded as an indicator of the accuracy of atomic numerical integration, where the deviation from zero is the integration error, and $L(\Omega) \leq$ c. 1.0×10^{-3} a.u. for the second- and third-row atoms of the periodic table and $L(\Omega) \leq$ c. 1.0×10^{-4} a.u. for hydrogen atoms are acceptable.

3.2.6.5 Total Atomic Energy

Dividing the total energy of a molecule into the sum of individual atomic energies is not a simple problem, but fortunately can be solved by Bader. The kinetic energy density can be written as:

$$K(\mathbf{r}) = -\frac{\hbar^2}{4m} N \int d\tau' [\Psi \nabla^2 \Psi^* + \Psi^* \nabla^2 \Psi]. \tag{3.25}$$

Comparison with Eq. (3.8) gives

$$K(\mathbf{r}) = G(\mathbf{r}) - \frac{\hbar^2}{4m} \nabla^2 \rho(\mathbf{r}), \tag{3.26}$$

where integrating the above equation over the volume of ω often yields different values of $K(\mathbf{r})$ and $G(\mathbf{r})$, because the integral of the Laplacian over the volume ω will not always be zero. Kinetic energy can be well defined only when the integral of the Laplacian is zero; therefore, this integral is performed over the whole system or the atomic basin enclosed by a zero-flux surface. Integrating Eq. (3.26) over ω gives:

$$K(\omega) = G(\omega) - \frac{\hbar^2}{4m} N \int_\omega d\mathbf{r}\, \nabla \cdot \nabla \rho. \tag{3.27}$$

According to the Gauss divergence theorem (the area integral of a physical quantity over a control surface is equal to the volume integral of the divergence of the physical quantity over the control volume enclosed by the control surface, i.e. $\iint_S \rho \boldsymbol{v} \cdot \hat{\boldsymbol{n}} \mathrm{d}S = \iiint_V \nabla \cdot (\rho \boldsymbol{v})\, \mathrm{d}V$), and Eq. (3.27) can be converted into an area integral:

$$K(\omega) = G(\omega) - \frac{\hbar^2}{4m} N \int_\omega \mathrm{d}S(\omega, \boldsymbol{r}) \nabla \rho \cdot \boldsymbol{n}(\boldsymbol{r}). \tag{3.28}$$

Obviously, the second term on the right-hand side of the above equation will be zero only in the system enclosed by the zero-flux surface or in the whole infinite space system (the integral of the Laplacian over the infinite space of the whole system is zero). Therefore, only under this condition can the kinetic energy have a definite value in the whole system or in a suitable subsystem, with $K(\Omega) = G(\Omega) = T(\Omega)$ in such a system. Because the integral of the Laplacian over the Ω space is zero, the virial theorem in Eq. (3.7) can be applied to a single atom to obtain the atomic virial theorem as follows:

$$-2T(\Omega) = \mathcal{V}(\Omega), \tag{3.29}$$

where $\mathcal{V}(\Omega)$ is the atomic virial.

The electron energy of an atom can be expressed as:

$$E_e(\Omega) = T(\Omega) + \mathcal{V}(\Omega). \tag{3.30}$$

For a system in equilibrium, there is no Hellmann–Feynman force on the nucleus; therefore, the virial is the mean potential energy of the molecule, i.e. $\mathcal{V} = V$. Under this condition, Eq. (3.29) becomes:

$$-2T(\Omega) = V(\Omega), \tag{3.31}$$

where $V(\Omega)$ is the potential energy of the atom Ω. Accordingly, Eq. (3.30) becomes

$$E(\Omega) = E_e(\Omega) = T(\Omega) + V(\Omega) = -T(\Omega) = \frac{1}{2} V(\Omega), \tag{3.32}$$

where $E(\Omega)$ is the total energy of the atom Ω. Therefore, the total energy of an atom in an equilibrium molecule can be obtained by applying the virial theorem to the atom. The sum of the individual atomic energies will naturally be the energy of the whole molecule, and the additivity of atomic energies can be expressed as:

$$E_{\text{total}} = \sum_\Omega E(\Omega). \tag{3.33}$$

Atomic energies are calculated separately and then summed to obtain the energy of the molecule, which will often differ from the directly calculated energy of the whole molecule. The integral calculation will generally be considered accurate if the difference does not exceed 1 Kcal/mol.

The above discussion is based on the assumption that the calculated molecular wave function will strictly conform to the virial theorem, and the molecular virial ratio $-V/T = 2$ strictly holds. In practical applications, the true value will often deviate slightly from this value, with possible reasons including the truncation effect of the basis set, residual force on the nucleus, and limited convergence accuracy.

3.2.6.6 Atomic Dipole Moment

The atomic dipole moment is also known as the first atomic electrostatic moment, which is the mean value of the electron position vector in the atomic space and is a 3D vector. It can be defined as:

$$\boldsymbol{\mu}(\Omega) = \begin{pmatrix} \mu_x \\ \mu_y \\ \mu_z \end{pmatrix} = \begin{pmatrix} -e\int_\Omega x\rho(\boldsymbol{r})\mathrm{d}\boldsymbol{r} \\ -e\int_\Omega y\rho(\boldsymbol{r})\mathrm{d}\boldsymbol{r} \\ -e\int_\Omega z\rho(\boldsymbol{r})\mathrm{d}\boldsymbol{r} \end{pmatrix} \equiv -e\int_\Omega \mathbf{r}_\Omega \rho(\boldsymbol{r})\mathrm{d}\boldsymbol{r}, \tag{3.34}$$

$$|\boldsymbol{\mu}(\Omega)| = \sqrt{\mu_x{}^2 + \mu_y{}^2 + \mu_z{}^2}, \tag{3.35}$$

where $|\boldsymbol{\mu}(\Omega)|$ is the magnitude of the dipole moment, and the origin of the vector $\mathbf{r}_\Omega$ is at the nuclear position of the atom Ω, i.e. $\boldsymbol{r}_\Omega = \boldsymbol{r} - \boldsymbol{R}_\Omega$, where $\boldsymbol{r}$ denotes the electron coordinates and $\boldsymbol{R}_\Omega$ the nuclear coordinates of the atom Ω. The first electrostatic moment describes the polarizability of the charge density, i.e. the deviation from spherical symmetry.

3.2.6.7 Atomic Quadrupole Moment

The atomic quadrupole moment tensor, also known as the second atomic electrostatic moment, is a traceless (sum of the main diagonal elements equal to zero) symmetric tensor and can be defined as:

$$\boldsymbol{Q}(\Omega) = \begin{pmatrix} Q_{xx} & Q_{xy} & Q_{xz} \\ Q_{yx} & Q_{yy} & Q_{yz} \\ Q_{zx} & Q_{zy} & Q_{zz} \end{pmatrix}$$

$$= -\frac{e}{2}\begin{pmatrix} \int_\Omega (3x^2{}_\Omega - r_\Omega^2)\,\rho(\boldsymbol{r})\,\mathrm{d}\boldsymbol{r} & 3\int_\Omega x_\Omega y_\Omega \rho(\boldsymbol{r})\,\mathrm{d}\boldsymbol{r} & 3\int_\Omega x_\Omega z_\Omega \rho(\boldsymbol{r})\,\mathrm{d}\boldsymbol{r} \\ 3\int_\Omega y_\Omega x_\Omega \rho(\boldsymbol{r})\,\mathrm{d}\boldsymbol{r} & \int_\Omega (3y^2{}_\Omega - r_\Omega^2)\,\rho(\boldsymbol{r})\,\mathrm{d}\boldsymbol{r} & \int_\Omega y_\Omega z_\Omega \rho(\boldsymbol{r})\,\mathrm{d}\boldsymbol{r} \\ 3\int_\Omega z_\Omega x_\Omega \rho(\boldsymbol{r})\,\mathrm{d}\boldsymbol{r} & \int_\Omega z_\Omega y_\Omega \rho(\boldsymbol{r})\,\mathrm{d}\boldsymbol{r} & \int_\Omega (3z^2{}_\Omega - r_\Omega^2)\,\rho(\boldsymbol{r})\,\mathrm{d}\boldsymbol{r} \end{pmatrix}. \tag{3.36}$$

Like the atomic dipole moment, the origin of the atomic quadrupole moment is also at the nucleus. If the electron density of an atom has a spherical symmetry, then $\int_\Omega x_\Omega^2 \rho(\boldsymbol{r})\,\mathrm{d}\boldsymbol{r} = \int_\Omega y_\Omega^2 \rho(\boldsymbol{r})\,\mathrm{d}\boldsymbol{r} = \int_\Omega z_\Omega^2 \rho(\boldsymbol{r})\,\mathrm{d}\boldsymbol{r} = \frac{1}{3}\int_\Omega \boldsymbol{r}_\Omega^2 \rho(\boldsymbol{r})\,\mathrm{d}\boldsymbol{r}$ and $Q_{xx} = Q_{yy} = Q_{zz} = 0$. Therefore, the quadrupole moment is another quantity that describes the deviation of the atomic electron density from spherical symmetry. For example, if the diagonal elements are less than zero, the electron density will accumulate along the axial direction. By rotating the original coordinate system, a suitable coordinate system can always be found to diagonalize the quadrupole moment tensor $\boldsymbol{Q}(\Omega)$ into a diagonal matrix $\boldsymbol{\Lambda}(\Omega)$:

$$\boldsymbol{\Lambda}(\Omega) = \begin{pmatrix} \Lambda_{x'x'} & 0 & 0 \\ 0 & \Lambda_{y'y'} & 0 \\ 0 & 0 & \Lambda_{z'z'} \end{pmatrix}, \tag{3.37}$$

where $\Lambda_{x'x'}$, $\Lambda_{y'y'}$, and $\Lambda_{z'z'}$ are the principal values of the quadrupole moment, corresponding to principal axes x', y', and z', respectively. The principal axes are also the symmetry axes if the electron density is distributed axisymmetrically.

The traceless property of the tensor in Eq. (3.36), or its diagonalized tensor, can be attributed to the fact that $\boldsymbol{r}_{\Omega}^2 = x_{\Omega}^2 + y_{\Omega}^2 + z_{\Omega}^2$ holds in any coordinate system; hence, $Q_{xx} + Q_{yy} + Q_{zz} = \Lambda_{xx} + \Lambda_{yy} + \Lambda_{zz} = 0$. There are only five independent matrix elements in the original coordinate system, with only two independent matrix elements in the diagonalized matrix. The modulus of the quadrupole moment can be defined as:

$$|\boldsymbol{Q}| = \sqrt{\frac{2}{3}\left(Q_{xx}^2 + Q_{yy}^2 + Q_{zz}^2\right)} = \sqrt{\frac{2}{3}\left(\Lambda_{xx}^2 + \Lambda_{yy}^2 + \Lambda_{zz}^2\right)}. \tag{3.38}$$

3.2.6.8 Atomic Information Entropy

The Shannon information entropy of the electron density $\rho(\boldsymbol{r})$ in coordinate space, $S_{\boldsymbol{r}}$, can be defined as: [11]

$$S_{\boldsymbol{r}} = -\int \rho(\boldsymbol{r}) \ln \rho(\boldsymbol{r}) \mathrm{d}\boldsymbol{r}. \tag{3.39}$$

Correspondingly, the information entropy of the electron density in momentum space, $S_{\boldsymbol{p}}$, can be defined as:

$$S_{\boldsymbol{p}} = -\int \gamma(\mathbf{p}) \ln \gamma(\mathbf{p}) \mathrm{d}\mathbf{p}, \tag{3.40}$$

where $\gamma(\mathbf{p})$ is the momentum density. The atomic information entropy can be obtained by integrating $S_{\boldsymbol{r}}$ or $S_{\boldsymbol{p}}$ over the atomic basin, which measures the uncertainty of the information regarding the electron density distribution in the atomic basin in coordinate or momentum space. A localized distribution of electron density implies a small information entropy, and according to the Heisenberg uncertainty principle, a localized distribution of electron density in coordinate space will correspond to a delocalized distribution of electron density in momentum space.

3.3 Chemical Interaction Analysis

3.3.1 Source Function

Bader and Gatti were the first to introduce the concept of the source function [12], which can represent the electron density at any point in a molecule equivalently as the sum of the contributions of individual atoms. Accordingly, we can imagine that the electron density of an atom in a molecule will depend not only on the atom itself but also on other atoms in the molecule. Thus, source functions can be used to analyze the electron density properties at certain typical locations (e.g. BCPs). Analysis of the contribution of each point on the bond path to the source function of the BCP will show which locations play a role in the accumulation or depletion of charge at the BCP. The local form of the source function (local source) can be further expressed as the sum of the contributions of electron kinetic energy density and potential energy density; thus, revealing the fingerprint information of the bond interactions. Source functions are very helpful for studying the portability of certain functional groups. For example, the characteristic electron density at the BCP of the C–H bond of a hydrocarbon chain end group requires that the contribution of other atoms in the hydrocarbon chain remain constant, independent of the length of the hydrocarbon chain. In addition, source functions can be used to classify hydrogen bonds.

Bader and Gatti noted that for a closed system with boundaries at infinity, the electron density at any point $\boldsymbol{r}$ in the atomic basin Ω can be expressed as:

$$\rho(\boldsymbol{r}) = \int \mathrm{LS}(\boldsymbol{r}, \boldsymbol{r}')\mathrm{d}\boldsymbol{r}' = \int_{\Omega} \mathrm{LS}(\boldsymbol{r}, \boldsymbol{r}')\mathrm{d}\boldsymbol{r}' + \sum_{\Omega' \neq \Omega} \int_{\Omega'} \mathrm{LS}(\boldsymbol{r}, \boldsymbol{r}')\mathrm{d}\boldsymbol{r}'. \tag{3.41}$$

The right-hand side of the equation is the integral over the whole space of the local sources contributing to the electron density at $\boldsymbol{r}$, which can be decomposed as the sum of the contributions from the basin Ω and the other basins Ω' in the system. Local sources can be defined as:

$$\mathrm{LS}(\boldsymbol{r}, \boldsymbol{r}') = -(1/4\pi)\nabla^2\rho(\boldsymbol{r}')/|\boldsymbol{r} - \boldsymbol{r}'|, \tag{3.42}$$

where the Green function $(4\pi|\boldsymbol{r} - \boldsymbol{r}'|)^{-1}$ can be considered the influence function of the source $\nabla^2\rho(\boldsymbol{r}')$ on $\rho(\boldsymbol{r})$. The influence of the Laplacian $\nabla^2\rho(\boldsymbol{r}')$ will be related to the distance between the source $\boldsymbol{r}'$ and $\boldsymbol{r}$, and the integral of the local source over the basin Ω can be defined as the source function $S(\boldsymbol{r}, \Omega)$, namely,

$$\int_{\Omega} \mathrm{LS}(\boldsymbol{r}, \boldsymbol{r}')\mathrm{d}\boldsymbol{r}' \equiv S(\boldsymbol{r}, \Omega). \tag{3.43}$$

Thus, the electron density at $\boldsymbol{r}$ within an atom can be written as the sum of the source function of this atom and the source functions of the other atoms in the system, according to:

$$\rho(\boldsymbol{r}) = S(\boldsymbol{r}, \Omega) + \sum_{\Omega' \neq \Omega} S(\boldsymbol{r}, \Omega'). \tag{3.44}$$

Therefore, the source function describes the importance of an atom or other atoms in the system for the electron density at any point.

In addition, the Laplacian also appears in the local expression of the virial theorem:

$$\nabla^2\rho(\boldsymbol{r}) = (4m/\hbar^2)[2G(\boldsymbol{r}) + V(\boldsymbol{r})]. \tag{3.45}$$

Therefore, the local source will be related to the electron kinetic energy density G and the electron potential energy density V (or virial) and can also be written as

$$\mathrm{LS}(\boldsymbol{r}, \boldsymbol{r}') = -(m/\pi\hbar^2)[2G(\boldsymbol{r}) + V(\boldsymbol{r})]/|\boldsymbol{r} - \boldsymbol{r}'|. \tag{3.46}$$

In addition, the local source can be expressed as the sum of the local electron kinetic energy contribution LG and the local electron potential energy contribution LV:

$$\begin{aligned} \mathrm{LS}(\boldsymbol{r},\boldsymbol{r}') &= LG(\boldsymbol{r},\boldsymbol{r}') + \mathrm{LV}(\boldsymbol{r},\boldsymbol{r}') \\ LG(\boldsymbol{r},\boldsymbol{r}') &= -(m/\pi\hbar^2)2G(\mathbf{r})/|\boldsymbol{r}-\boldsymbol{r}'|. \\ \mathrm{LV}(\boldsymbol{r},\boldsymbol{r}') &= -(m/\pi\hbar^2)V(\mathbf{r})/|\boldsymbol{r}-\boldsymbol{r}'| \end{aligned} \tag{3.47}$$

Because G is always positive, the contribution of LG to the electron density at $\boldsymbol{r}$ will always be negative. By contrast, because V is always negative, the contribution of LV to the electron density at $\boldsymbol{r}$ will always be positive. Therefore, the electron accumulation region ($\nabla^2\rho(\boldsymbol{r}) < 0$) in the system, where the electron potential energy dominates, will be the source region of the electron density at $\boldsymbol{r}'$. By contrast, the electron-depleted region ($\nabla^2\rho(\boldsymbol{r}) > 0$) in the system, where the electron kinetic energy dominates, will be the dissipation region of the electron density at $\boldsymbol{r}'$ [13].

3.3.2 Electron Localization Function

ELF was first introduced by Becke and Edgecombe [14] and has been widely used in chemical bond analysis [15]. Fermi hole is a key concept in understanding the behavior of electron localization and is a corollary of the Pauli repulsion principle. Obviously, assuming only Coulomb repulsion between electrons, it is impossible for any pair of electrons to be simultaneously present in a defined region of space. We understand the possible electron localization behavior as follows. Localization of an electron pair implies a high probability of finding two opposite-spin electrons in a defined region at the same time, where the probability of pair exchanges with electrons outside that region is low. For an electron in space, its motion will be linked to the motion of its corresponding Fermi hole. For an electron at a defined position, the corresponding Fermi hole will be a distribution function that can be used to describe the reduction in the probability of finding another same-spin electron at a position in space, and this reduction in probability will be due to the Pauli exclusion principle. The Fermi hole depends on the position $\boldsymbol{r}_1$ of the associated electron and the position $\boldsymbol{r}_2$ of another electron. For a given spin, the diagonal element (at $\boldsymbol{r}_1 = \boldsymbol{r}_2$) of the Fermi hole should be equal to the total density of the given spin at this position; thus, the probability of finding another same-spin electron at this position will be zero. For a reference electron fixed at position $\boldsymbol{r}_1$, the integral of the Fermi hole over $\boldsymbol{r}_2$ will be another same-spin electron that is repelled. Therefore, the Fermi hole describes the delocalization behavior of electrons, which can only move near the corresponding Fermi hole, and where the Fermi hole is, where the electron is localized. For example, a nucleus has a strong attractive potential for a nearby electron of definite spin, making it very difficult for the electron to overcome this potential barrier; thus, the Fermi hole of the electron will be strongly localized in this region. Assuming that its value reaches the maximum (i.e. the total density of a given spin) anywhere in the region where the Fermi hole is located, same-spin electrons will be completely repelled from this region. An electron with another spin that happens to be in this region would also have the same result. Thus, electron pairs with different spins will be confined to move in a particular region, while other electrons will be repelled regardless of their spins. Because Coulombic repulsion between electrons is exactly the opposite of this effect, electron localization cannot be maximized. Of note, the Fermi hole effect does not generate attractions between electrons of different spins, but rather exerts an additional repulsion between the electron pair and other electrons [16]. In general, this statement holds true for the inner $1s^2$ electron pair, as well as for the core electrons of any system.

In the study of electron correlation effects, it has long been noticed that the motions of same-spin electrons are more strongly correlated than opposite-spin electrons due to the Pauli repulsion principle. Therefore, it may be more convenient to separately study the densities of same-spin and opposite-spin electron pairs. The electron pair density $\rho_2^{\sigma\sigma'}(\boldsymbol{r}_1, \boldsymbol{r}_2)$ gives the probability of finding an electron with spin-σ at $\boldsymbol{r}_1$, as well as also finding another electron with spin-σ' at $\boldsymbol{r}_2$. Because the electron–electron interaction depends only on the distance between electrons and is independent of the orientation angle, it is convenient to transform the coordinate system into $\boldsymbol{r} = 1/2(\boldsymbol{r}_1 + \boldsymbol{r}_2)$ and $\boldsymbol{s} = \boldsymbol{r}_1 - \boldsymbol{r}_2$. One advantage of using this new coordinate system is that the number of variables for electron–electron interactions will depend on changes from six ($\boldsymbol{r}_1$ and $\boldsymbol{r}_2$) to four ($\boldsymbol{r}$ and s); thus, it is only necessary to study the spherically averaged electron pair density $\rho_{2,\text{av}}^{\sigma\sigma'}(\boldsymbol{r}_1, s)$, which can be defined as:

$$\rho_{2,\text{av}}^{\sigma\sigma'}(\mathbf{r}_1, s) = \frac{1}{4\pi} \int \rho_2^{\sigma\sigma'}(\boldsymbol{r}_1, s) \mathrm{d}\Omega_s, \tag{3.48}$$

where the integration is performed over the angle of the s vector.

Becke and Edgecombe preferred to use the conditional electron pair density for same-spin electrons to describe the probability of finding a same-spin electron at $\boldsymbol{r}_2$ under the condition that an electron is already at $\boldsymbol{r}_1$, given by the following equation:

$$P^{\sigma\sigma}(\boldsymbol{r}, s) = \frac{\rho_{2,\text{av}}^{\sigma\sigma'}(\boldsymbol{r}, s)}{\rho_\sigma(\boldsymbol{r})}, \tag{3.49}$$

where $\rho_\sigma(\boldsymbol{r})$ is the density of electrons with spin σ and, under the Kohn–Sham approximation, can be expressed as:

$$\rho_\sigma(\boldsymbol{r}) = \sum_i^\sigma \phi_i(\boldsymbol{r})^2, \tag{3.50}$$

where $\{\phi_i\}$ is a set of Kohn–Sham orbitals and the summation is performed on all occupied orbitals with spin σ.

A Taylor expansion of the spherically averaged conditional electron pair density in the domain of $s = 0$ will yield the short-range behavior of electrons at $\boldsymbol{r}_2$ near $\boldsymbol{r}_1$. The main terms of the Taylor expansion are given by:

$$P^{\sigma\sigma}(\boldsymbol{r}, s) = \frac{1}{3}\left(t_\sigma - \frac{1}{4}\frac{(\nabla\rho_\sigma)^2}{\rho_\sigma}\right) s^2 + \cdots, \tag{3.51}$$

where t_σ is the kinetic energy density and always positive, with the following definition:

$$t_\sigma = \sum_i^\sigma (\nabla\phi_i)^2. \tag{3.52}$$

The Taylor expansion (3.51) contains the localization information of all electrons, where the smaller the probability of finding a second electron at $\boldsymbol{r}$, the more localized the reference electron. Therefore, the degree of localization of electrons can be directly related to the quantity in the parentheses on the right-hand side of Eq. (3.51). A nonnegative quantity D_σ can be defined by:

$$D_\sigma = t_\sigma - \frac{1}{4}\frac{(\nabla\rho_\sigma)^2}{\rho_\sigma}. \tag{3.53}$$

Hence, there is a high probability of finding localized electrons or electron pairs in regions with small D_σ values. However, D_σ may be very high in other locations. For convenience, Becke and Edgecombe proposed two additional normalization rules. Referring to the kinetic energy density $D_\sigma^0 = (3/5)(6\pi^2)^{2/3}\rho^{5/3}$ of a uniform electron gas, this will limit the value of the function between 0 and 1, giving the definition of ELF (η):

$$\eta = \left(1 + \chi_\sigma^2\right)^{-1}, \tag{3.54}$$

where

$$\chi_\sigma = D_\sigma / D_\sigma^0. \tag{3.55}$$

In this way, the ELF satisfies the condition $0 \leq \text{ELF} \leq 1$. The probability of electron localization will be high when ELF is close to 1, and locations with ELF close to 0.5 will correspond to uniform electron gas behavior.

As shown in Eqs. (3.52)–(3.55), in order to calculate the ELF, it is necessary to obtain the wave function of the system, which is inconvenient for analyzing the experimental electronic structure because it is difficult to obtain the experimental electronic wave function in most cases. For a closed-shell system, Eq. (3.54) defining ELF can be rewritten, and the spin symbol σ can be omitted, as follows:

$$\eta = \left[1 + (D_\text{P}/D_0)^2\right]^{-1}, \tag{3.56}$$

where D_p is the Pauli kinetic energy, according to:

$$D_\text{p} = t(\boldsymbol{r}) - \frac{1}{8}\frac{|\nabla\rho(\boldsymbol{r})|^2}{\rho(\boldsymbol{r})}, \tag{3.57}$$

where $t(\boldsymbol{r})$ is the kinetic energy density (with a new factor of 1/2 compared to Eq. (3.52)):

$$t(\boldsymbol{r}) = \frac{1}{2}\sum_i (\nabla\phi_i)^2, \tag{3.58}$$

where D_0 is the kinetic energy density of a uniform electron gas:

$$D_0 = \frac{3}{10}(3\pi^2)^{2/3}\rho(\boldsymbol{r})^{5/3}. \tag{3.59}$$

To avoid using one-electron wave functions in the calculation of ELF, with the Kirzhnits approximation, Eq. (3.58) can be replaced by the second-order gradient expansion of the kinetic energy density [17]:

$$t(\boldsymbol{r}) = \frac{3}{10}(3\pi^2)^{2/3}\rho(\boldsymbol{r})^{5/3} + \frac{1}{72}\frac{|\nabla\rho(\boldsymbol{r})|^2}{\rho(\boldsymbol{r})} + \frac{1}{6}\nabla^2\rho(\boldsymbol{r}), \tag{3.60}$$

hence,

$$D_P = \frac{3}{10}(3\pi^2)^{2/3}\rho(\mathbf{r})^{5/3} - \frac{1}{9}\frac{|\nabla\rho(\mathbf{r})|^2}{\rho(\mathbf{r})} + \frac{1}{6}\nabla^2\rho(\mathbf{r}). \tag{3.61}$$

In this way, the ELF can be calculated directly from the electron density and its various orders of derivatives. In addition to the above equation, other approximate methods exist for calculating the ELF from the electron density.

3.3.3 Reduced Density Gradient

To study the noncovalent interactions (e.g. van der Waals forces, hydrogen bond, and intermolecular steric repulsion) that widely exist in chemistry and biology, Johnson et al. proposed reduced density gradient (RDG, denoted as s), a dimensionless quantity based on electron density and its gradient, to describe the characteristics of a uniform electron gas with electron drift in the system [18]. It can be defined as follows:

$$s = \frac{1}{2(3\pi^2)^{1/3}}\frac{|\nabla\rho|}{\rho^{4/3}}. \tag{3.62}$$

The value of s will be very large in regions far from the molecule and very small in regions of covalent or noncovalent interactions. The weak interaction region can be determined according to the plot of the variation of s with ρ. Because there was a saddle point with a large ρ but $\nabla\rho = 0$ for covalent bonds, a characteristic of covalent bonds in the ρ-s plot will indicate the presence of a region with $s = 0$ in locations where ρ is large. For noncovalent interactions, there will be one or more regions with very small ρ and $\nabla\rho$ in the ρ-s plot.

Although low electron density and its gradient can be used to find weak noncovalent interaction regions, the exact interaction type, such as van der Waals forces, hydrogen bonds, πbonds, or steric hindrance, cannot be obtained based solely on the electron density. The Laplacian $\nabla^2\rho = \lambda_1 + \lambda_2 + \lambda_3$, $(\lambda_1 \leq \lambda_2 \leq \lambda_3)$ has been widely used to distinguish different interaction types. Erin R. Johnson et al. further proposed that the sign$(\lambda_2)\rho$–s diagram can be used to describe the intensity and type of non-valent interactions in a system, where sign(λ_2) indicates the sign of the second largest Eigenvalue λ_2 in the Hessian matrix. In the sign$(\lambda_2)\rho$–s diagram, the density gradient peak on the negative sign$(\lambda_2)\rho$ side denotes bonding interactions, where a larger negative sign$(\lambda_2)\rho$ peak indicates attraction (e.g. hydrogen bond and dipole–dipole interactions), a close-to-zero negative sign$(\lambda_2)\rho$ peak indicates weak interactions (e.g. van der Waals forces), and a peak on the positive sign$(\lambda_2)\rho$ side indicates the nonbonding region (e.g. steric hindrance).

The distribution characteristics of non-covalent interactions in real space can be understood through the iso-surface plot of a specific RDG value (e.g. $s = 0.5$). If only noncovalent interactions need to be shown, the electron density will have to be truncated, and the covalent interactions can be excluded by only taking part with a smaller electron density (e.g. $\rho < 0.05$ a.u.). The sign$(\lambda_2)\rho$ is a good indicator of interaction strength, and the RDG iso-surface can be colored based on the sign$(\lambda_2)\rho$ value, providing an abundant visualization of non-covalent interactions.

Lu Tian et al. further promoted the RDG definition by proposing the interaction region indicator (IRI): [19] $\mathrm{IRI}(\boldsymbol{r}) = \frac{|\nabla\rho(\boldsymbol{r})|}{\rho(\boldsymbol{r})^{a}}$. For most systems, a can take the best empirical value, i.e. 1.1. Similar to RDG analysis, the IRI iso-surface can be colored based on $\mathrm{sign}(\lambda_2)\rho$. Compared to RDG, the IRI is advantageous in characterizing both covalent and non-covalent interactions.

3.4 Coarse Graining and Energy Partition of the Density Matrix

All operations described here are based on the partition of the corresponding physical quantities in 3D space, where the partition method should be as independent as possible from the accuracy of the calculation and the form of the wave function. In general, the space partition can either generate the corresponding objects for subsequent analysis or produce new distributions of physical quantities for further partition. We can consider two main partition schemes: ① the partition of space into various atomic regions, which are coarse-grained and also known as interacting quantum atoms (IQA) [20], and ② the partition of space into an infinite number of small, variable-sized units that possess a certain invariant quantum mechanical property. This section describes the first partition scheme.

3.4.1 Partition of the Density Matrix in Real Space

A weight function $w_A(\boldsymbol{r})$ can be assigned to a chemical object A (e.g. an atom) such that the sum of the weights of all objects will be 1 for each point in the space, according to:

$$\sum_{A} w_A(\boldsymbol{r}) = 1(\boldsymbol{r}). \tag{3.63}$$

This is a very universal condition because it unifies fuzzy and rigorous partition schemes. The fuzzy partition can be described using a continuous weight function, while the rigorous partition scheme will follow $w_A(\boldsymbol{r} \in \Omega_A) = 1$, $w_A(\boldsymbol{r} \notin \Omega_A) = 0$. The latter includes all partition methods based on the topological analysis of scalar fields, such as QTAIM and ELF partition methods, while the former includes all fuzzy partition methods, such as stockholder's or Hirshfeld atomic partitions [21], Li and Parr's minimum self-energy atoms [22], and Becke's decomposition [23].

The unit partition method can be used to decompose the density matrix. We can consider a first-order non-diagonal density matrix $\gamma^{(1)}(\boldsymbol{r}'; \boldsymbol{r})$ and assume that for any subsystem A, we will obtain:

$$\left.\frac{\hat{T}\gamma^{(1)}(\boldsymbol{r}'; \boldsymbol{r})}{\gamma^{(1)}(\boldsymbol{r}'; \boldsymbol{r})}\right|_{\boldsymbol{r}'\to\boldsymbol{r}} = \left.\frac{\hat{T}\gamma_A^{(1)}(\boldsymbol{r}'; \boldsymbol{r})}{\gamma_A^{(1)}(\boldsymbol{r}'; \boldsymbol{r})}\right|_{\boldsymbol{r}'\to\boldsymbol{r}}, \tag{3.64}$$

where $\hat{T}$ is the kinetic energy operator, and the kinetic energy density of each electron will not depend on whether we consider the electron as part of the whole system or a subsystem A. Thus:

$$\gamma_A^{(1)}(\boldsymbol{r}';\boldsymbol{r}) = w_A(\boldsymbol{r}')\gamma^{(1)}(\boldsymbol{r}';\boldsymbol{r}), \tag{3.65}$$

or

$$\gamma^{(1)}(\boldsymbol{r}';\boldsymbol{r}) = \sum_A \gamma_A^{(1)}(\boldsymbol{r}';\boldsymbol{r}) = 1(\boldsymbol{r}')\gamma^{(1)}(\boldsymbol{r}';\boldsymbol{r}). \tag{3.66}$$

The above equation gives us the common density partition $\rho(\boldsymbol{r}) = \sum_A \rho_A(\boldsymbol{r})$. Similarly, the two-electron density matrix $\gamma_{AB}^2\left(\boldsymbol{r}_1', \boldsymbol{r}_2'; \boldsymbol{r}_1, \boldsymbol{r}_2\right)$ can be written as:

$$\gamma_{AB}^{(2)}\left(\boldsymbol{r}_1', \boldsymbol{r}_2'; \boldsymbol{r}_1, \boldsymbol{r}_2\right) = w_A\left(\boldsymbol{r}_1'\right) w_B\left(\boldsymbol{r}_2'\right) \gamma^{(2)}\left(\boldsymbol{r}_1', \boldsymbol{r}_2'; \boldsymbol{r}_1, \boldsymbol{r}_2\right). \tag{3.67}$$

The density matrix of the whole system can be obtained by summing the density matrices of all atomic pairs A and B according to:

$$\begin{aligned} \gamma^{(2)}\left(\mathbf{r}_1', \mathbf{r}_2'; \mathbf{r}_1, \mathbf{r}_2\right) &= \sum_A \sum_B \gamma_{AB}^{(2)}\left(\boldsymbol{r}_1', \boldsymbol{r}_2'; \boldsymbol{r}_1, \boldsymbol{r}_2\right) \\ &= 1\left(\boldsymbol{r}_1'\right) 1\left(\boldsymbol{r}_2'\right) \gamma^{(2)}\left(\boldsymbol{r}_1', \boldsymbol{r}_2'; \boldsymbol{r}_1, \boldsymbol{r}_2\right) \end{aligned} . \tag{3.68}$$

The diagonal decomposition of the density matrix, $\gamma^{(2)}(\boldsymbol{r}_1, \boldsymbol{r}_2) = \sum_A \sum_B \gamma_{AB}^{(2)}(\boldsymbol{r}_1, \boldsymbol{r}_2)$, allows for the total electron repulsion energy to be written as the sum of the contributions from the monoatomic region ($A = B$) and the diatomic region ($A \neq B$). This idea can be immediately generalized to a density matrix of order p:

$$\gamma^{(p)}\left(\boldsymbol{r}_1', \ldots, \boldsymbol{r}_p'; \boldsymbol{r}_1, \ldots \boldsymbol{r}_p\right) = 1\left(\boldsymbol{r}_1'\right) \ldots 1\left(\boldsymbol{r}_p'\right) \gamma^{(p)}\left(\boldsymbol{r}_1', \ldots, \boldsymbol{r}_p'; \boldsymbol{r}_1, \ldots \boldsymbol{r}_p\right), \tag{3.69}$$

where

$$1\left(\boldsymbol{r}_1'\right) \ldots 1\left(\boldsymbol{r}_p'\right) = \sum_A \cdots \sum_P w_A\left(\boldsymbol{r}_1'\right) \ldots w_P\left(\boldsymbol{r}_p'\right). \tag{3.70}$$

Therefore, we can obtain:

$$\gamma^{(p)}\left(\boldsymbol{r}_1', \ldots, \boldsymbol{r}_p'; \boldsymbol{r}_1, \ldots \boldsymbol{r}_p\right) = \sum_A \cdots \sum_P \gamma_{A \ldots P}^{(p)}\left(\boldsymbol{r}_1', \ldots, \boldsymbol{r}_p'; \boldsymbol{r}_1, \ldots \boldsymbol{r}_p\right), \tag{3.71}$$

where

$$\gamma_{A \ldots P}^{(p)}\left(\boldsymbol{r}_1', \ldots, \boldsymbol{r}_p'; \boldsymbol{r}_1, \ldots \boldsymbol{r}_p\right) = w_A\left(\boldsymbol{r}_1'\right) \ldots w_P\left(\boldsymbol{r}_p'\right) \gamma^{(p)}\left(\boldsymbol{r}_1', \ldots, \boldsymbol{r}_p'; \boldsymbol{r}_1, \ldots \boldsymbol{r}_p\right). \tag{3.72}$$

It should be noted that the atoms $A, \ldots, P$ in the above three equations will correspond to the coordinates $\boldsymbol{r}_1, \ldots, \boldsymbol{r}_p$, respectively.

Coarse-grained density matrices (CGDM) can be obtained by region-weighted averaging [24]. Considering an N-electron system with wavefunction Ψ, which can be divided into m parts by unit partition, we can define a coarse-grained or regional average of an electron j in the region Ω_k as $N \int \mathrm{d}\boldsymbol{r}_j w_k(\boldsymbol{r}_j)\Psi^*\Psi$, which is a function of $\boldsymbol{r}_1, \ldots \boldsymbol{r}_{j-1}, \boldsymbol{r}_{j+1}, \ldots, \boldsymbol{r}_N$ with the factor N taking into account all cases of the N-electron system. Similarly, for the two-electron case, the coarse-graining of electrons j and l in the region Ω_k can be defined as:

$$N(N-1) \int \mathrm{d}\boldsymbol{r}_j \mathrm{d}\boldsymbol{r}_l w_k(\boldsymbol{r}_j) w_k'(\boldsymbol{r}_l) \Psi^* \Psi \quad (k \neq k')$$

or

$$\frac{N(N-1)}{2} \int \mathrm{d}\boldsymbol{r}_j \mathrm{d}\boldsymbol{r}_l w_k(\boldsymbol{r}_j) w'_k(\boldsymbol{r}_l) \Psi^* \Psi \quad (k = k'),$$

which is a function of the coordinates of all electrons except j and l. Similarly, we can define the coarse-graining process of c_1 electrons in the region Ω_1, c_2 electrons in the region Ω_2, …, and the c_m electrons in the region Ω_m. The number of electrons that are coarse-grained will be less than or equal to N, $c_1 + \cdots + c_m = \Lambda \leq N$, with the remaining $i = N - \Lambda \geq 0$ electrons known as free or non-coarsely-grained electrons. Without the loss of generality, we can consider electrons from 1 to i as free electrons and electrons from $i+1$ to $i+c_1$ as coarse-grained electrons on Ω_1. A coarse-grained C can be understood as the decomposition of coarse-grained electrons and represented as an ordered set $C = (c_1, \ldots, c_m)$. If all N electrons are also decomposed in this way, it is called a resonance structure S in real space. Considering the indistinguishability of electrons, we can define a coarse-grained density matrix of order i corresponding to C by the following:

$$\gamma^{(i)}(\boldsymbol{r}_1, \ldots, \boldsymbol{r}_i)[C] = I_C \int \mathrm{d}\boldsymbol{r}_{i+1} \ldots \mathrm{d}\boldsymbol{r}_N w_C \Psi^* \Psi, \tag{3.73}$$

where $I_C = N!/(i!c_1! \cdots c_m!) = \binom{N}{i} \Lambda!/(c_1! \cdots c_m!)$ corrects for all equivalent electron arrangements, and $w_C = w_C(\boldsymbol{r}_{i+1}, \ldots, \boldsymbol{r}_N)$ represents the product of Λ weighting factors, i.e., where weight w_1 is used for electrons from $i+1$ to $i+c_1$ using, weight w_2 for electrons from $i+c_1+1$ to $i+c_1+c_2$, and so on, until the appropriate weights have been used for all coarse-grained electrons. Certain special examples of Eq. (3.73) are familiar objects in the study of quantification, such as when $\Lambda = 0$ and $i = N$, $\gamma^{(N)}(\boldsymbol{r}_1, \ldots, \boldsymbol{r}_N)[] = \Psi^*\Psi$, where [] indicates that no electrons are coarse-grained, and when $\Lambda = N$, $i = 0$, $m = 1$, $\gamma^{(0)}()[C]$, it indicates the normalization of the wave function:

$$\gamma^{(0)}()[C; m = 1] = \int \mathrm{d}\boldsymbol{r}_1 \ldots \mathrm{d}\boldsymbol{r}_N \Psi^* \Psi = 1, \tag{3.74}$$

where () means that no electron is free. When $\Lambda = N - 1$ and $m = 1$, $\gamma^{(i)}(\boldsymbol{r}_1, \ldots, \boldsymbol{r}_i)$ $[C; m = 1]$ will be $\gamma^{(i)}(\boldsymbol{r}_1, \ldots, \boldsymbol{r}_i)$, with the diagonal element of the density matrix of order i. We can use the following property:

$$\begin{aligned} 1(\boldsymbol{r}_{i+1}) \cdots 1(\boldsymbol{r}_{i+\Lambda}) &= \sum_{A_1} \cdots \sum_{A_\Lambda} w_{A_1}(\boldsymbol{r}_{i+1}) \cdots w_{A_\Lambda}(\boldsymbol{r}_{i+\Lambda}) \\ &= \sum_{\{C\}} \frac{\Lambda!}{c_1! \cdots c_m!} w_C = \binom{i}{N} \sum_{\{C\}} I_C w_C \end{aligned} \tag{3.75}$$

The $\{C\}$ under the summation symbol indicates all possibilities of satisfying $c_1 + \cdots + c_m = \Lambda$.

The diagonal elements of any conventional density matrix of order i can be obtained by summing the coarse-grained density matrices of all possible Λ-electron coarse-grained sets $\{C\}$, according to

$$\gamma^{(i)}(\boldsymbol{r}_1, \ldots, \boldsymbol{r}_i) = \binom{N}{i} \int \mathrm{d}\boldsymbol{r}_{i+1} \ldots \mathrm{d}\boldsymbol{r}_N \Psi^* \Psi = \sum_{\{C\}} \gamma^{(i)}(\boldsymbol{r}_1, \ldots, \boldsymbol{r}_i)[C]. \tag{3.76}$$

Coarse-grained density matrices provide a region-based decomposition approach for any relevant quantum-mechanical expectation values. For a given resonance structure $S=(n_1, \ldots, n_m)$, the probability of finding n_1 electrons in Ω_1, n_2 electrons in $\Omega_2, \ldots$, and n_m electrons in Ω_m will be $p(S) = p(n_1, \ldots, n_m) = \gamma^{(0)}()[S]$.

When $i=0$, Eq. (3.76) becomes:

$$\gamma^{(0)}() = \int \mathrm{d}\boldsymbol{r}_1 \ldots \mathrm{d}\boldsymbol{r}_N \Psi^* \Psi = \sum_{\{S\}} \gamma^{(0)}()[S] = \sum_{\{S\}} p(S) = 1. \tag{3.77}$$

The above equation is reasonable because the sum of the probabilities of all resonance structures must be one.

If we are only interested in k regions ($\alpha, \ldots, \kappa$) in m, $k<m$, we can consider a subsystem containing $k+1$ regions, with the last region containing a weight function satisfying $w_\omega = 1 - \sum_{i=1}^{k} w_i$. In this case, the probability of finding $n_\alpha, \ldots, n_\kappa$ (satisfying $\Lambda = n_\alpha + \cdots + n_\kappa \leq N$) electrons in regions $\alpha, \ldots, \kappa$, respectively, will be $p(n_\alpha, \ldots, n_\kappa) = p(n_\alpha, \ldots, n_\kappa, (N-\Lambda)_\omega)$.

We can construct a conditional density matrix based on the coarse-grained objects:

$$\gamma_A^{(1)}(\boldsymbol{r}_1 \mid S) = w_A(\boldsymbol{r}_1) \frac{\gamma^{(1)}(\boldsymbol{r}_1)[C]}{p(S)}. \tag{3.78}$$

This equation describes the probability of finding an electron in the region A under the condition of the resonance structure S, where C is the coarse-grained set formed by removing an electron from the region A in S, satisfying $n_A = c_A + 1$. The number of electrons in the region A of structure S will be $\int \mathrm{d}\boldsymbol{r}_1 \gamma_A^{(1)}(\boldsymbol{r}_1 \mid S) = n_A$.

3.4.2 Energy Partition

When a coarse-graining scheme is determined, any observable of a molecule can be decomposed into the sum of individual regional contributions. This can be carried out at the regional level, where single-particle observations can be decomposed into the cumulative sum of regional contributions, and the two-particle observations will be in the form of a sum of regional pairs. A finer decomposition at the resonance structure level is also possible. At the regional level, under the Coulomb Hamiltonian assumption and the Born–Oppenheimer approximation, the electron energy can be expressed in terms of the first- and second-order density matrices as follows:

$$\begin{aligned} E_\mathrm{e} = &\int \mathrm{d}\boldsymbol{r}_1 \left(\hat{t} - \sum_\alpha \frac{Z_\alpha}{r_{1\alpha}}\right) 1\left(\boldsymbol{r}_1'\right) \gamma^{(1)}\left(\boldsymbol{r}_1'; \boldsymbol{r}_1\right) \\ &+ \frac{1}{2} \int \mathrm{d}\boldsymbol{r}_1 \mathrm{d}\boldsymbol{r}_2 \frac{1}{r_{12}} 1(\boldsymbol{r}_1) 1(\boldsymbol{r}_2) \gamma^{(2)}(\boldsymbol{r}_1, \boldsymbol{r}_2) \end{aligned}, \tag{3.79}$$

where the kinetic energy and electron-nucleus potential energy become the sum of the single-region contributions, and the electron repulsion energy will be the sum of the two-region (or basin) terms. Using Eq. (3.63), we can obtain:

$$E_\mathrm{e} = \sum_A \left(T^A + \sum_\alpha V_\mathrm{en}^{A\alpha}\right) + \sum_A V_\mathrm{ee}^{AA} + \sum_{A>B} V_\mathrm{ee}^{AB}. \tag{3.80}$$

We can use a simple notation $T^A = T(\Omega_A)$ for the kinetic energy of the atom and $V_{\text{en}}^{A\alpha}$ for the interaction energy between the electron and the nucleus α of atom A, noting that the V_{en} and V_{ee} terms are well-defined in any unit decomposition; however, only the QTAIM partition will have a well-defined regional kinetic energy. In practical applications, a special one-electron kinetic energy operator in the form of $\hat{t} = \hat{g} = (1/4)\nabla \cdot \nabla'$ is often used. When the unit partition is of all-atom type, each region A will correspond to a defined nucleus α, and the total energy of the molecule can be written as the sum of intra- and inter-atomic interaction terms:

$$E = \sum_{A} \left(T^A + V_{\text{en}}^{AA} + V_{\text{ee}}^{AA}\right) + \sum_{A>B} \left(V_{\text{en}}^{AB} + V_{\text{en}}^{BA} + V_{\text{ee}}^{AB} + V_{\text{nn}}^{AB}\right), \tag{3.81}$$

where the standard nuclear repulsion energy $V_{\text{nn}}^{AB} = Z_A Z_B / R_{AB}$ is added.

The atomic interaction energy decomposition method separates the intra-atomic contribution, accounting for the main contribution, and the inter-atomic interaction term, which has only a minor contribution. The self-energy of atom A can be defined as:

$$E_{\text{self}}^{A} = T^A + V_{\text{en}}^{AA} + V_{\text{ee}}^{AA}. \tag{3.82}$$

If the interaction term vanishes completely, the self-energy is the atomic energy in a vacuum. The interaction energy between A and B can be defined as:

$$E_{\text{int}}^{AB} = V_{\text{en}}^{AB} + V_{\text{en}}^{BA} + V_{\text{ee}}^{AB} + V_{\text{nn}}^{AB}. \tag{3.83}$$

Then, we can obtain:

$$E = \sum_{A} E_{\text{self}}^{A} + \sum_{A>B} E_{\text{int}}^{AB}. \tag{3.84}$$

In this sense, atoms (or regions) behave like regular physical systems in real space, where the self-energy includes single-particle energy contributions and interactions between internal particles, and the interaction energy includes all two-particle interactions (two particles belonging to two different systems). Therefore, if a region contains several atoms forming an atomic group, the total energy can be expressed as the sum of the contributions of the self-energy of the atomic group and the interaction energy between the groups.

All terms in the IQA partition have a clear physical meaning, although chemically, they may not necessarily be required. We all know that chemical bonds can be understood in terms of covalency and ionicity rather than through electron-nucleus attractions or electron–electron repulsion interactions. IQA can effectively separate the former from the latter, and a second-order density matrix can always be written as the sum of classical (or Coulomb) and non-classical (or exchange-correlation) terms:

$$\gamma^{(2)}(\boldsymbol{r}_1, \boldsymbol{r}_2) = \rho_2(\boldsymbol{r}_1, \boldsymbol{r}_2) = \rho(\boldsymbol{r}_1)\rho(\boldsymbol{r}_2) + \gamma_{\text{xc}}^{(2)}(\boldsymbol{r}_1, \boldsymbol{r}_2). \tag{3.85}$$

Substituting this equation into Eq. (3.79), the electron repulsion between regions A and B can be written as the sum of the Coulomb interaction V_{C}^{AB} and the exchange–correlation interaction V_{xc}^{AB}:

$$V_{\text{ee}}^{AB} = V_{\text{C}}^{AB} + V_{\text{xc}}^{AB}. \tag{3.86}$$

Considering all classical interaction terms, V_{en}, V_{nn}, and V_{c}, the classical interaction energy V_{cl}^{AB} can be written as:

$$V_{\text{cl}}^{AB} = V_{\text{C}}^{AB} + V_{\text{en}}^{AB} + V_{\text{en}}^{BA} + V_{\text{nn}}^{AB}. \tag{3.87}$$

For electrically neutral atoms, many of these terms will cancel each other out, reducing the total interaction energy by approximately an order of magnitude. The interaction energy between two atoms (or regions) includes a classical part V_{cl}^{AB} and the nonclassical part V_{xc}^{AB}:

$$E_{\text{int}}^{AB} = V_{\text{cl}}^{AB} + V_{\text{xc}}^{AB}. \tag{3.88}$$

These are related to the ionicity and covalency of the chemical interactions of A and B. Usually, the value of self-energy E_{self}^{A} itself is of little significance, and we will be more concerned with its difference from the reference state energy E_0^A. The deformation of atomic energy can be defined as follows:

$$E_{\text{def}}^{A} = E_{\text{self}}^{A} - E_0^A. \tag{3.89}$$

This is the second source of energy cancellation so that both the deformation atomic energy and the interaction energy will enter the energy scale in a chemical sense. The deformation atomic energy is always positive, and the interaction energy between atoms is always negative. We can define the binding energy E_{bind} as the balance between the deformation atomic energy and the interaction energy:

$$E_{\text{bind}} = \sum_{A} E_{\text{def}}^{A} + \sum_{A>B} E_{\text{int}}^{AB}. \tag{3.90}$$

This is very close to the traditional chemical scenario, where atoms must be raised to a high energy state to interact (chemical reactions and bonding) with each other, and the total deformation atomic energy $\left(E_{\text{def}} = \sum_A E_{\text{def}}^A\right)$ and binding energy will be several orders of magnitude lower than both the total energy E and atomic reference energy $\left(E_0^A\right)$.

Energy decomposition can also be performed at the resonance structure level. Here, we can introduce a unit decomposition for each electron coordinate:

$$E_{\text{e}} = \sum_{\{S\}} \int d\boldsymbol{r}_1 \dots \int d\boldsymbol{r}_N w_S \Psi^* \hat{H}_{\text{e}} \Psi \tag{3.91}$$

where w_S is the product of the N weight functions of the resonance structure S, and S contains m regions, i.e. $S = (n_A, \dots, n_M)$. According to the symmetry of electron permutation, the m^N terms summed in Eq. (3.91) will contain $n_S = N!/(n_A!\,n_B!\dots n_M!)$ sets of identical terms. We can define the energy contribution of each resonance structure as $\langle H_{\text{e}} \rangle_S$, and then $E_{\text{e}} = \sum_{\{S\}} \langle H_{\text{e}} \rangle_S$.

Using coarse-grained density matrices, $\langle H_{\text{e}} \rangle_S$ can be written as:

$$\begin{aligned}\langle H_{\text{e}} \rangle_S = &\sum_A \int d\boldsymbol{r} w_A(\boldsymbol{r}') \hat{h} \gamma^{(1)}(\boldsymbol{r}'; \boldsymbol{r})[C_A] \\ &+ \sum_{A,B} \int d\boldsymbol{r}_1 \int d\boldsymbol{r}_2 w_A(\boldsymbol{r}_1) w_B(\boldsymbol{r}_2) r_{12}^{-1} \gamma^{(2)}(\boldsymbol{r}_1, \boldsymbol{r}_2)[C_{AB}]\end{aligned}, \tag{3.92}$$

where $\widehat{h}$ is the one-electron Hamiltonian, and C_A and C_{AB} are coarse-grained objects of S, with one electron in C_A corresponding to atom A and two electrons in C_{AB} corresponding to atoms A and B, respectively.

Considering all of the above equations, the energy of any resonance structure can be written in IAQ form, according to:

$$\begin{aligned}\langle H_{\mathrm{e}}\rangle_S = &\sum_A \{T^A(S) + V_{\mathrm{ee}}^{AA}(S) + V_{\mathrm{en}}^{AA}(S)\} \\ &+ \sum_{A>B} \{V_{\mathrm{ee}}^{AB}(S) + V_{\mathrm{en}}^{AB}(S) + V_{\mathrm{en}}^{BA}(S)\}\end{aligned}. \tag{3.93}$$

If the probability of finding a resonance structure S is defined as $p(S)$, and the repulsion energy between atoms belonging to S is $V_{nm}^{AB}(S) = p(S)Z_AZ_B/R_{AB}$, then:

$$V_{\mathrm{int}}^{AB}(S) = V_{\mathrm{ee}}^{AB}(S) + V_{\mathrm{en}}^{AB}(S) + V_{\mathrm{en}}^{BA}(S) + V_{\mathrm{nn}}^{AB}(S). \tag{3.94}$$

Furthermore, a normalized energy contribution can be established for each resonance structure. For example, a quantity $A(S)$ can be normalized to $A(S)= A(S)/p(S)$. The total energy of the system can be written as the weighted average of the normalized energies of the resonance structures:

$$E = \sum_{\{S\}} p(S)\widetilde{E}(S), \tag{3.95}$$

where the normalized energy of the resonance structure is:

$$\widetilde{E}(S) = \sum_A \widetilde{E}_{\mathrm{self}}^{A}(S) + \sum_{A>B} \widetilde{E}_{\mathrm{int}}^{AB}(S). \tag{3.96}$$

The interaction energy of each resonance structure can be further decomposed into classical and nonclassical terms. Thus, a unit decomposition can describe the energy of an equilibrium system as an ensemble average of the energies of the resonance structures.

3.4.3 Electron Population Statistics

As a good starting point, we can begin with a generalized single-determinant population analysis based on the idempotency of the HF first-order density matrix, i.e.

$$Tr(\boldsymbol{PS})^i = N,$$

where $\boldsymbol{P}$ is the SCF electron density expressed in terms of spin orbitals (the BO matrix expressed in terms of spin-orbital bases), $\boldsymbol{S}$ is the overlap matrix, and i is any positive integer. This expression divides the N-electron system of order i into i-fold summation by:

$$N^{(i)} = \sum_{\underbrace{\alpha, \beta \ldots, l}_{i}} (PS)_{\alpha\beta}(PS)_{\beta\gamma} \ldots (PS)_{il}. \tag{3.97}$$

The original sets with subscripts $\alpha \ldots l$ can be categorized using a Mulliken-like method to decompose the electron population contributions into single, double, …, and i-central terms:

$$N^{(i)} = \sum_A N_A^{(i)} + \sum_{A>B} N_{AB}^{(i)} + \cdots + \sum_{A>B>\ldots>I} N_{AB\ldots I}^{(i)}, \tag{3.98}$$

where $N_A^{(i)}$ is called the ith-order localization index, $N_{AB\ldots K}^{(i)}$ is the ith-order, k-center delocalization index, $N_A^{(1)}$ is the Mulliken population of the A center, and $N_{AB}^{(2)}$ is the Wiberg BO. The multicenter delocalization index provides a rich description for multicenter bonds. For a single-determinant wave function, Eq. (3.97) can be written as:

$$N^{(i)} = \int \mathrm{d}\boldsymbol{r}_1 \int \mathrm{d}\boldsymbol{r}_2 \ldots \int \mathrm{d}\boldsymbol{r}_i \gamma^{(1)}(\boldsymbol{r}_1;\boldsymbol{r}_2)\gamma^{(1)}(\boldsymbol{r}_2;\boldsymbol{r}_3)\ldots\gamma^{(1)}(\boldsymbol{r}_i;\boldsymbol{r}_1). \tag{3.99}$$

In fact, for single-determinant wave functions, there are some non-equivalent electron permutations in Eq. (3.99) for $i > 3$; thus, the correct form must contain the average of all these nonequivalent terms, i.e.

$$N^{(i)} = \frac{1}{i!}\sum_P \hat{P} \int \mathrm{d}\boldsymbol{r}_1 \ldots \mathrm{d}\boldsymbol{r}_i \gamma^{(1)}(\boldsymbol{r}_1;\boldsymbol{r}_2)\gamma^{(1)}(\boldsymbol{r}_2;\boldsymbol{r}_3)\ldots\gamma^{(1)}(\boldsymbol{r}_i;\boldsymbol{r}_1), \tag{3.100}$$

where $\hat{P}$ is the permutation operator acting on the electron coordinates. This form is universal; however, it is only suitable for single-determinant wave functions. Here, it is extended to the fully correlated wave function by using the generalized density matrix. For example, for a single-determinant wave function, $\gamma^{(1)}(\boldsymbol{r}_1;\boldsymbol{r}_2)\gamma^{(1)}(\boldsymbol{r}_2;\boldsymbol{r}_1) = \rho(\boldsymbol{r}_1)\rho(\boldsymbol{r}_2) - \gamma^{(2)}(\boldsymbol{r}_1;\boldsymbol{r}_2) = -\gamma_{\mathrm{xc}}^{(2)}(\boldsymbol{r}_1,\boldsymbol{r}_2)$, which is known as the irreducible component of the second-order density matrix. Similarly,

$$\begin{aligned}\gamma_{irr}^{(3)}(1,2,3) &= \gamma^{(1)}(1)\gamma^{(1)}(2)\gamma^{(1)}(3) - \frac{1}{2}\gamma^{(1)}(1)\gamma^{(2)}(2,3) - \frac{1}{2}\gamma^{(1)}(2)\gamma^{(2)}(1,3)\\ &\quad - \frac{1}{2}\gamma^{(1)}(3)\gamma^{(2)}(1,2) + \frac{1}{2}\gamma^{(3)}(1,2,3).\end{aligned} \tag{3.101}$$

The simple notation for electron coordinates can be used here, noting that only the diagonal terms of the density matrix appear on the right-hand side. Equation (3.98) can be expressed in the form of the i-fold integration of irreducible components of $\gamma_{\mathrm{irr}}^{(i)}$:

$$N^{(i)} = \int \mathrm{d}\boldsymbol{r}_1 \ldots \mathrm{d}\boldsymbol{r}_i 1(\boldsymbol{r}_1)\ldots 1(\boldsymbol{r}_i)\gamma_{\mathrm{irr}}^{(i)}(\boldsymbol{r}_1,\ldots,\boldsymbol{r}_i). \tag{3.102}$$

The electron number operator $\hat{n}_A$ for atom A can be expressed as:

$$\hat{n}_A = \int \mathrm{d}\boldsymbol{r} w_A(\boldsymbol{r}) \sum_i^N \delta(\boldsymbol{r} - \boldsymbol{r}_i). \tag{3.103}$$

Setting $i = 2$ in Eq. (3.102), $N_{AB}^{(2)}$ in Eq. (3.98) will be $-2\langle(\hat{n}_A - \overline{n}_A)(\hat{n}_B - \overline{n}_B)\rangle$, which measure the covariance of the electron number distributions of A and B. The standard two-center delocalization index $N_{AB}^{(2)}$ can usually be written as $\delta(\Omega_A, \Omega_B)$ or δ^{AB}, which can be interpreted as the shared electron pair of atoms A and B. In addition, we can obtain:

$$N_{AB\ldots I}^{(i)} \equiv i(-1)^{i-1}\langle(\hat{n}_A - \overline{n}_A)(\hat{n}_B - \overline{n}_B)\ldots(\hat{n}_I - \overline{n}_I)\rangle. \tag{3.104}$$

Therefore, multicenter bonds will be related to multicenter loops of electron population fluctuations. It should be noted that to obtain the multicenter index for $i > 2$, we must calculate the full wave function or higher-order density matrix, which is much more complicated than conventional energy calculations, and electron correlations may play a very important role.

Based on the probability $p(S)$ of all possible resonance structures S in a unit partition, the electron population distribution function (EDF) can be derived, and through the population fluctuation relationship, any localization and delocalization indices can be obtained, such as

$$\begin{aligned} \hat{n}_A &= \overline{n}_A = N_A^{(1)} \\ \overline{n}_A - \left\langle (\hat{n}_A - \overline{n}_A)^2 \right\rangle &= \lambda^A = N_A^{(2)} \\ -2\langle (\hat{n}_A - \overline{n}_A)(\hat{n}_B - \overline{n}_B) \rangle &= \delta^{AB} = N_{AB}^{(2)}, \end{aligned} \tag{3.105}$$

where $\lambda(\Omega_A) = \lambda^A$ is the standard localization index. Accordingly,

$$\begin{aligned} N_A^{(1)} &= \sum_{n_A=0}^{N} n_A p(n_A) \\ N_A^{(2)} &= N_A^{(1)} - \sum_{n_A=0}^{N} (n_A - \overline{n}_A)^2 p(n_A) \\ N_{AB}^{(2)} &= -2 \sum_{n_A,n_B=0}^{N} (n_A - \overline{n}_A)(n_B - \overline{n}_B) p(n_A, n_B) \end{aligned} \tag{3.106}$$

3.5 Restricted Space Partition

In the previous section, we discussed that a system can be partitioned into separate spatial regions that extend roughly the same distance as between the atoms (coarse-grained decomposition). Another method for space partitioning is presented here, which produces a large number of extremely small regions, and we can choose a property and calculate this quantity in each region; thus, obtaining the high-density distribution of this property. The separate regions produced by this method are not intended to exhibit certain chemically relevant quantities, but rather generate new distributions that facilitate the further decomposition of the system into meaningful parts.

3.5.1 ω-Restricted Partition

To gain a better understanding of a system, we can examine the specific properties of samples with an equal number of specific attributes, such as samples with the same mass, volume, or velocity, and then compare their properties. This approach can also be used for continuous distributions, where both the sample size and properties can be determined by integral functions. In particular, for a sample of μ_i in a region with a volume of V_i near coordinate $\boldsymbol{a}_i$, the extent of the region can be determined by integrating a control function over this region to a fixed value ω, a rule that gives a large degree of freedom in partitioning the space. Therefore, the sampling area describing the system must be as local as possible and the coordinates within the sample volume cannot be shifted too far from the center; thus, the sample area should be as compact as possible. Clearly, under these conditions, a single sample will be a spherical region around a location in which the integration of the control function will yield a value ω. However, here, we use a different partition approach.

In contrast to the scheme of individually probing different locations, this partitioning method allows us to divide the entire system into a very large number of samples that form a compact nonoverlapping set occupying the entire space, with each sample region satisfying the constraint ω and the sum of all sample volumes is the volume of the system. For this type of partition, it is impossible for all regions to be spherical, and each region will take the form that achieves the most compact effect with neighboring regions. Despite the reduction in the freedom of sample shapes (due to the need to achieve compactness), there will still be an infinite number of partitioning approaches for any constraint ω. This differs from the previous single partition scheme determined by the scalar field gradient.

This compact space partition significantly depends on the control function $f_c(\boldsymbol{r})$, and also on the integral value of the control function over the entire system, $F_c = \int f_c(\boldsymbol{r})d\boldsymbol{r}$. The following general laws apply to the number of partitioned regions and the allowable ω:

1. when $F_c = \pm\infty$, there will be an infinite number of regions μ_i for any real number ω;
2. when $F_c = 0$, only $\omega = 0$ is allowed because $F_c = \kappa\omega$, where κ is the total number of regions. The number of partitioned compact regions, μ_i, cannot be determined, because it can be the whole system or any region that encloses the positive regions (where the control function is positive) and negative regions (where the control function is negative).
3. When $F_c \neq 0$, there will be a finite number ($\kappa = F_c/\omega$) of compact regions for any non-zero ω.

For a relatively simple case of the first situation, the control function will be a nonzero constant, i.e. $f_c \equiv \xi$, and all compact regions μ_i will have the same volume $V_i = \omega/\xi$ and similar sample shapes. If the control function is not constant, the density of the compact regions in the system will vary.

For the second situation, $F_c = 0$, the integral of the control function over the whole system will be zero, which will be different from the case of $f_c \equiv 0$, such as the Laplacian of a scalar field. Therefore, the whole system space can be partitioned into different atomic basins, and the integral of the Laplacian over each of them will be zero. As mentioned before, these conditions will not be sufficient to restrict the whole system to a single partition, because of other possible partitioning methods that can also make the integral of the control function zero.

Interestingly, the integral of the control function over the whole region will be a nonzero value, as the control function will not change the sign in the system volume because this allows for no mutual cancellation of the control function values within the sample region to satisfy the ω-restricted condition. The number of sample regions can be uniquely determined, and the local properties of the samples can be scaled by the ratio of the regional volume to ω, for example. For smooth control functions with a very small ω, the sample volume V_i will remain almost constant for small shifts in the center $\boldsymbol{a}_i$ of the sample region μ_i. Hence, even if the shape and volume of the sample region due to space partitioning are not unique, for a sufficiently small ω, the density of the sample region will be approximately constant among all possible partition schemes satisfying the ω condition. Therefore, the ω-restricted space partition (ωRSP) can be defined as a method of partitioning the system space into refined non-overlapping regions such that the integral of the control function over each region will be a fixed value of ω. For a sufficiently small ω, the ωRSP regions will also be very small and denoted as microregions. The integral of the control function over a microregion can safely be replaced by the first nonzero term in the Taylor expansion.

We can apply the ωRSP idea in the following context, where the electron density $\rho_1(\boldsymbol{r})$ is used as a control function, where the integral of which over the entire molecular region will be the total number of electrons N. If the system can be decomposed into non-overlapping regions μ_i, then each will contain a fixed charge:

$$q = \int_{\mu_i} \rho_1(\boldsymbol{r})\mathrm{d}\boldsymbol{r} = \omega. \tag{3.107}$$

This will result in $\kappa = N/q$ for such regions. If ω is sufficiently small, the charge in each microregion with a volume of V_i will be $q \approx \rho_1(\boldsymbol{a}_i)V_i$, and the volume of the microregion will be inversely proportional to the density, i.e. $V_i \approx q/\rho_1(\boldsymbol{a}_i)$, and this inverse correlation will hold for any positive control function normalized to $F_c = \int f_c(\boldsymbol{r})\mathrm{d}\boldsymbol{r}$ under the ω restriction:

$$V_i \approx \omega/f_c(\boldsymbol{a}_i); \quad F_c = \kappa\omega. \tag{3.108}$$

Even for $f_c(\boldsymbol{a}_i) = 0$, the volume V_i of a microregion will be strictly defined and may need to be approximated using a higher-order Taylor expansion. Thus, ωRSP can also be performed for the control function $f_c(\boldsymbol{r}_1, \ldots, \boldsymbol{r}_n)$ with n coordinates. For each microregion, we can obtain:

$$\omega = \int_{\mu_i} \mathrm{d}\boldsymbol{r}_1 \ldots \int_{\mu_i} f_c(\boldsymbol{r}_1, \ldots \boldsymbol{r}_n)\mathrm{d}\boldsymbol{r}_n. \tag{3.109}$$

Here, all coordinates will be restricted to the microregion μ_i, yielding a fixed value of ω. Unlike the case of single coordinates, the number of microregions κ cannot be inferred simply by normalizing the control function to F_c. For example, the space can be partitioned by using the pair density distribution function $\rho_2(\boldsymbol{r}_1, \boldsymbol{r}_2)$ as the control function, and the integral over each microregion will be fixed to D electron pairs. For any region Ω,

$$D = \iint_{\Omega} \rho_2(\boldsymbol{r}_1, \boldsymbol{r}_2)\mathrm{d}\boldsymbol{r}_1\mathrm{d}\boldsymbol{r}_2, \tag{3.110}$$

where D is a positive value. For a given ω, the number of microregions κ will satisfy $\kappa\omega \leq N(N-1)/2$. This is mainly based on the fact that the total number of electron pairs in all microregions cannot exceed the total number of electron pairs in the system. This is also restricted by a sufficiently small ω, where the volume V_i of the microregion μ_i is scaled using ω, and the actual inverse proportional expression can be obtained by performing a Taylor expansion of the control function at the center $\boldsymbol{a}_i$ of the microregion.

3.5.2 Restricted Electron Population Analysis

Another function f_s, the so-called sampled property function, can be evaluated in each microregion resulting from ωRSP. The idea behind microregion sampling is that the expected value of an operator $\hat{A}$ acting on a wave function Ψ can be expressed as an integral over the full space:

$$\langle \hat{A} \rangle = \int \Psi^* \hat{A} \Psi \mathrm{d}V. \tag{3.111}$$

If $\hat{A}$ is a one-particle operator, the above equation can be simplified (without considering the spin) as:

$$\langle \hat{A} \rangle = \int \hat{A} \gamma^{(1)}(\boldsymbol{r}'; \boldsymbol{r}) \mathrm{d}\boldsymbol{r} = \sum_{i=1}^{K} \int_{\mu_i} \hat{A} \gamma^{(1)}(\boldsymbol{r}'; \boldsymbol{r}) \mathrm{d}\boldsymbol{r}. \tag{3.112}$$

In this case, the sum of all property values $f_s(\boldsymbol{r}) = \hat{A}\gamma^{(1)}(\boldsymbol{r}'; \boldsymbol{r})\,|_{r' \to r}$ of the microregion μ_i will be the expected value of the operator. For the two-particle operator $f_s(\mathbf{r}_1, \mathbf{r}_2) = \hat{A}\gamma^{(2)}(\boldsymbol{r}_1', \boldsymbol{r}_2'; \boldsymbol{r}_1, \boldsymbol{r}_2)\,|_{r' \to r}$, the sum of the integrals over the sampled properties will give only a fraction of the expected value. The integration of the sampled properties over the microregion will yield a distribution of the discrete values $\{\zeta_i\}$ of the system. Both this sample distribution and the sampled values will depend strongly on the control function f_c that determines the space partition and the property function f_s.

Here, we present two special examples, as follows.

1. When $f_c \equiv f_s$, the control function is the same as the property function. In this case, the sampling of f_s will only yield a fixed value ω, i.e. the result will be a constant distribution. The density of the sampled values will vary with the location of the system, which will be the same as the density of the partitioned microregions, with a low density of samples at locations where the f_c value is high.
2. When f_c is constant, all of the microregions will have the same volume and will be proportional to ω. When ω is sufficiently small, $\{\zeta_i\}$ will have a discrete distribution simulating the sampled property f_s and will be uniformly distributed throughout the system.

An example of a sampled property is the diagonal element of the first-order density matrix $\gamma^{(1)}(\boldsymbol{r}'; \boldsymbol{r})$, i.e. the electron density $\rho_1(\boldsymbol{r})$, which can be normalized for N electrons by integrating over the entire system. The integration of $\rho_1(\boldsymbol{r})$ over the microregion μ_i will yield a charge population distribution that satisfies $\sum q_i = N$. Another example is the diagonal element of the second-order density matrix $\gamma^{(2)}\left(\boldsymbol{r}_1', \boldsymbol{r}_2'; \boldsymbol{r}_1, \boldsymbol{r}_2\right)$, i.e. the electron pair density $\rho_2(\boldsymbol{r}_1, \boldsymbol{r}_2)$, the integral of which, as a sampling function over the whole system, will be the number of electron pairs

$N(N-1)/2$. The individual integration of $\rho_2(\boldsymbol{r}_1, \boldsymbol{r}_2)$ over each microregion will produce an electron pair population distribution $\{D_i\}$. Obviously, the sum of the number of electron pairs in all microregions will not be total number of electron pairs in the system, as it will be necessary to add the electron pair population D_{ij} of different microregions. The actual electron population $\{q_i\}$ and electron pair population $\{D_i\}$ will depend strongly on the control function. The discrete distribution resulting from the integral value of the nth-order electron density $\rho_n(\boldsymbol{r}_1, \ldots, \boldsymbol{r}_n)$ over the ωRSP microregion is known as the ω-restricted population.

3.5.3 Quasi-continuous Distribution

Sampling of the property function f_s in the microregion μ_i will yield a discrete distribution of $\{\zeta_i\}$. For a suitable control function, the sample density (number of samples per unit volume) will increase with decreasing ω. Clearly, the individual sampled property ζ_i will decrease at this point, causing inconvenience in some cases, especially when the sum of all samples approaches zero. Therefore, it is necessary to extract information that is not significantly affected by ω. This can be achieved by scaling the sample values appropriately, with a scale that depends only on ω and with a sample distribution that converges to a single function in the limit of $\omega \to 0$. This scale can be obtained by analyzing the behavior of the integral value ζ_i. A Taylor expansion of the control function $f_c(\boldsymbol{r}_1, \ldots, \boldsymbol{r}_m)$ with m coordinates can be performed at the center $\boldsymbol{a}_i$ of the microregion, and the ω-restricted integral over the microregion μ_i can be written as:

$$\omega = \int_{\mu_i} \mathrm{d}\boldsymbol{r}_1 \ldots \int_{\mu_i} f_c(\boldsymbol{r}_1, \ldots, \boldsymbol{r}_m) \mathrm{d}\boldsymbol{r}_m = t_c(\boldsymbol{a}_i) V_i^{\vartheta_c} + \varepsilon_c(\boldsymbol{a}_i), \tag{3.113}$$

where t_c and ϑ_c are the function and parameter, respectively, determined by the Taylor expansion of the control function, and V_i is the volume of the microregion μ_i, noting that $t_c(\boldsymbol{a}_i)$ is the value at $\boldsymbol{a}_i$ and ϑ_c is applicable to the entire distribution. In addition, $\varepsilon_c(\boldsymbol{a}_i)$ is the higher-order correction term in the Taylor expansion and can be gradually ignored when the volume of the microregion becomes very small. The volume V_i can be expressed as:

$$V_i = \left[\frac{\omega - \varepsilon_c(\boldsymbol{a}_i)}{t_c(\boldsymbol{a}_i)} \right]^{1/\vartheta_c}. \tag{3.114}$$

The same processing flow can be applied to the sampling of a property function $f_s(\boldsymbol{r}_1, \ldots, \boldsymbol{r}_n)$ with n coordinates. The sample value ζ_i can be expressed as:

$$\zeta_i = \int_{\mu_i} \mathrm{d}\boldsymbol{r}_1 \ldots \int_{\mu_i} f_s(\boldsymbol{r}_1, \ldots, \boldsymbol{r}_n) \mathrm{d}\boldsymbol{r}_n = t_s(\boldsymbol{a}_i) V_i^{\vartheta_c} + \varepsilon_s(\boldsymbol{a}_i). \tag{3.115}$$

Substitution of the volume V_i of Eq. (3.114) yields:

$$\zeta_i = t_s(\boldsymbol{a}_i) \left[\frac{\omega - \varepsilon_c(\boldsymbol{a}_i)}{t_c(\boldsymbol{a}_i)} \right]^{\vartheta_s/\vartheta_c} + \varepsilon_s(\boldsymbol{a}_i) = t_s(\boldsymbol{a}_i) \left[\frac{\omega}{t_c(\boldsymbol{a}_i)} \right]^{\vartheta_s/\vartheta_c} + \varepsilon(\boldsymbol{a}_i), \tag{3.116}$$

where all correction terms are concentrated in $\varepsilon(\boldsymbol{a}_i)$. We can rescale $\{\zeta_i\}$ to $\{\zeta_i/\omega^{\vartheta_s/\vartheta_c}\}$, and as ω decreases, the value of the rescaled discrete distribution will become $\{t_s(\boldsymbol{a}_i)/t_c(\boldsymbol{a}_i)^{\vartheta_s/\vartheta_c}\}$ in the limit of $\omega \to 0$, i.e.

$$\lim_{\omega\to 0}\{\zeta_i/\omega^{\vartheta_s/\vartheta_c}\} = t_s(\boldsymbol{r})\left[\frac{1}{t_c(\boldsymbol{r})}\right]^{\vartheta_s/\vartheta_c} = t_s(\boldsymbol{r})\widetilde{V}^{\vartheta_s}(\boldsymbol{r}), \tag{3.117}$$

where $\widetilde{V}(\boldsymbol{r})$ is the limit of the volume of the rescaled microregion, known as the volume function. Of note, restricted by a nonzero ω, the set $\{\zeta_i/\omega^{\vartheta_s/\vartheta_c}\}$ will be a discrete function distribution; however, its limit will be a continuous function, called a quasi-continuous distribution.

3.5.4 Electron Localization Indicators (ELI)

The chemical bonding information of the N-electron system will be contained in the Nth-order density matrix $\Gamma^{(N)}$. Because the energy is determined by one- and two-particle operators, it can be assumed that the second-order density matrix contains the main information regarding chemical bonding. If the electron density $\rho_1(\boldsymbol{r})$ and electron pair density $\rho_2(\boldsymbol{r}_1, \boldsymbol{r}_2)$ are used as the control and sampling functions, respectively, the resulting ω-restricted population can be viewed as a charge distribution (D-restricted distribution) with fixed electron pair density or an electron pair population (q-restricted population) with fixed electron density.

If the q-restricted population is used, each microregion will have the same charge of $q = \omega$. The difference from q^2 can actually be revealed by sampling the electron pair density, and the integral of the electron pair density can be written as:

$$\frac{1}{2}\int\int_{\mu_i} \rho_1(\boldsymbol{r}_1)\rho_1(\boldsymbol{r}_2)\left[1 + f_{XC}(\boldsymbol{r}_1, \boldsymbol{r}_2)\right] d\boldsymbol{r}_1 d\boldsymbol{r}_2 = \frac{1}{2}\left[q^2 + F_i\right], \tag{3.118}$$

where $f_{XC}(\boldsymbol{r}_1, \boldsymbol{r}_2)$ is the correlation factor, and F_i can be regarded as a manifestation of the correlation of electron motions in the microregion μ_i.

Because the electron density is a one-particle property, the integral over the microregion can be expanded by Taylor series and the same step size can be employed regardless of whether the total density $\rho_1(\boldsymbol{r})$ or spin density $\rho_1^{\sigma}(\boldsymbol{r})$ ($\sigma = \alpha$ or β) is used. However, this is not the case for the electron pair density, which is a two-particle property. According to the Pauli principle, $\rho_2^{\sigma\sigma}(\boldsymbol{r}, \boldsymbol{r}) = 0$ while $\rho_2^{\alpha\beta}(\boldsymbol{r}, \boldsymbol{r}) \neq 0$. These properties not only control the first nonzero term of the Taylor expansion, but also show dominance by opposite-spin electron pairs in sufficiently small microregions. The information of same-spin electron pairs can be analyzed only when opposite-spin electron pairs are not considered. Thus, the charge and electron pair population in a microregion can be correlated by the rejection of electron sharing in the same location, and D-restricted and q-restricted populations can reflect the localization of electrons.

The ELI is the distribution obtained by partitioning and rescaling the infinitesimal ωRSP with the electron density and electron pair density as the control and sampling functions, respectively. The electron pair population (ELI-q, Υ_q) can be obtained if the electron density is the control function, and the electron density population (ELI-D, Υ_q) can be generated by using the electron pair density as the control function, where Y_q denotes the sampling of the α-spin density in the $\alpha\alpha$ electron pair density-restricted microregion μ:

$$\begin{aligned} Y_D^{\alpha}(\mu) &= \frac{1}{\omega^{3/8}} \int_{\mu} \rho_1^{\alpha}(\mathbf{r})\mathrm{d}\mathbf{r} \\ \omega &= \iint_{\mu_i} \rho_2^{\alpha\alpha}(\mathbf{r}_1, \mathbf{r}_2)\mathrm{d}\mathbf{r}_1\mathrm{d}\mathbf{r}_2 = const \end{aligned}, \tag{3.119}$$

where the exponent of ω generated after rescaling.

3.5.4.1 Same-spin Electron Pairs

The same-spin electron pair density $\rho_2^{\sigma\sigma}(\mathbf{r}_1, \mathbf{r}_2)$ can be used for sampling or ω-restriction in the ωRSP microregion. In both cases, the integral over the corresponding microregion μ_i should be calculated, for which Taylor expansion can be performed at the center $\mathbf{a}_i$ of the microregion and, due to the Pauli rule and cusp condition, only the quadratic terms have to be retained:

$$D_i^{\sigma\sigma} \approx \frac{1}{2} \int \mathrm{d}\mathbf{r}_1 \int_{\mu} (\mathbf{s}_2 \cdot \nabla_{\mathbf{r}_2})^2 \rho_2^{\sigma\sigma}(\mathbf{r}_1, \mathbf{r}_2) \,|_{r\to a} \mathrm{d}\mathbf{r}_2 , \tag{3.120}$$

where $\mathbf{s}_2 = \mathbf{r}_2 - \mathbf{a}$. After the expansion of $\mathbf{r}_2$, both $\mathbf{r}_1$ and $\mathbf{r}_2$ should be set to $\mathbf{a}$. Of note, Eq. (3.120) retains only the second-order Taylor expansion terms, and the zeroth-order term of $D_i^{\sigma\sigma}$ at $\mathbf{a}$ will be zero. In addition, due to the cusp condition, the first-order derivative of $D_i^{\sigma\sigma}$ will be proportional to itself, and the first-order Taylor expansion term will also be zero [25, 26].

Expanding $\rho_2^{\sigma\sigma}(\mathbf{r}_1, \mathbf{r}_2)$ by the Slater determinant, we can obtain:

$$\rho_2^{\sigma\sigma}(\mathbf{r}_1, \mathbf{r}_2) = \frac{1}{2} \sum_{i<j}^{\sigma} \sum_{k<l}^{\sigma} P_{ij,kl} |\phi_i(\mathbf{r}_1)\phi_j(\mathbf{r}_2)||\phi_k(\mathbf{r}_1)\phi_l(\mathbf{r}_2)|. \tag{3.121}$$

After substitution into Eq. (3.120) and integration, the electron pair population of the microregion μ_i will be [25]:

$$D_i^{\sigma\sigma} \approx \frac{1}{12} V_i^{8/3} g^{\sigma}(\mathbf{a}_i), \tag{3.122}$$

where V_i is the volume of the microregion μ_i, and $g^{\sigma}(\mathbf{a}_i)$ is the Fermi hole curvature at the center of the microregion:

$$\begin{aligned} g^{\sigma}(\mathbf{a}_i) = \sum_{i<j}^{\sigma} \sum_{k<l}^{\sigma} & P_{ij,kl} [\phi_i(\mathbf{a}_i)\nabla\phi_j(\mathbf{a}_i) - \phi_j(\mathbf{a}_i)\nabla\phi_i(\mathbf{a}_i)] \\ & \times [\phi_k(\mathbf{a}_i)\nabla\phi_l(\mathbf{a}_i) - \phi_l(\mathbf{a}_i)\nabla\phi_k(\mathbf{a}_i)] \end{aligned}. \tag{3.123}$$

If ωRSP is controlled by an infinitesimal $\omega = D^{\sigma\sigma}$, and the population of same-spin electron pairs is fixed, the volume V_i of the microregion μ_i can be expressed as follows, according to Eq. (3.122),

$$\begin{aligned} V_i &\approx \left[\frac{12\omega}{g^{\sigma}(\mathbf{a}_i)} \right]^{3/8} \\ \widetilde{V}_D(\mathbf{r}) &= \lim_{\omega\to 0} \left\{ \frac{1}{\omega^{3/8}} V_i \right\} = \left[\frac{12}{g^{\sigma}(\mathbf{r})} \right]^{3/8}, \end{aligned} \tag{3.124}$$

where $\widetilde{V}_D(\boldsymbol{r})$ is the electron-pair volume function by taking the limit of the volume of the rescaled microregion. Sampling the σ-spin electron density in such a microregion μ_i will yield ELI-D:

$$Y_D^\sigma(\mu_i) = \frac{1}{\omega^{3/8}} \int_\mu \rho_1^\sigma(\boldsymbol{r})\mathrm{d}\boldsymbol{r} = Y_D^\sigma(\boldsymbol{a}_i) \approx \frac{1}{\omega^{3/8}} \rho_1^\sigma(\boldsymbol{a}_i) V_i. \tag{3.125}$$

After rescaling, we can obtain:

$$\widetilde{Y}_D^\sigma(\boldsymbol{r}) = \lim_{\omega\to 0} \left\{Y_D^\sigma(\mu_i)\right\} = \rho_1^\sigma(\boldsymbol{r})\widetilde{V}_D(\boldsymbol{r}). \tag{3.126}$$

Because the scale factor $\omega^{3/8}$ is derived from the Taylor expansion, a strict ELI-D will depend on the choice of ω. Nevertheless, the rescaling operation demonstrates the limiting behavior of the ELI-D distribution.

3.5.4.2 Singlet and Triplet Electron Pairs

In the case of spin polarization, both same-spin ELI-Ds must be analyzed. The second-order density matrices $\gamma^{(2)}\left(\boldsymbol{r}_1', \boldsymbol{r}'; \boldsymbol{r}_1, \boldsymbol{r}_2\right)$ can be used to form symmetric and antisymmetric combinations, according to:

$$\begin{aligned} \gamma^{(s)}\left(\boldsymbol{r}_1', \boldsymbol{r}_2'; \boldsymbol{r}_1, \boldsymbol{r}_2\right) &= \frac{1}{2}\left[\gamma^{(2)}\left(\boldsymbol{r}_1', \boldsymbol{r}_2'; \boldsymbol{r}_1, \boldsymbol{r}_2\right) + \gamma^{(2)}\left(\boldsymbol{r}_1', \boldsymbol{r}_2'; \boldsymbol{r}_2, \boldsymbol{r}_1\right)\right] \\ \gamma^{(t)}\left(\boldsymbol{r}_1', \boldsymbol{r}_2'; \boldsymbol{r}_1, \boldsymbol{r}_2\right) &= \frac{1}{2}\left[\gamma^{(2)}\left(\boldsymbol{r}_1', \boldsymbol{r}_2'; \boldsymbol{r}_1, \boldsymbol{r}_2\right) - \gamma^{(2)}\left(\boldsymbol{r}_1', \boldsymbol{r}_2'; \boldsymbol{r}_2, \boldsymbol{r}_1\right)\right] \end{aligned}. \tag{3.127}$$

The diagonal elements of these matrices will be the singlet electron pair density $\rho_2^{(s)}(\boldsymbol{r}_1, \boldsymbol{r}_2)$ and triplet electron pair density $\rho_2^{(t)}(\boldsymbol{r}_1, \boldsymbol{r}_2)$, respectively, which can be integrated over specific regions to obtain the number of singlet electron pairs $D^{(s)}$ and the number of triplet electron pairs $D^{(t)}$ for these regions, respectively. The same-spin electron pairs will contribute only to the triplet state, while the opposite-spin electron pairs will contribute to both the singlet and triplet states. Operating on the electron pair densities $\rho_2^{(s)}(\boldsymbol{r}_1, \boldsymbol{r}_2)$ and $\rho_2^{(t)}(\boldsymbol{r}_1, \boldsymbol{r}_2)$ will result in the electron densities $\rho_1^{(s)}(\boldsymbol{r})$ and $\rho_1^{(t)}(\boldsymbol{r})$ contributing to the singlet and triplet states, respectively.

The triplet electron density distribution $\left\{q_i^{(t)}\right\}$ can be obtained by integrating $\rho_1^{(t)}(\boldsymbol{r})$ over the microregion μ_i partitioned by the ωRSP with a fixed number of triplet electron pairs ($\omega = D^{(t)}$) and, after rescaling, the ELI-D ($Y_D^{(t)}$) for triplet coupled electrons can be achieved.

After taking its limit, we can obtain:

$$\widetilde{Y}_D^{(t)}(\boldsymbol{r}) = \lim_{\omega\to 0} \left\{Y_D^{(t)}(\mu_i)\right\} = \rho_1^{(t)}(\boldsymbol{r})\widetilde{V}_{D^{(t)}}(\boldsymbol{r}). \tag{3.128}$$

Similar to the case of same-spin electron pairs, the electron pair volume function $\widetilde{V}_{D^{(t)}}$ can also be calculated from the triplet electron density and Fermi hole curvature. For a single-determinant closed-shell wave function, $Y_D^{(t)} \propto Y_D^\alpha$. For a singlet coupled case, the number of singlet-coupled electrons will be fixed ($\omega = q^{(s)}$) in the ωRSP partition, i.e. q-restricted partition. With this choice, a high value of ELI will correspond to a high singlet electron pair population, and a singlet-like electron pair population distribution $\{D_i^{(s)}\}$ can be obtained by integrating $\rho_2^{(s)}(\boldsymbol{r}_1, \boldsymbol{r}_2)$ over the microregion. After rescaling, the ELI-q ($\Upsilon_q^{(s)}$) of the singlet coupled electrons will be generated, with the exponent of the scale factor obtained by performing a Taylor expansion at the center $\boldsymbol{a}_i$ of the microregion μ_i. The singlet electron pair density includes the opposite-spin electron pair density, where $\rho_2^{(s)}(\boldsymbol{r}, \boldsymbol{r})$ dominates after the expansion. The singlet electron pair density can be approximately expressed by:

$$D_i^{(s)} \approx \rho_2^{(s)}(\boldsymbol{a}_i, \boldsymbol{a}_i)V_i^2 \approx \rho_2^{(s)}(\boldsymbol{a}_i, \boldsymbol{a}_i)\left[\frac{\omega}{\rho_1^{(s)}(\boldsymbol{a}_i)}\right]^2. \tag{3.129}$$

The volume of the microregion can be approximated as $V_i \approx q^{(s)}/\rho_1^{(s)}$. After re-scaling $\Upsilon_q^{(s)}$ with ω^2 and taking the limit, we can obtain:

$$\widetilde{\Upsilon}_q^{(s)}(\boldsymbol{r}) = \lim_{\omega\to 0}\left\{\Upsilon_q^{(s)}(\mu_i)\right\} = \frac{1}{2}\left[\frac{N+2}{N-1}\right]^2 \frac{\rho_2^{(s)}(\boldsymbol{r}, \boldsymbol{r})}{\left[\rho_1^{(s)}(\boldsymbol{r})\right]^2}, \tag{3.130}$$

where N is the total number of electrons.

Regarding the ELI of the correlated wave function, the expressions in the previous two sections will remain almost unchanged, with only the correlated density matrix substituted into them. The topology of the same-spin and triplet coupled spin ELI-D distribution can be principle determined by the Fermi correlation, where its characteristics have already been manifested at the HF level. The Coulomb correlation will have a small effect on the ELI-D, and its effect on the electron pair volume will be greater than on the electron density. The effect of Coulomb correlation on the ELI-D of the whole molecule will not be pronounced, except for some systems (e.g. F_2 molecule) where the correlation plays a major role.

$\Upsilon_q^{(s)}$ will increase if the correlated wave function is used. Thus, the number of opposite-spin pairs will deviate significantly from the simple product of the numbers of the two spin components due to the correlation of the motion of the opposite-spin electrons. In addition, $\Upsilon_q^{(s)}$ will reflect how many singlet electron pairs form for a fixed number of singlet coupled electrons, and the extent to which this correlation decreases $\Upsilon_q^{(s)}$ will depend on the location of the microregion in the system and how many correlations are considered.

3.5.4.3 ELI in Momentum Space

The derivation of ELI is based on the ω-restricted partition in real space and is also suitable in momentum space if the density matrix in momentum space is used. The momentum space can likewise be partitioned into many microregions with the same electron or electron pair population, and the microregions will be sampled using electron momentum $\boldsymbol{P}$. In momentum space, the Fermi and Coulomb correlation behaviors of the electrons will be the same as the position representation, where at the same momentum, the same-spin electron pair density will be zero and the opposite-spin electron pair density will have a non-zero pole, and the scale factor will be the same as in real space. We can take the limit of the rescaled same-spin ELI-D in momentum space to have the following definition:

$$\widetilde{\Upsilon}_D^{\sigma}(\boldsymbol{p}) = \lim_{\omega\to 0}\left\{\Upsilon_D^{\sigma}(\mu_i)\right\} = \pi_1^{\sigma}(\boldsymbol{p})\widetilde{V}_D(\boldsymbol{p}), \tag{3.131}$$

where $\pi_1^\sigma(\boldsymbol{p})$ is the σ-spin momentum density and $\widetilde{V}_D$ is the momentum electron pair volume. Similarly, the ELI-D of the triplet coupled electrons in momentum space can be calculated using the calculation method for the real space.

ELI-D is the integral distribution of the electron density in a microregion. The electron density can formally be decomposed into any component of the electron density, and ELI-D can also be written as the sum of the contributions of these components. Based on the orbital density, ELI-D can be written as

$$Y_D(\mu) = \frac{1}{\omega^{3/8}} \int_{\mu_i} \pi(\boldsymbol{p})\mathrm{d}\boldsymbol{p} = \frac{1}{\omega^{3/8}} \int_{\mu_i} \sum_k \pi_k(\boldsymbol{p})\mathrm{d}\boldsymbol{p} = \sum_k Y_D(\mu_i \mid \phi_k). \tag{3.132}$$

The momentum density can be derived from the sum of the contributions of all orbitals ϕ_k, i.e. $\pi = \sum_k \pi_k$. $Y_D(\mu_i|\phi_k)$ can be referred to as partial ELI-D.

3.6 Intermolecular Interaction Energy

3.6.1 Interaction Energy of Experimental Electron Density

In molecular compounds, the forces between structural units, molecules, or ions originate mainly from the classical electrostatic force between them, and are the main part of the binding energy of molecular crystals, with anisotropic electrostatic forces playing a very important role in the formation of the final crystal structure. Advances in experimental electron density measurements provide new avenues for the qualitative and quantitative studies of molecular forces.

From a quantum mechanical point of view, the interaction energy between molecules A, B, C, … can be defined as the difference between the compound energy $E_{\mathrm{ABC}}\ldots$ and the sum of the monomer energies $E_\mathrm{A}, E_\mathrm{B}, E_\mathrm{C}, \ldots$:

$$E_{\mathrm{int}}(\boldsymbol{R}) = E_{\mathrm{ABC}\ldots}(\boldsymbol{R}) - [E_\mathrm{A} + E_\mathrm{B} + E_\mathrm{C} + \ldots]. \tag{3.133}$$

The internal structure of monomers can generally be assumed to be unchanged. For monomers with fixed structures, the interactions will depend only on the relative positions and orientations between these monomers. This statement also assumes that coupling between the electron and nuclear motions can be ignored, i.e. it is consistent with the Born–Oppenheimer approximation. The total energy in Eq. (3.133) can be derived from the solution of the electron Schrödinger equation for fixed nucleus and monomer structures.

Intermolecular interaction energies can be obtained using difference or perturbation methods. Using the difference method, as shown in Eq. (3.133), it can be calculated by subtracting the energy of the compound from the total energy of all monomers. In general, the exact interaction energy will be 4 to 7 orders of magnitude smaller than the energy terms in the subtraction. For multi-electron systems, the error in the total energy from first-principles calculation will often be much larger than the intermolecular interaction energy. Hence, the reliability of the difference method will depend on how much the errors of individual energy terms will cancel each other out during subtraction. Furthermore, this method fails to explain any physical nature of interactions.

The perturbation theory method is effective in that the interaction energy can be calculated directly rather than obtained by subtracting it from the total energy. The perturbation method decomposes the total energy into terms with clear physical meanings, such as the electrostatic, induction, dispersion, and exchange-repulsion interaction energies. In the perturbation method, the electrostatic interaction energy E_{es} represents the Coulomb interactions between the charge densities $\rho_{\text{tot}}^{A}(\boldsymbol{r})$ and $\rho_{\text{tot}}^{B}(\boldsymbol{r})$ of the unperturbed monomers:

$$E_{\text{es}}^{AB}(\boldsymbol{R}) = \int_{A}\int_{B} \frac{\rho_{\text{tot}}^{A}(\boldsymbol{r}_1)\rho_{\text{tot}}^{B}(\boldsymbol{r}_2)}{|\boldsymbol{r}_1 - \boldsymbol{r}_2|} \mathrm{d}\boldsymbol{r}_1 \mathrm{d}\boldsymbol{r}_2. \tag{3.134}$$

The electron densities of monomers A and B include contributions from both nuclei and electrons, according to:

$$\rho_{\text{tot}}^{A}(\boldsymbol{r}_1) = \sum_{i \in A} Z_i \delta(\boldsymbol{r}_1 - \boldsymbol{R}_i) - \rho_{\text{elec}}^{A}(\boldsymbol{r}_1). \tag{3.135}$$

where the first term on the right-hand side of the equation is the positive charge of the nucleus i at position $\boldsymbol{R}_i$, and $\rho_{\text{elec}}^{A}(\boldsymbol{r})$ is the electron density.

Induction energy, sometimes referred to as polarization energy, is always be mutually attractive between molecules and arises from the polarization of one molecule in an electrostatic field generated by other non-perturbed molecules. Dispersion energy, which is also always mutually attractive, arises from the instantaneous correlation resulting from fluctuations in the electron density of non-perturbed molecules. Exchange-repulsion energy comes from the Pauli rule, which states that same-spin electrons will repel each other in overlapping regions, with exchange energy by which molecules attract each other.

Quantitative estimation of interaction energies from experimental data is also possible due to the availability of quantitative electron density information with current experimental techniques. The pseudo-atomic model proposed by Stewart is the most used experimental electron density model, where the electron density can be expanded by spherical harmonic functions at the center of the atom, according to:

$$\rho_i(\boldsymbol{r}) = \sum_{\ell=0}^{\ell_{\max}} \sum_{m=0}^{\ell} \sum_{p} P_{\ell mp} N_{\ell mp} \rho_{\ell}(r) \mathcal{Y}_{\ell mp}(\theta, \phi), \tag{3.136}$$

where $\rho_{\ell}(r)$ is the radial function, $P_{\ell mp}$ is the electron population parameter, $N_{\ell mp}$ is the normalization factor, $\mathcal{Y}_{\ell mp}(\theta, \phi)$ is the real spherical harmonic function, commonly known as the multipole moment, and p can be positive or negative when m is not zero.

The electron density of a crystal can be written as a superposition of pseudoatomic electron densities, according to:

$$\rho_{\text{elec}}^{A}(\boldsymbol{r}) = \sum_{i} \rho_i(\boldsymbol{r} - \boldsymbol{R}_i), \tag{3.137}$$

where $\boldsymbol{R}_i$ represents the position of the atom i in the unit cell.

Among the many variants of the pseudoatomic model, the most widely used is the Hansen–Coppens multipole model, in which the pseudo-atomic electron density can be defined as:

$$\begin{aligned}\rho_i(\boldsymbol{r}) &= P_c\rho_{\text{core}}(\boldsymbol{r}) + P_v\kappa^3\rho_{\text{valence}}(\kappa\boldsymbol{r}) \\ &+ \sum_{\ell=0}^{\ell_{\max}}\kappa'^3 R_\ell(\kappa'\zeta r)\sum_{m=0}^{m=\ell}\sum_{p} P_{\ell mp} d_{\ell mp}(\theta,\phi)\end{aligned}, \tag{3.138}$$

where $\rho_{\text{core}}(\boldsymbol{r})$ and $\rho_{\text{valence}}(\boldsymbol{r})$ are the spherically symmetric electron densities of the core and valence shells of the free atom, respectively, and both can be normalized using the full integral value of one electron, where $R_\ell(\kappa'\zeta r)$ is a Slater-type radial function and $d_{\ell mp}(\theta, \phi)$ is a density-normalized spherical harmonic function. The valence electron population parameters P_v and $P_{\ell mp}$, as well as the dimensionless parameters κ and κ', can be used to refine the experimental data, during which the core electron population parameter P_c is usually fixed.

The experimentally obtained crystal electron density should reflect all many-body interaction effects such as polarization, charge transfer, electron correlation, and relativistic effects. Due to intermolecular interactions, the electron density perturbations can be observed in the experimental electron densities obtained by the highly refined diffraction experiments and rigorous multipole refinement. To calculate intermolecular interaction energy, the electron density of a crystal can be divided into the sum of the electron densities of several molecules under the necessary assumption that the molecules keep themselves constant during the formation of the crystal. When using the multipole model, the molecular density can usually be obtained by superposing the electron densities of the atoms that make up the molecule, a process that is valid only if there is no overlap between the densities of the individual pseudoatoms. Abramov argued that the overlap error in the experimental basis set could be substantially reduced by fixing κ' to the κ' obtained from the theoretical crystal structure factor during the refinement process.

3.6.2 Pseudoatomic Representation of Electrostatic Interactions

3.6.2.1 Multipole Expansion Approximation

In practical application, the exact Eq. (3.134) will not be suitable for calculating the electrostatic energy because it is a six-dimensional integral. One way to avoid the complicated integral calculation is to approximate Eq. (3.134) by multipole moment expansion and to perform the Taylor expansion of $|\boldsymbol{r}_1 - \boldsymbol{r}_2|^{-1}$. If the expansion is carried out in a rectangular coordinate system, the electrostatic interaction energy can be written as the Buckingham traceless multipole moment equation:

$$\begin{aligned}E_{\text{esMM}}(\boldsymbol{R}) &= Tq^A q^B + T_\alpha\left(\mu_\alpha^A q^B - q^A\mu_\alpha^B\right) \\ &+ T_{\alpha\beta}\left(\frac{1}{3}\Theta_{\alpha\beta}^A q^B + \frac{1}{3}q^A\Theta_{\alpha\beta}^B - \mu_\alpha^A\mu_\beta^B\right) \\ &+ T_{\alpha\beta\gamma}\left(\frac{1}{15}q^A\Omega_{\alpha\beta\gamma}^B - \frac{1}{15}\Omega_{\alpha\beta\gamma}^A q^B - \frac{1}{3}\mu_\alpha^A\Theta_{\beta\gamma}^B + \frac{1}{3}\Theta_{\beta\gamma}^A\mu_\alpha^B\right) \\ &+ \frac{1}{9}\Theta_{\alpha\beta}^A T_{\alpha\beta\gamma\delta}\Theta_{\gamma\delta}^B + \dots\end{aligned}, \tag{3.139}$$

where q, μ, Θ, … denote the permanent molecular electrostatic moments (monopole, dipole, and quadrupole, …), $T_{\alpha\beta\gamma\ldots} = \nabla_\alpha \nabla_\beta \nabla_\gamma \ldots (R^{AB})^{-1}$ is the Cartesian tensor, α, β, γ, … represents the radial vector components, and the Einstein summation is used. In Eq. (3.139), it is not necessary to use the details of electron densities $\rho^A_{\text{tot}}(\boldsymbol{r})$ and $\rho^B_{\text{tot}}(\boldsymbol{r})$, and only the permanent multipole moments are needed.

The multipole expansion of electrostatic interaction energy can only be used if the interacting atoms and molecules do not overlap. Strictly speaking, the multipole expansion will converge only when the difference between the coordinates describing the electron densities of atoms i and j is smaller than the distance between the two atoms, i.e. $|\boldsymbol{r}_1 - \boldsymbol{r}_2| < |\boldsymbol{R}_{ij}|$. Due to the fuzzy boundary division, the electron densities of the molecules and atoms will extend to infinity. Therefore, by definition, their multipole expansions will not converge in principle; however, due to the use of finite basis sets in practice, it is possible to achieve convergence.

3.6.2.2 Exact Potential and Multipole Moment (EPMM) Model

The EPMM method uses the exact electrostatic potential (EP) for the short-range pseudoatom–pseudoatom interactions in the calculation and uses the Buckingham multipole approximation for the numerous long-range interactions. The direct electrostatic interactions between the two pseudoatomic electron densities $\rho_{\text{tot},i}(\boldsymbol{r})$ and $\rho_{\text{tot},j}(\boldsymbol{r})$ can be written as:

$$
\begin{aligned}
E_{\text{esEP},ij} &= \int_i \rho_{\text{tot},i}(\boldsymbol{r}_1) V_{\text{tot},j}(\boldsymbol{r}_1) \mathrm{d}\boldsymbol{r}_1 = \int_j \rho_{\text{tot},j}(\boldsymbol{r}_2) V_{\text{tot},i}(\boldsymbol{r}_2) \mathrm{d}\boldsymbol{r}_2 \\
&= \frac{Z_i Z_j}{R_{ij}} + \int_i \rho_i(\boldsymbol{r}_1) V_{\text{nuc},j}(\boldsymbol{r}_1) \mathrm{d}\boldsymbol{r}_1 + \int_j \rho_j(\boldsymbol{r}_2) V_{\text{nuc},i}(\boldsymbol{r}_2) \mathrm{d}\boldsymbol{r}_2, \\
&\quad + \int_j \rho_j(\boldsymbol{r}_2) V_{\text{elec},i}(\boldsymbol{r}_2) \mathrm{d}\boldsymbol{r}_2
\end{aligned}
\tag{3.140}
$$

where $V_{\text{tot},i}$ and $V_{\text{tot},j}$ denote the total electrostatic potentials (including the potentials generated by nuclei and electrons) of pseudoatoms i and j, respectively, Z_i and Z_j indicate the nuclear charges, R_{ij} is the distance between the two atoms, $\rho_i(\boldsymbol{r}_1)$ and $\rho_i(\boldsymbol{r}_2)$ are the electron densities of the two atoms, $V_{\text{nuc},i}$ and $V_{\text{nuc},j}$ is the nuclear potential of the two atoms, and $V_{\text{elec},i}$ is the electron potential of the pseudo-atom i.

In general, the electrostatic potential can be obtained using a variety of methods, where the EPMM method requires that the user define a critical atomic distance ($\boldsymbol{R}_{\text{crit}}$) as the dividing line between the EP and MM calculations. Compared to the full calculation of each pair of atoms in Eq. (3.140), the EPMM method can give accurate results while significantly reducing the computational effort.

3.6.2.3 Promolecular Approximation

Spackman proposed a different approximation of the electrostatic energy by dividing the electron density of each pseudo-atom into a spherical atomic term $\rho_{\text{atomic}}(\boldsymbol{r})$ (free atom density) and a deformation term $\Delta\rho_{\text{elec}}(\boldsymbol{r})$:

$$
\rho_{\text{pseudoatom}}(\boldsymbol{r}) = \rho_{\text{atomic}}(\boldsymbol{r}) + \Delta\rho_{\text{elec}}(\boldsymbol{r}).
$$

Thus, the integrand of Eq. (3.134) can be written as:

$$\rho_{\text{tot}}^{A}(\boldsymbol{r}_1)\rho_{\text{tot}}^{B}(\boldsymbol{r}_2) = \sum_{i\in A}\sum_{j\in B}\left[\rho_{\text{atomic},i}^{A}(\boldsymbol{r}_1)\rho_{\text{atomic},j}^{B}(\boldsymbol{r}_2) + \rho_{\text{atomic},i}^{A}(\boldsymbol{r}_1)\Delta\rho_{\text{elec},j}^{B}(\boldsymbol{r}_2)\right.$$
$$\left. + \Delta\rho_{\text{elec},i}^{A}(\boldsymbol{r}_1)\rho_{\text{atomic},j}^{B}(\boldsymbol{r}_2) + \Delta\rho_{\text{elec},i}^{A}(\boldsymbol{r}_1)\Delta\rho_{\text{elec},j}^{B}(\boldsymbol{r}_2)\right].$$

The spherical terms of pseudo-atoms belonging to a molecule will form a promolecule. Therefore, the Spackman electrostatic energy E_{esSP} can be written as the sum of promolecule–promolecule, where the promolecule–deformation and deformation–deformation contributions follow:

$$E_{\text{esSP}} = E_{\text{pro-pro}} + E_{\text{pro-def}} + E_{\text{def-def}}. \tag{3.141}$$

Using the multipole expansion approximation, only the term $E_{\text{def-def}}$ in Eq. (3.141) can be approximated because the multipole moments of spherical and neutral atoms will be zero, which would make the remaining two terms zero. For short-range interatomic distances, the multipole approximation introduces errors in $E_{\text{def-def}}$ due to the overlap of deformation charges. The first term in Eq. (3.141) contains the sum of electrostatic interactions of the electron densities of spherical atoms, and this energy can be exactly expressed in reciprocal space:

$$E_{\text{es,spherical},ij} = \frac{2}{\pi}\int_0^{\infty}[Z_i - f_i(s)][Z_j - f_j(s)]j_0(s\boldsymbol{R}_{ij})\text{d}s, \tag{3.142}$$

where Z_i is the nuclear charge of atom i, $f_i(s)$ is the scattering factor of atom i, as a function of the scattering vector $s = 4\pi\sin\theta/\lambda$, and $j_0(s\boldsymbol{R}_{ij})$ is the spherical Bessel function of order zero, in which $\boldsymbol{R}_{ij}$ is the internuclear distance. For computational convenience, the integrand can also be divided into four parts similar to Eq. (3.140), where the nuclear interaction term is replaced by Z_iZ_j/R_{ij}, while the remaining terms oscillate with s but converge quickly with an increase in s; hence, they can be easily obtained by a 1D integration strategy.

The term $E_{\text{pro-def}}$ in Eq. (3.141) contains the interaction between the spherical atomic density of one molecule ($\rho_{\text{atomic},i}$) and the deformation density term of the other molecule ($\Delta\rho_{\text{elec},i}$). The deformation term can be approximated by a multipole moment, which can similarly lead to inaccurate $E_{\text{pro-def}}$.

3.6.3 Non-electrostatic Interactions

Although electrostatic interactions are dominant, they fail to describe the full nature of intermolecular interaction energies. The Coulomb interactions calculated from experimental electron densities already contain some induction interactions in addition to classical electrostatic interactions. Regarding exchange repulsion and dispersion contributions, currently, it is not yet possible to extract them from the electron density represented by the pseudoatomic model.

To extract repulsion and dispersion interactions from experimental electron densities, researchers have suggested the use of the parametric isotropic atom-atom potential proposed by Williams or Spackman, where both parametric schemes can be derived from a model and the interactions between the AB dimers can be approximated as the sum of the repulsion energy, dispersion energy, and electrostatic energy, according to:

$$E_{\text{int}} = E_{\text{rep}} + E_{\text{disp}} + E_{\text{es}}. \tag{3.143}$$

The first two terms can be expressed as:

$$E_{\text{rep}} + E_{\text{disp}} = \sum_{i\in A,j\in B} b_i b_j \exp[-(c_i + c_j)r_{ij}] - \sum_{i\in A,j\in B} a_i a_j r_{ij}^{-6}, \tag{3.144}$$

where r_{ij} is the distance between the nonbonded atoms i and j, and a, b, c are the adjustable potential parameters for individual elements.

The third term is the electrostatic interaction between the monomers, which can be obtained by point-charge approximation (Cox–Williams model) or by multipole moment and electrostatic potential derivatives (Spackman).

In the Cox–Williams model, the electrostatic energy can be calculated from the Coulomb interaction of the q point charges:

$$E_{\text{es}} = -\sum_{i\in A}\sum_{j\in B} q_i q_j r_{ij}^{-1}. \tag{3.145}$$

The exponential parameter c in Eq. (3.144) can be estimated by theoretical calculations, with parameters a and b obtained by fitting the actual crystal structure.

Spackman proposed a set of non-bonded atom-atom potentials. Unlike those in the Cox–Williams model, the repulsion and dispersion parameters in the Spackman potential can be obtained entirely from ab initio calculations.

3.6.4 Lattice Energy

Lattice energy can be used as an overall assessment of the intermolecular interactions in the lattice and is usually defined as the energy at which isolated (gas) molecules form a crystal. In general, E_{latt} can be obtained by calculating the difference between the intermolecular interaction energy (binding energy) E_{bind} and the relaxation energy E_{rel}:

$$E_{\text{latt}} = E_{\text{bind}} - E_{\text{rel}}. \tag{3.146}$$

The crystal binding energy can be defined as:

$$E_{\text{bind}} = E_{\text{cryst}}/Z - E_{\text{mol}}(\text{cryst}), \tag{3.147}$$

where E_{cryst} is the energy per unit cell of the crystal, Z is the number of molecules per unit cell, and E_{mol}(cryst) is the energy of an individual molecule at the geometrical position of the crystal. Assuming additivity E_{bind} can be obtained by gradually accumulating the intermolecular interactions of neighboring unit cells until convergence:

$$E_{\text{bind}} = \frac{1}{2}\sum_{B} E_{\text{int}}^{AB}, \tag{3.148}$$

where E_{int}^{AB} is the interaction energy between molecules A and B. Because interaction exists between the two molecules, a factor of 1/2 can be used to eliminate double counting.

The relaxation energy E_{rel} can be defined as the difference between the energies of a molecule in the gaseous state and in the crystal, respectively:

$$E_{\text{rel}} = E_{\text{mol}}(\text{gas}) - E_{\text{mol}}(\text{cryst}). \tag{3.149}$$

The calculation of lattice energy includes the calculation of the interactions between all relevant structural units in the crystal. During the summation of the lattice, the Coulomb interaction will converge only under certain conditions, and the dispersion term will converge at a very slow rate, and certain special computational techniques, such as the Ewald method, are needed to obtain reliable results.

For a molecule with an electron density of $\rho_{\text{tot},A}(\boldsymbol{r})$ in a periodic crystal environment, the interaction energy between the molecule and the crystal environment will follow:

$$E_{\text{int}} = \int \mathrm{d}r \rho_{\text{tot},A}(\boldsymbol{r}) V_{\text{cryst}}(\boldsymbol{r}), \tag{3.150}$$

where $V_{\text{cryst}}(\boldsymbol{r})$ is the electrostatic potential of the crystal.

The conditional convergence of the lattice summation indicates that the summation calculation needs to be performed for finite samples of a certain size and shape. The Ewald summation method is currently the best-known method to accelerate the convergence of lattice summation, and essentially combines two conditionally convergent lattice summations so that the sum of the two can reach absolute convergence. In practice, the Coulomb kernel can be divided into two parts represented by an error function and a complementary error function, respectively. The contribution of the short-range error compensation function can be safely summed in direct space, while the contribution of the long-range error function will converge quickly in reciprocal space:

$$G(\boldsymbol{r}, \boldsymbol{r}') = \sum_{L} \frac{\operatorname{erfc}(\mu|\boldsymbol{r} - \boldsymbol{r}' + \boldsymbol{L}|)}{|\boldsymbol{r} - \boldsymbol{r}' + \boldsymbol{L}|} + \frac{4\pi}{\Omega} \sum_{k \neq 0} \frac{\mathrm{e}^{-k^2/4\mu^2}}{k^2} \mathrm{e}^{i\boldsymbol{k}\cdot(\boldsymbol{r}-\boldsymbol{r}')}, \tag{3.151}$$

where the parameter μ determines the relative weights of the summation in direct space and reciprocal space, and it should be chosen to facilitate the overall convergence.

Stewart's method can be used in most studies for the calculation of the electrostatic potential or interaction energy from experimental electron densities. In this method, the electron density can be decomposed into the sum of the spherical and neutral atomic density $\rho_{\text{IAM}}(\boldsymbol{r})$ and a multipole deformation term $\Delta\rho(\boldsymbol{r})$, i.e. where the total electrostatic potential of the crystal can be written as:

$$V_{\text{cryst}}(\boldsymbol{r}) = V_{\text{IAM}}(\boldsymbol{r}) + \Delta V(\boldsymbol{r}), \tag{3.152}$$

where $V_{\text{IAM}}(\boldsymbol{r})$ is the potential of the procrystal and $\Delta V(\boldsymbol{r})$ is the potential of the multipole deformation density term. Furthermore, $V_{\text{IAM}}(\boldsymbol{r})$ can be summed directly in direct space:

$$V_{\mathrm{IAM}}(\boldsymbol{r}) = \int \mathrm{d}\boldsymbol{r}' \frac{\rho_{\mathrm{IAM}}(\boldsymbol{r}')}{|\boldsymbol{r} - \boldsymbol{r}'|}. \tag{3.153}$$

In addition, the multipole deformation term contribution $\Delta V(\boldsymbol{r})$ will converge quickly in reciprocal space:

$$\Delta V(\boldsymbol{r}) = -\frac{1}{\pi\Omega} \sum_{\boldsymbol{k}(\neq 0)} \frac{F_{\mathrm{cryst}}(\boldsymbol{k}) - F_{\mathrm{IAM}}(\boldsymbol{k})}{k^2} \mathrm{e}^{-2\pi i\boldsymbol{k}\cdot\boldsymbol{r}}. \tag{3.154}$$

3.6.5 Interaction Energies Obtained from Experimental Charge Analysis

As an excellent analytical tool for analyzing the interactions between atoms or molecules from electron densities, the QTAIM method can be used to extract all topological and energetic properties from theoretically calculated wave functions; however, for experimental electron densities, it can only obtain this type of information from the multipole function $\rho(\boldsymbol{r})$. Abramov proposed a method for estimating the local electron kinetic energy density $G(\boldsymbol{r})$ of closed-shell interactions, based on the semiclassical Thomas-Fermi equation and gradient quantum correction and using the electron density $\rho(\boldsymbol{r})$, density gradient $\nabla\rho(\boldsymbol{r})$, and Laplacian $\nabla^2\rho(\boldsymbol{r})$, where $G(\boldsymbol{r})$ can be written as

$$G(\boldsymbol{r}) = \left(\frac{3}{10}\right)(3\pi^2)^{2/3}(\rho(\boldsymbol{r}))^{5/3} + \left(\frac{1}{72}\right)\left(\frac{(\nabla\rho(\boldsymbol{r}))^2}{\rho(\boldsymbol{r})}\right) + \left(\frac{1}{6}\right)\nabla^2\rho(\boldsymbol{r}). \tag{3.155}$$

The $G(\boldsymbol{r})$ estimated from Eq. (3.155) will be in good agreement with the HF theoretical calculations in the mid-range region (with an interatomic distance of 0.5–2.1 Å), resulting in a deviation of less than 4%. However, Eq. (3.155) will only be suitable for closed-shell interactions and will give only rough results for electron-sharing systems. The approach of calculating $G(\boldsymbol{r})$ opens up new possibilities for experimentalists to analyze intermolecular interactions, especially for analyzing molecular or atomic interactions. At a BCP of (3, −1), $\nabla\rho(\boldsymbol{r}_{\mathrm{BCP}})=0$, Eq. (3.155) can be simplified to

$$G = \left(\frac{3}{10}\right)(3\pi^2)^{2/3}\rho^{5/3} + \left(\frac{1}{6}\right)\nabla^2\rho. \tag{3.156}$$

Applying the local form of the virial theorem gives:

$$2G + V = \left(\frac{1}{4}\right)\nabla^2\rho. \tag{3.157}$$

The local electron potential V at the BCP and the total energy density H can be estimated as:

$$\begin{aligned} V &= \left(\frac{1}{4}\right)\nabla^2\rho - 2G \\ H &= G + V \end{aligned}. \tag{3.158}$$

References

1 Bader, R.F.W. (1990). *Atoms in Molecules: A Quantum Theory*. Oxford, U.K.: Oxford University Press.

2 Matta, C.F. and Boyd, R.J. (2007). *The Quantum Theory of Atoms in Molecules: From Solid State to DNA and Drug Design*. Weinheim: Wiley-VHC.

3 Coppens, P. (1997). *X-ray Charge Densities and Chemical Bonding*. New York: Oxford University Press, Inc.

4 Pendás, A.M., Kohout, M., Blanco, M.A., and Francisco, E. (2012). Beyond standard charge density topological analyses. In: *Modern Charge-Density Analysis* (ed. C. Gatti and P. Macchi), 303–358. Berlin: Springer.

5 Dominiak, P.M., Espinosa, E., and Angyan, J.G. (2012). Intermolecular interaction energies from experimental charge density studies. In: *Modern Charge-Density Analysis* (ed. C. Gatti and P. Macchi), 387–433. Berlin: Springer.

6 Abramov, Y.A. (1997). On the possibility of kinetic energy density evaluation from the experimental electron-density distribution. *Acta Cryst. A* 53: 264–272.

7 Cremer, D. and Kraka, E. (1984). Chemical bonds without bonding electron density — does the difference electron-density analysis suffice for a description of the chemical bond? *Angew. Chem. Int. Ed.* 23: 627–628.

8 Tsirelson, V.G., Stash, A.I., Potemkin, V.A. et al. (2006). Molecular and crystal properties of ethyl 4,6-dimethyl-2-thioxo-1,2,3,4-tetrahydropyrimidine-5-carboxylate from experimental and theoretical electron densities. *Acta Cryst. B* 62: 676–688.

9 Gatti, C. (2005). Chemical bonding in crystals: new directions. *Z. Kristallogr.* 220: 399–457.

10 Matta, C.F. and Hernández-Trujillo, J. (2003). Bonding in polycyclic aromatic hydrocarbons in terms of the electron density and of electron delocalization. *J. Phys. Chem. A* 107: 7496–7504. (Correction: J. Phys. Chem. A, 2005, 109: 10798).

11 Sen, K.D. (2005). Characteristic features of Shannon information entropy of confined atoms. *J. Chem. Phys.* 123: 074110.

12 Bader, R.F.W. and Gatti, C. (1998). A Green's function for the density. *Chem. Phys. Lett.* 287: 233–238.

13 Gatti, C. and Bertini, L. (2004). The local form of the source function as a fingerprint of strong and weak intra- and intermolecular interactions. *Acta Cryst. A* 60: 438–449.

14 Becke, A.D. and Edgecombe, K.E.J. (1990). A simple measure of electron localization in atomic and molecular systems. *Chem. Phys.* 92: 5397–5403.

15 Savin, A., Nesper, R., Wengert, S., and Fassler, T. (1997). ELF: The electron localization function. *Angew. Chem. Int. Ed. Engl.* 36: 1808–1832.

16 Fuentealba, P., Chamorro, E., and Santos, J.C. (2008). Understanding and using the electron localization function. In: *Theoretical Aspects of Chemical Reactivity*, Theoretical and Computational Chemistry, vol. 19 (ed. A. Toro-Labbé), 57–85. Elsevier B.V.

17 Tsirelson, V. and Stash, A. (2002). Determination of the electron localization function from electron density. *Chem. Phys. Lett.* 351: 142–148.

18 Johnson, E.R., Keinan, S., Mori-Sánchez, P. et al. (2010). Revealing noncovalent interactions. *J. Am. Chem. Soc.* 132: 6498–6506.

19 Tian, L. and Chen, Q. (2021). Interaction region indicator (IRI): a simple real space function clearly revealing both chemical bonds and weak interactions. *Chemistry–Methods* 1: 231–239.

20 Blanco, M.A., Martín Pendás, A., and Francisco, E. (2005). Interacting quantum atoms: a correlated energy decomposition scheme based on the quantum theory of atoms in molecules. *J. Chem. Theory Comput.* 1: 1096–1109.

21 Hirshfeld, F.L. (1977). Bonded-atom fragments for describing molecular charge densities. *Theor. Chim. Acta* 44: 129–138.

22 Li, L. and Parr, R.G. (1986). The atom in a molecule: a density matrix approach. *J. Chem. Phys.* 84: 1704–1711.

23 Becke, A.D. (1988). A multicenter numerical integration scheme for polyatomic molecules. *J. Chem. Phys.* 88: 2547–2553.

24 Martín Pendás, A., Francisco, E., and Blanco, M.A. (2007). Pauling resonant structures in real space through electron number probability distributions. *J. Phys. Chem. A* 111: 1084–1090.

25 Kohout, M., Pernal, K., Wagner, F.R., and Grin, Y. (2004). Electron localizability indicator for correlated wavefunctions. I. parallel-spin pairs. *Theor. Chem. Acc.* 112: 453–459.

26 Kimball, J.C. (1973). Short-range correlations and electron-gas response functions. *Phys. Rev. A* 7: 1648–1652.

4

Principles of Electronic Structure Measurement

4.1 Introduction

In 1915, Debye first realized the feasibility of using the X-ray diffraction technique to obtain the electron density distribution of solids. However, at that time, instrumental precision had not reached the point where accurate electron density could be obtained, and it was not until decades later that Debye's idea was implemented, benefiting from advances in the following areas:

(1) Improvements in diffraction data quality as a result of the development of automated diffractometer and high-precision detector.
(2) The emergence and development of high-brilliant light sources, such as synchrotron radiation and neutron sources; substantial enhancement of laboratory diffractometer intensity.
(3) The development of a theoretical electronic structure model for the reliable refinements.

When determining electronic structures based on experimental diffraction data, it is necessary to first establish a parameterized theoretical electronic structure model (pseudo-atomic model, density matrix model, or electronic wave function model), and theoretical structure factors can be calculated using these theoretical models. Then, the parameters of the electronic structure model can be adjusted (refined) to minimize the deviation (usually the square of the difference between theoretical and experimental values) between the model and experimentally calculated structure factor amplitude (or its square). The electronic structure can be confirmed when the deviation is minimized. However, because experimental diffraction data inevitably have errors, it is necessary to select an appropriate weighting method based on experimental errors [1].

Real atoms are in constant motion. Therefore, experimental electron density is an average result of atomic thermal vibrations. In conventional electron density analysis, the atom vibration can usually be described using a simple harmonic vibration model. The probability that atom j deviates from its equilibrium position $\boldsymbol{r}_j$ can be described using a probability distribution function $p_j(\boldsymbol{r})$, while the static electron density of atom j can be referred to as $\rho_j^{\mathrm{at,static}}$, and the mean electron density can be described as the convolution of the atomic static electron density and $p_j(\boldsymbol{r})$ (denoted

Electronic Structure Crystallography and Functional Motifs of Materials, First Edition.
Guo-Cong Guo and Xiao-Ming Jiang.

using *):

$$\rho_j^{\text{at}}(\boldsymbol{r}) = \rho_j^{\text{at,static}}(\boldsymbol{r}) * p_j(\boldsymbol{r}) = \int \rho_j^{\text{at,static}}(\boldsymbol{r}-\boldsymbol{r}') \cdot p_j(\boldsymbol{r}')\mathrm{d}\boldsymbol{r}'. \tag{4.1}$$

The unit cell density is the sum of the convolution of the electron density of each atom in a unit cell and the function δ, i.e.:

$$\rho_{\text{unit cell}}(\boldsymbol{r}) = \sum_j \rho_j^{\text{at}}(\boldsymbol{r}) * \delta(\boldsymbol{r}-\boldsymbol{r}_j). \tag{4.2}$$

Substituting (4.1) into (4.2) gives:

$$\rho_{\text{unit cell}}(\boldsymbol{r}) = \sum_j \left\{ \left[\rho_j^{\text{at,static}}(\boldsymbol{r}) * p_j(\boldsymbol{r})\right] * \delta(\boldsymbol{r}-\boldsymbol{r}_j)\right\}. \tag{4.3}$$

The electron density and structure factor $F(\boldsymbol{H})$ can be Fourier transformed, according to:

$$\begin{aligned} F(\boldsymbol{H}) = FT[\rho_{\text{unit cell}}(\boldsymbol{r})] &= \sum_j \left\{ FT\left[\rho_j^{\text{at,static}}(\boldsymbol{r}) * p_j(\boldsymbol{r})\right] \cdot FT[\delta\boldsymbol{r}-\boldsymbol{r}_j\} \\ &= \sum_j f_j(\boldsymbol{H}) \cdot T_j(\boldsymbol{H}) \cdot e^{2\pi i\boldsymbol{H}\cdot\boldsymbol{r}_j} \end{aligned} \tag{4.4}$$

where

$$\begin{aligned} f_j(\boldsymbol{H}) &= \int \rho_j^{\text{at,static}}(\boldsymbol{r})e^{2\pi i\boldsymbol{H}\cdot\boldsymbol{r}}\mathrm{d}\boldsymbol{r}; \\ T_j(\boldsymbol{H}) &= \int p_j(\boldsymbol{r}')e^{i\boldsymbol{H}\cdot\boldsymbol{r}'}\mathrm{d}\boldsymbol{r}'. \end{aligned} \tag{4.5}$$

The above equations use the Fourier transform convolution theorem. The Fourier transform of the convolution of the product of two functions is equal to the product of the Fourier transform of the two functions, where f_j is the form factor of the static atom, T_j is the atom temperature factor, usually referred to as the Debye–Waller factor. Different f_j and T_j forms and structure parameters can be used to build different electronic structure refinement models. The least square method is often used to minimize the residual density (difference between experimental and modeled density). In addition, other methods, such as the conjugate gradient method, are also commonly used. Parameters associated with f_j, such as the basis functions that describe the electron density and their expansion coefficients, will determine $\rho_j^{\text{at,static}}$. In the refinement of conventional crystal structure and most electron structures, the probability density function $p_j(\boldsymbol{r})$ describing the atom deviation from its equilibrium position can be approximated using the isotropic or anisotropic Gaussian function (referred to as simple harmonic motion approximation). The Fourier transform of a Gaussian function will still be a Gaussian function. Therefore, T_j is a Gaussian function and its exponential part can be represented by a symmetric 3×3 matrix. The isotropic and anisotropic temperature factors can be described by one and six independent atomic displacement parameters (ADP). Under certain complicated circumstances, the temperature factor can be described using higher order anharmonic approximation, e.g. the Gram Charlier coefficient. In this case, it is usually necessary to collect high-angle diffraction points for reliable refinement.

In the simplest circumstance, the atomic static electron density $\rho_j^{\text{at,static}}$ can be derived from the Hartree–Fock or Dirac–Fock function of a neutral spherically symmetric atom, which will only allow for the refinement of atomic coordinates and temperature factors. This is known as the independent atom model (IAM), which can be used to display the images of crystal structures via the least-square refinement method; however, it cannot be used to obtain electron density redistribution information as a result of atomic interaction (chemical bond). Improvements can be conducted for this model. One approach (the Kappa Model) involves dividing the atom into the core electron shell and valence shell, fixing the core electron shell density to be the same as the neutral atom. This approach also allows for density changes of electron shells, e.g. non-neutral transformation, expansion, and shrinkage, reflecting the actual bonding environment around an atom. Compared to the IAM model, the Kappa model will significantly improve the refinement results with only two extra parameters (i.e. the valence of the electron and the radial shrinkage coefficient κ of the valence electron) for each atom. Although the Kappa model can give a rough electron density distribution, it is rarely used alone in actual electron density analysis. Instead, it is generally used as an intermediate step in the entire electronic structure refinement. A more flexible multipole model (MM) is often used in electron density analysis, where the valence shell electron density of each atom can be expanded in multipole. The non-spherical symmetric behavior of valence electrons can be refined via the population number of the multipole moment and the shrinkage factor of the multipole function. Therefore, a more complicated chemical bonding environment can be simulated. In the pseudo-atomic model, this multipole architecture usually gives the best-fitting results. Because each atom usually contains over 20 refinement parameters, high data quality and structure resolution (obtained using Mo kα X-rays with a shorter wavelength and a higher diffraction angle) are required to obtain reliable refinement results. Although MM has become almost a standard method for experimental electron density analysis, it still suffers from a deficiency in flexibility, sometimes introducing systemic biases to the electron density results. Currently, researchers have proposed multiple ways to improve MM.

Experimentally obtained electron density results imply certain weak electronic effects, such as electron correlation and relativistic effect, which are difficult to calculate theoretically. In addition, they also contain the crystal field effect, where the effect of molecular crystallization on the molecular electron density can be obtained through periodic ab initio calculations. Studies have found that the electron density information of the mixed ground and excited states can also be obtained from experimental electron density analysis. In addition, the transient experimental electron density can be obtained through femtosecond X-ray diffraction. Despite these advantages, the electron density data quality, even at the equilibrium state, will be affected by many factors, such as the testing error of the structure factor, the flaws of the MM model, and the expanding method of the electron density basis function centered on the nucleus. These restrictions will hinder analysis of all two-electron terms, whereas the importance of two-electron terms can be represented in the extended electron density model or density matrix, where a single electron reduced density matrix can be constructed from a single electron atomic orbital. The density matrix can be obtained by jointly refining the scattering data in the position and momentum space, while the elastic and inelastic scattering can provide density information for the position and momentum space, respectively. In addition to pseudoatomic and density matrix models, another potential approach involves constructing electron wave function models, i.e. the so-called X-ray constrained wavefunction (XCW). The standard ab initio variational approach in this method minimizes both the wavefunction energy and the difference between the modeled and experimental structure factors. Obviously, in addition to providing an experimental single electron density matrix, the XCW method can also generate the electron density that matches the experiments.

Other methods, such as the X-ray atomic orbital (XAO) method, directly refines the electronic state of an atom based on the atomic orbital hypothesis, and are mainly used in the electro-density analysis of ionic solids. In this method, the coefficients of the orthogonalized atomic orbital can be calculated using the perturbation method and refined using the measured structure factor. The XAO is similar to the valence orbital model (VOM), which can be used in complexes if low overlap is assumed between the electron orbitals of the metal and the organic ligand. Therefore, the electron density of metals or organic ligands can be treated as an atomic density perturbation. This assumption can typically be used in the MM to determine orbital coefficients.

These methods reconstruct electron density from diffraction data based on atomic molecular density or orbital assumptions. In fact, other methods, such as the Monte Carlo wavefunction, can also be used. Many alternative methods for determining electron density are related to the maximum likelihood principle, the most prominent of which is the maximum entropy method (MEM), which provides a statistical thermodynamic approach to data processing. In addition, the charge-flipping method based on pseudo-thermodynamics can also be potentially used for electron density analysis, which deserves further attention.

4.2 Thermal Vibration Analysis

The biggest obstacle to comparing the experimental electron density with the theoretical results is the simulation of atomic thermal vibrations. The theoretically calculated electron density excludes the influence of thermal vibrations; however, the actual atoms in crystals always vibrate near the equilibrium position. Although the influence of thermal vibrations can be reduced by cooling, it cannot be completely eliminated. Theoretically, even at 0 K, vibrations (called zero vibrations) still exist.

Atomic vibrations have a significant influence on the diffracting intensity of diffraction experiments, where the diffraction of X-rays through atoms is an ultrafast process that lasts around 10^{-18} s and is much faster than atomic vibrations (10^{-13} s). Each image obtained from an X-ray scattering experiment can typically be exposed for a few seconds, during which it contains a large number of states. Stewart and Feil noted that the scattering average of the instantaneous distribution of electrons is equivalent to the time average of the electron density [2].

To create a comparison between the experimental and theoretical electron density, in experimental electron density analysis, static electron density must be studied and obtained by deconvolution of the atomic thermal vibrations. The simulation of atomic thermal vibrations is critical for two reasons, namely, (i) atomic vibrations provide information regarding the potential energy surface of molecules and atoms in crystals, and (ii) the electron density model and atomic vibration model are correlated with each other and often simulate the same set of diffraction data [3].

4.2.1 Lattice Dynamics

A crystal can be regarded as a giant molecule, where each atom in the crystal contains three degrees of freedom. Given that the crystal contains N cells and each cell contains n atoms, each crystal will have $3nN$ degrees of freedom. The vibrations of these atoms can be linked together, moving in waves throughout the crystal, known as phonons. Based on the periodic boundary conditions, the vibration of each atom can be described by:

$$\boldsymbol{u}(kl, t, \boldsymbol{q}) = \boldsymbol{U}(k|\boldsymbol{q}) \exp[i(\boldsymbol{q} \cdot \boldsymbol{r}(kl) - \omega(\boldsymbol{q})t + \phi(k|\boldsymbol{q}))], \tag{4.6}$$

where k refers to one atom in a cell l refers to a unit cell. The displacement of (k, l) atom from its equilibrium position $\boldsymbol{r}(kl)$, namely, $\boldsymbol{u}(kl, t, \boldsymbol{q})$, is related with the wave vector $\boldsymbol{q}$. $\boldsymbol{U}(k|\boldsymbol{q})$ describes the maximum vibration amplitude and direction of atom vibration. $\boldsymbol{U}$ will not be related to the unit cell l, as the equivalent atoms in different unit cells will have the same magnitude and direction and only differ in phase. This is known as the Bloch Theory, which dramatically simplifies the atomic vibration problem and focuses on the $3n$ equations of n atoms in a unit cell instead of $3nN$ equations in nN atoms in the crystal. The frequency ω in Eq. (4.6) is the continuous function of $\boldsymbol{q}$, where the ω dependence of $\boldsymbol{q}$ is called the dispersion relationship. For a known $\boldsymbol{q}$, Eq. (4.6) describes the $3n$ vibrational modes. Therefore, N different $\boldsymbol{q}$ vectors will be needed to describe the atomic vibration modes in the crystal. These vectors can be uniformly distributed in the first Brillouin zone (also called the Brillouin zone). The Brillouin zone is the area in reciprocal space centered at the origin and surrounded by the mid-vertical plane from the origin to the nearest reciprocal lattice point.

Under the simple harmonic vibrations and adiabatic approximation, the motion equation of an atom k in a unit cell l can be written as:

$$m(k)\ddot{\boldsymbol{u}}(kl, t) = -\sum_{k'l'} \boldsymbol{\Phi} \begin{pmatrix} k & k' \\ l & l' \end{pmatrix} \boldsymbol{u}(k'l', t), \tag{4.7}$$

where $\boldsymbol{\Phi}$ is the force constant matrix (second derivation of energy with respect to atomic displacement), and $\boldsymbol{\Phi}$ is related to crystal potential energy V:

$$2V = \sum_{kl\alpha} \sum_{k'l'\alpha'} \Phi_{\alpha\alpha'} \begin{pmatrix} k & k' \\ l & l' \end{pmatrix} u_\alpha(kl) u_{\alpha'}(k'l'). \tag{4.8}$$

A system of equations can be obtained by substituting Eq. (4.6) in (4.7), and the equations can be written in matrix form:

$$\omega^2 \boldsymbol{U}_m = \boldsymbol{D}\boldsymbol{U}_m, \tag{4.9}$$

where $\boldsymbol{U}_m$ is the 3n×1 column vector of the renormalized square root of mass:

$$\boldsymbol{U}_m = \boldsymbol{m}^{1/2}\boldsymbol{U}_0. \tag{4.10}$$

The mass matrix $\boldsymbol{m}$ is a 3n×3n diagonal matrix obtained by repeating the mass of each atom three times on the diagonal, where $\boldsymbol{U}_0$ is the column vector of the displacement of each atom in a unit cell, defined as:

$$\boldsymbol{U}_0(q) = \begin{pmatrix} U_1(1|\boldsymbol{q}) \\ U_2(1|\boldsymbol{q}) \\ U_3(1|\boldsymbol{q}) \\ U_1(2|\boldsymbol{q}) \\ \vdots \\ U_3(n|\boldsymbol{q}) \end{pmatrix}. \tag{4.11}$$

$\boldsymbol{D}(\boldsymbol{q})$ is a 3n×3n kinetic matrix after mass renormalization and can be defined as:

$$\boldsymbol{D} = \boldsymbol{m}^{-1/2}\boldsymbol{D}_0\boldsymbol{m}^{-1/2}, \tag{4.12}$$

where the $\boldsymbol{D}_0$ matrix can be obtained by calculating the force between atoms in different unit cells and can be seen as the Fourier transformation of the force constant matrix. The matrix element of the $\boldsymbol{D}$ matrix is given by:

$$D_{\alpha\alpha'}(kk'|\boldsymbol{q}) = (m(k)m(k'))^{-1/2}\sum_{l'}\Phi_{\alpha\alpha'}\begin{pmatrix} k & k' \\ 0 & l' \end{pmatrix}\exp[i\boldsymbol{q}\cdot\Delta\boldsymbol{r}(kk'l')], \tag{4.13}$$

where $\Phi_{\alpha\alpha'}$ is the force constant between atoms with coordinates of α and α', and $\Delta\boldsymbol{r}(kk'l')$ is the distance between the origins of the asymmetric units. The solution to the eigenproblem of Eq. (4.9) will be the eigenvector $\boldsymbol{e}(j\boldsymbol{q})$ of the $j\boldsymbol{q}$ mode shift, where the eigenvalue is the square $\omega_j(\boldsymbol{q})^2$ of the $j\boldsymbol{q}$ mode frequency. The force that builds $\boldsymbol{D}(\boldsymbol{q})$ can be calculated through the force field and ab initio calculations. Otherwise, the dispersion curve can be obtained using an empirical force model to fit the dispersion curve of the inelastic neutron scattering measurements, and the vibration frequency information can be obtained from the Raman and IR spectra. Once the vibration modes and frequencies of an atom are known, they can be used to calculate the contributions to crystal thermodynamic properties, such as entropy and thermal capacity. In addition, they can be used to calculate the mean squared displacement tensor and anisotropic displacement parameters of an atom.

The mean squared displacement tensor of an atom can be written as the sum of the contribution of 3nN vibration modes:

$$\boldsymbol{B}_{\text{atom}}(k) = \frac{1}{Nm_k}\sum_{j\boldsymbol{q}}\frac{E_j(\boldsymbol{q})}{\omega_j^2(\boldsymbol{q})}\boldsymbol{e}(k|j\boldsymbol{q})\boldsymbol{e}^*((k|j\boldsymbol{q}))^T, \tag{4.14}$$

where $\boldsymbol{e}(k\,|\,j\boldsymbol{q})$ is the component of the complex eigenvector $\boldsymbol{e}(j\boldsymbol{q})$ corresponding to the atom k, ω_j is the frequency of j mode, m_k is the mass of the atom k, and $E_j(\boldsymbol{q})$ is the energy of the vibration mode:

$$E_j(\boldsymbol{q}) = \hbar\omega_j(\boldsymbol{q})\left(\frac{1}{2} + \frac{1}{\exp(\hbar\omega_j(\boldsymbol{q})/k_BT) - 1}\right). \tag{4.15}$$

The $\boldsymbol{B}_{\text{atom}}(k)$ in Eq. (4.14) denotes the mean atom displacement tensor and is a 3×3 symmetric tensor equivalent to the ADP tensor. The $\boldsymbol{e}(j\boldsymbol{q})$ contains atom vibration-related information, which cannot be obtained from the scattering experiment under a single temperature. However, certain information can be obtained from experiments under multiple different temperature points with rigid body approximation.

4.2.2 Atomic Displacement Parameters

As described in the previous section, the lattice dynamics theory can describe the vibrations of atoms in a crystal as individual vibration modes. In actual electron density analysis and structure refinement, lattice dynamics methods are not used. Instead, the atoms can be treated as independent simple harmonic oscillators moving in the average field of surrounding atoms. In this method, the thermally averaged electron density can be written as the convolution of the probability density function (p.d.f.), describing how an atom k deviates from its equilibrium position $\boldsymbol{r}_{k0}$:

$$\langle \rho_k(\boldsymbol{r}) \rangle = \int \rho_k(\boldsymbol{r} - \boldsymbol{r}_k) p_k(\boldsymbol{r}_k - \boldsymbol{r}_{k0}) \mathrm{d}\boldsymbol{r}_k. \tag{4.16}$$

It should be noted that this model assumes that the static electron density will not be deformed during atomic vibrations. Electron density moves with the nucleus, and this approximation offers a rigid pseudo-atom approximation. It is difficult to validate this model; however, the atom displacement parameters obtained using this model were well consistent with inelastic neutron scattering results.

The X-ray form factor of the scattering vector $\boldsymbol{h}$ can be expressed as the Fourier transform of the mean electron density per cell:

$$F(\boldsymbol{h}) = \int < \rho(\boldsymbol{r}) > e^{2\pi i \boldsymbol{h} \cdot \boldsymbol{r}} \mathrm{d}\boldsymbol{r}. \tag{4.17}$$

The mean electron density per cell can be approximated as the sum of contributions from each atom:

$$< \rho(\boldsymbol{r}) >= \sum_{N}^{k=1} n_k \int \rho_k(\boldsymbol{r} - \boldsymbol{r}_k) p_k(\boldsymbol{r}_k - \boldsymbol{r}_{k0}) \mathrm{d}\boldsymbol{r}_k, \tag{4.18}$$

where n_k is the occupation factor of atom k.

By substituting Eq. (4.18) into (4.17), we can obtain the structure factor as the sum of contributions from each atom $F_k(\boldsymbol{h})$:

$$F(\boldsymbol{h}) = \sum_{N}^{k=1} n_k F_k(\boldsymbol{h}). \tag{4.19}$$

The contribution of each atom contains a product of the structure factor $f_k(\boldsymbol{h})$ with $T_k(\boldsymbol{h})$, where $f_k(\boldsymbol{h})$ and $T_k(\boldsymbol{h})$ are the Fourier transform of the static atom electron density and atom p.d.f., respectively:

$$F(\boldsymbol{h}) \approx \sum_{N}^{k=1} n_k f_k(\boldsymbol{h}) T_k(\boldsymbol{h}) \exp(2\pi i \boldsymbol{h} \cdot \boldsymbol{r}_{k0}), \tag{4.20}$$

where $\boldsymbol{v} = (\boldsymbol{r} - \boldsymbol{r}_k)$ and $\boldsymbol{u} = (\boldsymbol{r}_k - \boldsymbol{r}_{k0})$, and the scattering factor of the k atom, also known as the atomic form factor can be written as:

$$f_k(\boldsymbol{h}) = \int \rho_k(\boldsymbol{r}) \exp(2\pi i \boldsymbol{h} \cdot \boldsymbol{v}) \mathrm{d}\boldsymbol{v}. \tag{4.21}$$

The Fourier transform $T_k(\boldsymbol{h})$ of atomic p.d.f. is given by:

$$T_k(\boldsymbol{h}) = \int p_k(\boldsymbol{u}) \exp(2\pi i \boldsymbol{h} \cdot \boldsymbol{u}) \mathrm{d}\boldsymbol{u}. \tag{4.22}$$

Equation (4.22) describes the dependence of the structure factor on atomic displacement, and so far, no restrictions have been added to the format of p.d.f and $T_k(\boldsymbol{h})$. In regular structure refinement and most experimental electron density studies, the atomic p.d.f. can be approximated using an isotropic or anisotropic Gaussian function. According to statistical physics, with simple harmonic approximation, p.d.f. will be the Gaussian function, and the Fourier transform of the Gaussian function will be another Gaussian function. Without the atom subscript k, $T(\boldsymbol{h})$ can be written as:

$$T(\boldsymbol{h}) = \exp[-2\pi^2(\boldsymbol{h} \cdot \boldsymbol{u})^2]. \tag{4.23}$$

The exact format of Eq. (4.23) will depend on the scattering vector $\boldsymbol{h}$ and the displacement vector $\boldsymbol{u}$. Two critical basis vectors are often used, where the diffraction vector $\boldsymbol{h}$ uses reciprocal space lattice basis vector ($\boldsymbol{a}^*, \boldsymbol{b}^*, \boldsymbol{c}^*$), and the atom displacement vector $\boldsymbol{u}$ uses straight space basis vector ($\boldsymbol{a}$, $\boldsymbol{b}$, $\boldsymbol{c}$), which can be normalized. Therefore, the atomic displacement component has a length dimension:

$$\begin{aligned} \boldsymbol{h} &= h\boldsymbol{a}^* + k\boldsymbol{b}^* + l\boldsymbol{c}^* \\ \boldsymbol{u} &= \Delta\xi^1 a^* \boldsymbol{a} + \Delta\xi^2 b^* \boldsymbol{b} + \Delta\xi^3 c^* \boldsymbol{c}. \end{aligned} \tag{4.24}$$

T can be obtained from the above equations:

$$\begin{aligned} T &= \exp\left(-2\pi^2 \sum_{j=1}^{3} \sum_{l=1}^{3} h_j a^j < \Delta\xi^j \Delta\xi^l > a^l h_l\right), \\ &\equiv \exp\left(-2\pi^2 \sum_{j=1}^{3} \sum_{l=1}^{3} h_j a^j U^{jl} a^l h_l\right) \end{aligned} \tag{4.25}$$

where $U^{jl} = < \Delta\xi^j \Delta\xi^l >$, U^{jl} has an anisotropic perimetric form and square length dimensions. In addition, it can be directly related to the mean square displacement of an atom in a certain direction, and $\boldsymbol{U}$ is the 3×3 symmetric tensor. It is usually not represented by using Cartesian coordinates; however, certain complicated calculations can be avoided when the diffraction vector and atom displacement vector use the same Cartesian coordinates. Therefore, Cartesian coordinates is also often used in lattice dynamics studies and thermal vibration analysis. When $\boldsymbol{h}$ and $\boldsymbol{u}$ are both represented in the Cartesian coordinates, then:

$$\begin{aligned} \boldsymbol{h} &= h_1^C \boldsymbol{e}_1 + h_2^C \boldsymbol{e}_2 + h_3^C \boldsymbol{e}_3 \\ \boldsymbol{u} &= \xi_1^C \boldsymbol{e}_1 + \xi_2^C \boldsymbol{e}_2 + \xi_3^C \boldsymbol{e}_3. \end{aligned} \tag{4.26}$$

Subsequently, T can be written as:

$$T = \exp\left(-2\pi^2 \sum_{j=1}^{3}\sum_{l=1}^{3} h_j^C < \Delta\xi_j^C \Delta\xi_l^C > h_l^C\right)$$
$$\equiv \exp\left(-2\pi^2 \sum_{j=1}^{3}\sum_{l=1}^{3} h_j^C U_{jl}^C h_l^C\right), \tag{4.27}$$

where $U_{jl}^C =< \Delta\xi_j^C \Delta\xi_l^C >$ is the Debye–Waller factor, and U_{jl}^C can be directly related to the tensor $\boldsymbol{B}$ in Eq. (4.14), as both are expressions of the mean squared displacement tensor in the Cartesian coordinates. Furthermore, they can be directly compared under the same coordinates. Otherwise, a comparison can only be made after linear transformation.

The situation gets more complicated when the atomic density function is not a Gaussian function. The most popular way involves using the GramCharlier (GC) expansion method. This method describes a non-Gaussian distribution as a differential expansion of a Gaussian p.d.f., where $T_G(\boldsymbol{h})$ is represented by using the Gaussian Debye–Waller factor $T_G(\boldsymbol{h})$:

$$T_{GC}(\boldsymbol{h}) = T_G(\boldsymbol{h})[1 + (2\pi i)^3 \gamma^{jkl} h_j h_k h_l/3! + (2\pi i)^4 \delta^{jklm} h_j h_k h_l h_m/4! + \ldots], \tag{4.28}$$

where γ^{jkl}, δ^{jklm}, ... denote the third-order, fourth-order... anharmonic tensor coefficients, respectively, which are subjected to Einstein summation. Structural refinement using GC expansion requires high-resolution diffraction data, and the minimum resolution of the diffraction data based on the mean atomic square displacement $< u^2 >$ can be calculated as:

$$K_n = 2\left(\frac{n \ln 2}{\pi}\right)^{1/2} \frac{1}{< u^2 >^{1/2}}, \tag{4.29}$$

where n is the GC expanded order. For certain systems, the K_n calculated from (4.29) will only be meaningful theoretically.

Although anharmonic parameter refinement can be used in certain electron density analyses, the necessity of these parameters has rarely been examined. A notable indicator is the standard deviation of these parameters, and it is also required to check whether the anharmonic p.d.f. diagram is physically reasonable or not.

The mean atomic squared displacement tensor can usually be expressed by using the ellipsoid with fixed probability:

$$\boldsymbol{u}^T \boldsymbol{B}^{-1} \boldsymbol{u} = c^2, \tag{4.30}$$

where c is a constant. When c is 1.5382, there is a 50% probability of an atom falling in the ellipsoid, and a major direction of atom vibrations can be observed by the external shape of the ellipsoid.

Checking the equal probability ellipsoid of ADP during structure refinement can help to identify the problem. Oblate or prolate distortion ellipsoids will often result from static disorder. Displacement disorder is typical in crystals; however, static disorder rarely exists in electron density systems. Static disorder is irrelevant to temperature, which differs from thermal vibrations. Multiple temperature measurements can also be an effective method, which can distinguish contributions from different types of ADP. If the ellipsoids of all atoms are elongated in one direction, it will most likely be due to insufficient absorption correction.

In general, Hirshfeld rigid bond testing on ADP is required. Covalent bonds formed by atoms of close mass, such as atoms in the second row of the periodic table in organic molecules, must have a similar mean square displacement in the bond direction. The mean square displacement of atom k at the direction of the unit vector $\boldsymbol{v}$ will be $\boldsymbol{v}\boldsymbol{B}(k)\boldsymbol{v}^T$, where $\boldsymbol{B}(k)$ is the mean squared displacement tensor, $\boldsymbol{v}^T$ is the transposed matrix of $\boldsymbol{v}$, and a difference greater than 10^{-4} indicates certain problem. Rigid bond testing can be expanded to nonbonding atoms, where if the mean square displacement of the nonbonding atom in a molecule passes rigid bond testing, then the entire molecule or a part of it will be a rigid fragment.

4.2.3 Rigid Fragment Analysis

Rigid fragment analysis mainly studies the mean square displacement of a molecule as a rigid fragment without the influence of surrounding molecules in a crystal. Researchers found that ADP was influenced by the collective motion of atoms, and the most famous model is the TLS model built by Schomaker and Trueblook [4]. Intramolecular and intermolecular interaction strengths will vary greatly in molecular crystals, which means that vibrational modes can be divided into two parts, i.e. the low-frequency mode outside the molecule (rigid segment) and high-frequency mode (bond bending, stretching, and twisting) inside the molecule. Generally, the frequencies of the external low-frequency and internal high-frequency modes will vary considerably. For some rigid molecules, a combination of the two modes may exist. To understand the relative contribution of high/low-frequency modes on atomic mean square displacement, the vibration model can be treated as a harmonic oscillator, where the mean square displacement of the atom follows:

$$< u^2 >= \frac{h}{4\pi^2 m v}\left(\frac{1}{2} + \frac{1}{\exp(hv/k_B T) - 1}\right), \tag{4.31}$$

where m is the reduced mass, v is the frequency, and T is the absolute temperature. Equations (4.31) and (4.14) are closely related to Eq. (4.15), as every vibration model in the crystal can be treated as a harmonic oscillator. Because the frequency v has an inverse relationship with $< u^2 >$ in (4.31), the mean square displacement will be mainly contributed by the low-frequency vibrations of the rigid fragments, especially under high temperatures.

The mean square vector of an atom can be obtained by summing all vibration modes, and the rigid molecule can be treated as an independent object. Equation (4.14) can be written as:

$$\boldsymbol{B}_{\text{molecule}}(k) = \sum_{jq} \boldsymbol{e}_{\text{molecule}}(k|j\boldsymbol{q})\boldsymbol{e}^*_{\text{molecule}}((k|j\boldsymbol{q}))^T, \tag{4.32}$$

where $\boldsymbol{e}_{\text{molecule}}(k \mid j\boldsymbol{q})$ describes the translation and rotation displacement of the j vibration model of a rigid molecule k along the $\boldsymbol{q}$ vector direction:

$$\boldsymbol{e}_{\text{molecule}}(k|j\boldsymbol{q}) = \begin{bmatrix} U_1(k|j\boldsymbol{q}) \\ U_2(k|j\boldsymbol{q}) \\ U_3(k|j\boldsymbol{q}) \\ --- \\ \Theta_1(k|j\boldsymbol{q}) \\ \Theta_2(k|j\boldsymbol{q}) \\ \Theta_3(k|j\boldsymbol{q}) \end{bmatrix}, \tag{4.33}$$

where k is a molecule in the original cell, and summation can be conducted at the vibration model of the $6nN$ molecule, n is the number of molecules in a cell, U_i is the translational displacement along the i axis, and $\boldsymbol{\Phi}_I$ is the rotational displacement around this axis. In addition, the $\boldsymbol{B}_{\text{molecule}}$ is a 6×6 symmetric matrix, which can be divided into four 3×3 matrices and used to describe the translational displacement $\boldsymbol{T}$, rotational displacement $\boldsymbol{L}$, and their correlation $\boldsymbol{S}$:

$$\boldsymbol{B}_{\text{molecule}}(k) = \begin{bmatrix} \boldsymbol{T} & \vdots & \boldsymbol{S} \\ \cdots & \cdots & \cdots \\ (\boldsymbol{S}^*)^T & \vdots & \boldsymbol{L} \end{bmatrix}, \tag{4.34}$$

where:

$$\begin{aligned} \boldsymbol{T}(k) &= \sum_{jq} \boldsymbol{U}(k|j\boldsymbol{q})\boldsymbol{U}^*(k|j\boldsymbol{q})^T \\ \boldsymbol{L}(k) &= \sum_{jq} \Theta(k|j\boldsymbol{q})\Theta^*(k|j\boldsymbol{q})^T \\ \boldsymbol{S}(k) &= \sum_{jq} \boldsymbol{U}(k|j\boldsymbol{q})\Theta^*(k|j\boldsymbol{q})^T, \end{aligned} \tag{4.35}$$

where $\boldsymbol{T}$ is the equivalent expression of the rigid fragment of the mean squared displacement tensor of an atom, and the $\boldsymbol{T}, \boldsymbol{L}, \boldsymbol{S}$, of $\boldsymbol{B}_{\text{molecule}}$ can be obtained by conducting the least square fit on ADPs. H atoms have dramatic internal vibrations, which will not allow them to fit the rigid fragment conditions. Hence, they shall not be included in the refinement. The ADP used for rigid fragment analysis must be transformed to Cartesian coordinates used for the mean squared displacement tensor, where the relationship between the atom and the rigid fragment displacement follows:

$$\boldsymbol{B}_{\text{atom}}(k\alpha) = \boldsymbol{T}(k) + \boldsymbol{R}(k\alpha)\boldsymbol{L}(k)\boldsymbol{R}^T(k\alpha) + \boldsymbol{S}(k)\boldsymbol{R}^T(k\alpha) + \boldsymbol{R}(k\alpha)\boldsymbol{S}^T(k), \tag{4.36}$$

where $\boldsymbol{R}(k\alpha)$ is the antisymmetric matrix of the equilibrium position vector $\boldsymbol{r}(k\alpha)$ of the α atom in molecule k:

$$\boldsymbol{R}(k\alpha) = \begin{bmatrix} 0 & r_3 & -r_2 \\ -r_3 & 0 & r_1 \\ r_2 & -r_1 & 0 \end{bmatrix}. \tag{4.37}$$

Because the diagonal matrix element of $\boldsymbol{S}$ can be represented as the combination of $(S_{22} - S_{11})$, $(S_{33} - S_{22})$, and $(S_{11} - S_{33})$, only a part of the matrix element of the correlation tensor $\boldsymbol{S}$ can be confirmed, and the trace of $\boldsymbol{S}$ cannot be confirmed. However, adding a constant to each diagonal will not affect the observation equation, and the trace of $\boldsymbol{S}$ is often set as zero. The TLS model can be used to describe the quality of molecular vibrations, and the residual value can be calculated as:

$$R_w(U_{i,j}) = \sum \left(\Delta(wU_{i,j})^2\right) / \sum \left(wU_{i,j}^{obs}\right)^2,$$

where $\Delta(wU_{i,j})$ is the weighted difference between the experimental ADP and TLS model calculated ADP. For a real rigid molecule, this value will usually be around 5%. However, for a molecule with potential torsional vibrations, this value will usually be around 8–12%. In addition to the entire molecule, rigid fragment analysis can also be used in a part of the molecule. The low-frequency molecular vibration information obtained from the rigid segment analysis of ADP can be used to estimate the vibration entropy and heat capacity of molecular crystals.

4.2.4 Neutron Diffraction-assisted Analysis

Multipole refinement using the X-ray diffraction structure factor cannot guarantee the complete separation of thermal vibrations and electron density, as their parameters are linked together. Therefore, it is critical to obtain nuclear coordinates and atom vibration information from other experiments. The most common method involves adding neutron diffraction measurements in addition to X-ray diffraction experiments; however, neutron diffraction experiments are often difficult to perform. First, there are few places where neutron diffraction can be measured, and second, neutron diffraction usually requires a crystal with a big size, which is difficult to obtain. Lattice dynamics calculations based on empirical force fields or ab initio methods can provide significant information regarding atomic vibrations.

Neutron diffraction is based on the interactions between neutrons and nuclei, which can directly obtain information regarding the position and mean square displacement of the nucleus. All information is critical in experimental charge analysis, especially for systems containing hydrogen atoms. Because hydrogen atoms do not have core electrons, it is difficult to analyze the charge density based solely on X-ray diffraction.

However, it is difficult to combine data measured from different techniques because experimental conditions will rarely be the same. Significant systematic errors will often be found between the temperature factors obtained by X-ray diffraction and neutron diffraction. Spherical (ellipsoid) atomic approximation is commonly used in X-ray diffraction analysis for several reasons, as the anisotropic temperature factor not only fits the mean squared displacement of the atom, but also fits the non-spherical distortion of valence electron density. This effect can be effectively removed in multipole model refinement.

Certain significant experimental differences exist between X-ray diffraction and neutron diffraction.

4.2.4.1 Temperature

The mean square displacement as a result of thermal vibrations will have an approximately linear relationship with temperature, which can be obtained from Eq. (4.31) under low frequency and high temperatures ($h\nu \leq k_B T$):

$$\langle u^2 \rangle = h/4\pi^2 m\nu \left(\frac{1}{2} + k_B T/h\nu \right). \tag{4.38}$$

Temperature control in X-ray diffraction experiments is difficult because accurate temperature control will differ from using nitrogen or helium gas flow. Temperature adjustments can usually be achieved by using crystals with phase transitions. For example, the phase transition temperature of potassium dihydrogen phosphate is 122.4 K.

4.2.4.2 Absorption

The absorption effects in neutron diffraction and X-ray diffraction will be significantly different. Although the absorption coefficient of a crystal for neutrons will be relatively small, the incoherent scattering of hydrogen atoms can cause similar absorption problems. The crystal used for neutron scattering is usually much larger than the one used for X-ray diffraction (>1 mm^3), making the overall absorption effect much larger. Because the attenuation of low-angle diffraction spots caused by absorption will be larger than high-angle diffraction spots, incorrect absorption correction will make the temperature factor too small.

4.2.4.3 Extinction

Severe extinction problems exist in neutron diffraction due to the large sample size, and the extinction of X-ray diffraction will generally attenuate strong low-index diffraction spots. However, for neutron diffraction, extinction affects high-angle spots, as neutron diffraction intensity will not decrease with increasing diffraction angle. This is different from X-ray diffraction. Incorrect extinction correction can lead to a similar effect as incorrect absorption correction, that can result in excessively small mean square displacement. The most commonly used extinction correction method is the one proposed by Becker and Coppens [5].

4.2.4.4 Thermal Diffuse Scattering

Thermal diffuse scattering (TDS) is due to the energy exchange between scattered X-ray and low-frequency lattice vibration modes. Due to the influence of thermal vibrations, elastic Bragg scattering intensity will decrease with increasing temperature, while the inelastic TDS intensity will increase. Total X-ray scattering will be independent of temperature. Similar to Bragg scattering, the peak intensity of TDS will also appear at the reciprocal grid point. Because TDS is inelastic, compared to the Bragg peak, the TDS intensity peak will have a broader wavelength distribution and broader peak shape. The TDS peak can be estimated empirically by peak shape scanning, and the best way to avoid TDS is to perform the measurements under low temperatures.

4.2.4.5 Multiple Scattering

Multiple scattering effects will occur when several reciprocal lattice points are tangent to the Ewald sphere simultaneously. Under this circumstance, the diffracted beam can act as an incident beam, causing further diffraction, weakening the strong points, and strengthening the weak points. Multiple scattering can also lead to smaller ADP, which is similar to the effects caused by incorrect TDS, extinction, and absorption correction.

4.3 Scattering Experiments

4.3.1 X-ray Diffraction

X-rays play a significant role in the fields of chemistry and materials, where the X-rays can scatter or absorb when the X-rays interact with matter. Because the typical wavelength of X-rays (~1 Å) is close to the atomic distance in matter, X-rays can be used to obtain structural information. In addition, the photon energy of X-rays will be close to the binding energy of electrons in the inner atom-shell, and the X-ray absorption spectrum can be used to obtain information on the inner electron energy and characteristic absorption spectrum.

Total X-ray scattering contains elastic and inelastic components. Because electrons scatter X-rays to a larger extent than atomic nuclei, elastic strength will be mainly determined by electron density. This contrasts with neutron scattering, where the nuclei scatter neutrons to a larger extent than electrons, and the largest electron density is located at the nucleus position.

Actual atoms in crystals continuously thermally vibrate; therefore, the electron density will be thermally averaged in a unit cell, and known as the dynamic electron density. In an ideal case, the electron density is called static electron density when every atom is static. The coherent scattering (diffraction) of X-rays can be described using Laue's Equation or Bragg's Law, where the diffraction amplitude (F) is the Fourier transform (FT) of the thermally averaged electron density $\langle \rho(\boldsymbol{r}) \rangle$ in a unit cell:

$$F(\boldsymbol{H}) = \int_V \langle \rho(\boldsymbol{r}) \rangle e^{2\pi i \boldsymbol{H} \bullet \boldsymbol{r}} \mathrm{d}\boldsymbol{r} = FT\langle \rho(\boldsymbol{r}) \rangle, \tag{4.39}$$

where $\boldsymbol{H} = h\boldsymbol{a}^* + k\boldsymbol{b}^* + l\boldsymbol{c}^*$ is the scattering vector in the reciprocal space, and V is the volume of a unit cell. The Fourier transform of the static electron density in a unit cell, F_{hk}, is known as the structure factor, indicating the sum of the scattering amplitude of the electron density scattered by X-rays at the grid point in an reciprocal space hkl. The structure factor will only be defined at some discrete points in the reciprocal space, and the unit cell (static) electron density can be obtained by the reverse Fourier transform of the structure factor:

$$\rho(\boldsymbol{r}) = 1/V \sum_{h,k,l} F_{hkl} e^{-2\pi i(hx+ky+lz)}. \tag{4.40}$$

This is the sum of infinite hkl, where x, y, z is the factorial coordinate of a unit cell in a real space, defined as $\boldsymbol{r} = x\boldsymbol{a} + y\boldsymbol{b} + z\boldsymbol{c}$. In principle, the accurate electron density of a crystal can be obtained through Eq. (4.40). However, actual accurate electrons will be difficult to obtain mainly due to the following. (i) The X-ray wavelength (λ) is not infinitely small, and the measurement instrument is restricted geometrically. Therefore, the actual measurable diffraction points will be limited, i.e.

$|\boldsymbol{H}|_{\max} = 2\sin(\theta_{\max})/\lambda$. (ii) The X-ray diffraction intensity will decrease dramatically as the diffraction angle increases, causing the high-angle diffraction to be too weak to be detected. (iii) Under dynamic approximation, only the $F(\boldsymbol{H})$ intensity can be directly measured using experiments, and the phase position cannot be obtained experimentally, however, it is needed to rebuild $\rho(\boldsymbol{r})$.

4.3.2 Polarized Neutron Diffraction

Polarized neutron diffraction (PND) is the most widely used technique to study magnetization density. This technique can be used to obtain the magnetization density of any point in a unit cell, and can be applied to different types of magnetic crystal materials. The molecular magnetic material is a critical field where PND can be used, and this type of material is rich in magnetism. Compared to transition metals or rare-earth compounds, the spin density in molecular-based magnetic materials will be delocalized at individual atoms. Experimental magnetization density analysis is an effective method for studying the electronic structure and magnetic interactions of open-shell molecular systems [6].

The PND experiment measures the flipping ratio $R(\boldsymbol{k})$, i.e. the diffraction intensity ratio $\frac{I_+(\boldsymbol{k})}{I_-(\boldsymbol{k})}$, with the polarization vector $\boldsymbol{P}$ parallel to the crystal magnetization direction and inverse parallel of neutron magnetization:

$$R(\boldsymbol{k}) = \frac{I_+(\boldsymbol{k})}{I_-(\boldsymbol{k})} = \frac{F_N^2 + 2\boldsymbol{P}\cdot\left(F'_N\boldsymbol{Q}'_\perp + F''_N\boldsymbol{Q}''_\perp\right) + Q_\perp^2}{F_N^2 - 2\boldsymbol{P}\cdot\left(F'_N\boldsymbol{Q}'_\perp + F''_N\boldsymbol{Q}''_\perp\right) + Q_\perp^2}, \tag{4.41}$$

where $F_N(\boldsymbol{k})$ is the structure factor of the nuclei, $\boldsymbol{Q}_\perp(\boldsymbol{k})$ is the magnetic interaction vector, $\boldsymbol{Q}_\perp(\boldsymbol{k}) = \widehat{\boldsymbol{k}} \times \boldsymbol{F_M}(\boldsymbol{k}) \times \widehat{\boldsymbol{k}}$, and $\widehat{\boldsymbol{k}}$ is the unit vector parallel to the diffraction vector $\boldsymbol{k}$. Both $F_N(\boldsymbol{k})$ and $\boldsymbol{Q}_\perp(\boldsymbol{k})$ are complex numbers, and signs 'and" denote their real and imaginary parts, respectively. The magnetic structure factor $\boldsymbol{F}_M(\boldsymbol{k})$ is a vector, i.e. the sum of the atomic spins and orbital magnetic moments in a unit cell, where its direction is $\boldsymbol{\mu}$, and its amplitude is the Fourier transform of the electron spin density $m(\boldsymbol{r})$:

$$\boldsymbol{F}_M(\boldsymbol{k}) = \boldsymbol{\mu}\int_{\text{cell}} m(\boldsymbol{r})e^{\mathrm{i}\boldsymbol{\kappa}\boldsymbol{r}}\mathrm{d}\boldsymbol{r}, \tag{4.42}$$

where $m(\boldsymbol{r})$ is the sum of spin magnetization density $s(\boldsymbol{r})$ and the orbital magnetization density $L(\boldsymbol{r})$:

$$m(\boldsymbol{r}) = s(\boldsymbol{r}) + L(\boldsymbol{r}). \tag{4.43}$$

When the atomic magnetic moment μ_i is collinear with the external magnetic field, the magnetic structure factor $\boldsymbol{F}_M(\boldsymbol{k})$ will be parallel to the magnetic field, and the corresponding rotation ratio is given by:

$$R(\boldsymbol{k}) = \frac{F_N^2 + 2Pq^2\left(F'_NF'_M + F''_NF''_M\right) + q^2F_M^2}{F_N^2 - 2Peq^2\left(F'_NF'_M + F''_NF''_M\right) + q^2F_M^2}, \tag{4.44}$$

where e is the rotation efficiency, $q^2 = (\sin\alpha)^2$ and α is the angle between the perpendicular magnetization direction and the scattering vector. For crystals centrosymmetric space group, F''_N and F''_M will be both zero and the magnetic structure factor can be obtained from the rotation ratio. In compounds with strong anisotropic paramagnetic properties, there may be noncollinear magnetic moments, where the general expression Eq. (4.41) has to be used.

4.3.3 Compton Scattering

The Compton effect is a well-known inelastic scattering phenomenon, which refers to the phenomenon where the X-ray wavelength elongates due to energy loss as a result of the interactions between X-ray and electrons. This can often be described by using the wavelength shift $\Delta\lambda$, and determined using the scattering angle ϕ, Planck's constant h, static electron mass m, and vacuum light speed c:

$$\Delta\lambda = \frac{h}{mc}(1 - \cos\phi). \tag{4.45}$$

To understand how the electron motion affects such a phenomenon, it can be expressed by the photon energy formula:

$$E_2 = \frac{E_1}{[1 + (E_1/mc^2)(1 - \cos\phi)]}. \tag{4.46}$$

Compton scattering caused by a freely moving electron can be understood as the Doppler effect, where the electrons moving in the X-ray scattering direction vector cause a frequency (or energy) shift of the X-rays. The frequency shift will be proportional to the electron momentum (velocity) projection in the direction of the X-ray scattering vector $\boldsymbol{K} = \boldsymbol{k}_1 - \boldsymbol{k}_2$. If this direction is defined as the z-axis of the Cartesian coordinate system, the scattered energy E_2 can be related to the component of the electron momentum along the scattering vector p_z:

$$\frac{p_z}{mc} = \frac{(E_2 - E_1) + E_1E_2(1 - \cos\varphi)/mc^2}{\left(E_1^2 + E_2^2 - 2E_1E_2\cos\varphi\right)^{\frac{1}{2}}}. \tag{4.47}$$

This result, if expressed using the function of electron momentum p_z, will be symmetric on both sides of $p_z = 0$, i.e. with projection of the electron momentum distribution $n(\mathbf{p})$ along the scattering vector. This is known as the Compton profile and is usually denoted as $J(p_z)$. Therefore, the Compton scattering can be used to detect the electron momentum density $n(\mathbf{p})$, and the relationship between $J(p_z)$ and $n(\mathbf{p})$ can be defined by:

$$J(p_z) = \iint n(\mathbf{p})dp_x dp_y. \tag{4.48}$$

Under impulse approximation, i.e. assuming that the transferred energy exceeds the electron binding energy, the electrons can be considered as moving at the same speed. Thus, the scattering cross-section can be written as:

$$\begin{aligned}\frac{d^2\sigma}{d\Omega dE_2} = {} & N\left(\frac{e^2}{mc^2}\right)\left(\frac{E_2}{E_1}\right)\left(\frac{m}{\hbar^2 K}\right)\\ & \times \Big\{[1 + \cos^2\varphi + P_1\sin^2\varphi][J(p_Z)]\\ & + 2\frac{E_1}{mc^2}[(\cos\varphi - 1)P_c\hat{\boldsymbol{\sigma}}\cdot(k_1\cos\varphi + k_2)][J_{\text{mag}}(p_Z)]\Big\},\end{aligned} \tag{4.49}$$

where P_l and P_c are the linear polarization and circular polarization of the incident photon, respectively, and $\hat{\sigma}$ is the unit vector along the electron magnetic matrix direction. The last term $J_{\text{mag}}(p_z)$ in the above equation denotes the magnetic Compton profile, which can be defined as the difference between the spin-up momentum density and spin-down momentum density:

$$J_{\text{mag}}(p_z) = \iint (n(\mathbf{p}\uparrow) - n(\mathbf{p}\downarrow))\mathrm{d}p_x\mathrm{d}p_y. \tag{4.50}$$

Under impulse approximation, orbital magnetization can be ignored, as orbital motion cannot exert influence during instantaneous interactions. In this aspect, the magnetic Compton scattering will be different from neutron scattering, where the former can only detect the spin magnetic moment, while the latter can probe the total magnetic moment of the spin and orbit.

4.4 Refinement Algorithm for Experimental Electronic Structure

4.4.1 Least-square Method

The amount of data from diffraction experiments will be enormous and far above the number of refinement parameters in an ordinary electronic structure model. Therefore, the least-square refinement method can be used to study experimental electronic structures [7]. In the least square method, the error function S can be defined as the square of the sum of observed and experimental values, and the error function can be minimized by adjusting the refinement parameters.

4.4.1.1 Mathematical

If we have n experiment observations, $f_1, f_2, \cdots, f_n$, each of them relies on m parameters $x_1, \cdots, x_m$, and $m \leq n$. Every f_i contains random error ε_i, and the relationship between the observed value and parameters can be written as:

$$\begin{aligned} f_1 &= a_{11}x_1 + a_{12}x_2 + \cdots + a_{1m}x_m + \varepsilon_1 \\ f_2 &= a_{21}x_1 + a_{22}x_2 + \cdots + a_{2m}x_m + \varepsilon_2 \\ &\cdots\cdots\cdots\cdots\cdots\cdots\cdots \\ fn &= a_{n1}x_1 + a_{n2}x_2 + \cdots + a_{nm}x_m + \varepsilon_n, \end{aligned} \tag{4.51}$$

or in matrix form as

$$\boldsymbol{F}_{n,1} = \boldsymbol{A}_{n,m}\boldsymbol{X}_{m,1} + \boldsymbol{E}_{n,1}, \tag{4.52}$$

where the matrix element of the coefficient matrix $\boldsymbol{A}$ is a partial derivative: $\frac{\partial f_i}{\partial x_j} = a_{ij}$.

Given n observations, the goal is to find the optimum $\boldsymbol{X}$. This can be achieved by minimizing the square sum of the errors, i.e. error function S. If a weight is attached to each observation, then S can be defined as:

$$S = \sum_{i=1}^{n} w_i(f_i - \hat{f}_i)^2 = \sum_{i=1}^{n} w_i\Delta_i^2, \tag{4.53}$$

where $\hat{f}_i$ is the calculated value based on $\boldsymbol{X}$.

Assuming a correlation coefficient γ_{ij} exists between the i and j observed values, then the variance and covariance of the observed values can be given by the variance-covariance matrix $\boldsymbol{M}_f$. The matrix element is $\sigma_i\sigma_j\gamma_{ij}$, where σ_i is the standard deviation of the i observed value, and the corresponding error function follows:

$$S = \boldsymbol{V}^T\boldsymbol{M}_f^{-1}\boldsymbol{V}, \tag{4.54}$$

where $\boldsymbol{V}$ is the column vector, and the vector element is $f_i - \hat{f}_i$, i.e. $\boldsymbol{V} \equiv \boldsymbol{F} - \hat{\boldsymbol{F}} \equiv \boldsymbol{F} - \boldsymbol{A}\hat{\boldsymbol{X}}$; therefore,

$$\begin{aligned} S &= \boldsymbol{V}^T\boldsymbol{M}_f^{-1}\boldsymbol{V} = (\boldsymbol{F} - \boldsymbol{A}\hat{\boldsymbol{X}})^T\boldsymbol{M}_f^{-1}(F - \boldsymbol{A}\hat{\boldsymbol{X}}) \\ &= \mathbf{F}^T\boldsymbol{M}_f^{-1}\boldsymbol{F} + \hat{\boldsymbol{X}}^T\boldsymbol{A}^T\boldsymbol{M}_f^{-1}\boldsymbol{A}\hat{\boldsymbol{X}} - \boldsymbol{F}'\boldsymbol{M}_f^{-1}\boldsymbol{A}\hat{\boldsymbol{X}} - \hat{\boldsymbol{X}}^T\boldsymbol{A}^T\boldsymbol{M}_f^{-1}\boldsymbol{F} \\ &= \mathbf{F}^T\boldsymbol{M}_f^{-1}\boldsymbol{F} + \hat{\boldsymbol{X}}^T\boldsymbol{A}^T\boldsymbol{M}_f^{-1}\boldsymbol{A}\hat{\boldsymbol{X}} - 2\boldsymbol{F}^T\boldsymbol{M}_f^{-1}\boldsymbol{A}\hat{\boldsymbol{X}}. \end{aligned} \tag{4.55}$$

For each position parameter, when $\frac{\partial S}{\partial x_i} = 0$, the best parameter estimation can be obtained. With the matrix symbol and differential operator δ, these conditions can be written as:

$$\delta S = \delta\left(\boldsymbol{V}^T\boldsymbol{M}_f^{-1}\boldsymbol{V}\right) = 0. \tag{4.56}$$

Substituting (4.55) into above equation gives:

$$\delta\left(\boldsymbol{V}^T\boldsymbol{M}_f^{-1}\boldsymbol{V}\right) = 2(\delta\hat{\boldsymbol{X}})^T\left(\boldsymbol{A}^T\boldsymbol{M}_f^{-1}\boldsymbol{A}\hat{\boldsymbol{X}} - \boldsymbol{A}^T\boldsymbol{M}_f^{-1}\boldsymbol{F}\right) = 0, \tag{4.57}$$

where the solution to this equation can be written as a normal equation:

$$\left(\boldsymbol{A}^T\boldsymbol{M}_f^{-1}\boldsymbol{A}\right)\hat{\boldsymbol{X}} = \boldsymbol{A}^T\boldsymbol{M}_f^{-1}\boldsymbol{F}, \tag{4.58}$$

and with $\boldsymbol{B} \equiv \boldsymbol{A}^T\boldsymbol{M}_f^{-1}\boldsymbol{A}$, Equation (4.58) can be written as:

$$\boldsymbol{B}\hat{\boldsymbol{X}} = \boldsymbol{A}^T\boldsymbol{M}_f^{-1}\boldsymbol{F}. \tag{4.59}$$

Without considering the observed conditions, $\boldsymbol{M}_f^{-1}$ will be the diagonalized weighted matrix:

$$\boldsymbol{B} = \sum_{i=1}^{n} w_i \frac{\partial f_i}{\partial x_j}\frac{\partial f_i}{\partial x_k}. \tag{4.60}$$

According to (4.59), the best estimation of X is given by:

$$\hat{\boldsymbol{X}} = \boldsymbol{B}^{-1}\boldsymbol{A}^T\boldsymbol{M}_f^{-1}\boldsymbol{F}. \tag{4.61}$$

4.4.1.2 Least-square Refinement of Structure Factors

In the least square refinement of diffraction intensity data, the measured f_i is often defined as the square of the structure factor F^2, or the structure factor F itself. The deviation between the observed and calculated values follows:

$$\Delta = \left|F_{\text{obs}} - k|F_{\text{calc}}|\right|, \tag{4.62}$$

$$\Delta = \left|F_{\text{obs}}^2 - k^2|F_{\text{calc}}|^2\right|, \tag{4.63}$$

where the scale factor k is used to scale the calculated value and the observed value of the structure factor to the same level, and the obs and calc subscripts of the structure factor denote the calculated observed value and the calculated value, respectively.

Unlike the processing methods introduced above, the relationship between the structure factor and the unknown x_j will no longer be the linear function described by (4.51), and the above methods will only be applicable when the partial derivatives of the second or higher order are assumed to be zero. This assumption may make these equations inaccurate. Therefore, this single calculation will not minimize the error function, and if the error function is far from linear, its minimum value can be achieved with many more iteration steps.

To linearize the equation, the slope S_j' of the error function S can be written as the Taylor expansion of the unknown value x_j. For the j unknown value:

$$S_j' = S_j'(\boldsymbol{X}_0) + \sum_k \frac{\partial S_j'(\boldsymbol{X}_0)}{\partial x_k}\delta x_k + \frac{1}{2}\sum_k \sum_p \frac{\partial^2 S_j'(\boldsymbol{X}_0)}{\partial x_k \partial x_p}\delta x_k \delta x_p + \cdots, \tag{4.64}$$

where $k, p = 1, \cdots, m$ denotes m unknown values, and $\boldsymbol{X}_0$ is the current value of the unknown values. The first term indicates the error function slope at $\boldsymbol{X}_0$ along the x_j, which can be calculated by Eq. (4.53):

$$S_j'(\boldsymbol{X}_0) = \frac{\partial S}{\partial x_j} = 2\sum_i^n w_i \Delta_i \frac{\partial \Delta_i}{\partial x_j}. \tag{4.65}$$

The linearization process removes all terms in Eq. (4.64), except for the first-order partial derivatives and previous terms:

$$S_j' = S_j'(\boldsymbol{X}_0) + \sum_k \frac{\partial S_j'(\boldsymbol{X}_0)}{\partial x_k}\delta x_k. \tag{4.66}$$

The processing of the structure factor is based on Eq. (4.62):

$$\frac{\partial \Delta_i}{\partial x_j} = -\frac{\partial |kF_{i,c}|}{\partial x_j}\frac{(F_{\text{obs}} - |kF_{\text{calc}}|)}{\left|F_{\text{obs}} - |kF_{\text{calc}}|\right|}, \tag{4.67}$$

and substituting this equation into (4.65) gives:

$$S_j'(\boldsymbol{X}_0) = -2\sum_{i=1}^n w_i (F_{i,o} - |kF_{i,c}|)\frac{\partial |kF_{i,c}|}{\partial x_j}. \tag{4.68}$$

Equation (4.66) can be written as:

$$S_j' = -2\sum_{i=1}^n w_i (F_{i,o} - |kF_{i,c}|)\frac{\partial |kF_{i,c}|}{\partial x_j} + 2\sum_k \left\{ \sum_{i=1}^n w_i \left(\frac{\partial |kF_{i,c}|}{\partial x_j}\right)\left(\frac{\partial |kF_{i,c}|}{\partial x_k}\right) - \sum_{i=1}^n w_i \Delta_i \frac{\partial^2 |kF_{i,c}|}{\partial x_k \partial x_j}\delta x_k \right\}. \tag{4.69}$$

The last term in Eq. (4.69) includes the second-order derivative, and the m equations can be obtained based on $S_j' = 0(j = 1, \cdots, m)$:

$$\sum_k \left\{ \left[\sum_{i=1}^{n} w_i \left(\frac{\partial |kF_{i,c}|}{\partial x_j} \right) \left(\frac{\partial |kF_{i,c}|}{\partial x_k} \right) \right] \delta x_k \right\} = \sum_{i=1}^{n} w_i \{F_{i,o} - |kF_{i,c}|\} \frac{\partial |kF_{i,c}|}{\partial x_j}. \tag{4.70}$$

This equation will be equivalent to Eq. (4.59) $\boldsymbol{B}\widehat{\boldsymbol{X}} = \boldsymbol{A}^T\boldsymbol{M}_f^{-1}\boldsymbol{F}$, which contains a diagonal variance-covariance matrix $\boldsymbol{M}_f$. The matrix element of $\boldsymbol{A}(nm)$ is $A_{ij} = \partial \frac{|kF_{i,c}|}{\partial x_j}$, while the vector $\boldsymbol{F}(n1)$ corresponds to $F_i = w_i|F_{i,0} - |kF_{i,c}||$, and the matrix element of the symmetric matrix $\boldsymbol{B}(nn)$ is given by:

$$B_{jk} = \sum_{i=1}^{n} w_i \left(\frac{\partial |kF_{i,c}|}{\partial x_j} \right) \left(\frac{\partial |kF_{i,c}|}{\partial x_k} \right).$$

The main steps of the least square iteration include calculating matrices $\boldsymbol{A}, \boldsymbol{B}, \boldsymbol{B}^{-1}$ and matrix multiplication. As indicated by Eq. (4.70), the vector $\boldsymbol{X}$ in Eq. (4.59) can be defined as $\boldsymbol{X}_i = \delta x_i$. Therefore, the solution for Equation (4.61) will be the deviation of the unknown value instead of the unknown parameter itself. Because the truncation of Eq. (4.64) is an approximation, this deviation can be overestimated or underestimated. When this deviation is added to the original parameter value. i.e. $\boldsymbol{X}_2 = \boldsymbol{X}_1 + \delta \boldsymbol{X}$, the matrix elements of $\boldsymbol{A}$, which is dependent on $\boldsymbol{X}$, will change and need multiple steps to convergence.

4.4.1.3 Parameter-estimated variance and covariance

The matrix $\boldsymbol{M}_x$ describes the variance and covariance of the optimal estimated $\boldsymbol{X}$ value:

$$\boldsymbol{M}_x = \varepsilon\{(\widehat{\boldsymbol{X}} - \boldsymbol{X}^0)(\widehat{\boldsymbol{X}} - \boldsymbol{X}^0)^T\}, \tag{4.71}$$

where the superscript 0 denotes the real value, and ε denotes an estimation. Substituting Eq. (4.61) into the above equation generates:

$$\begin{aligned} \boldsymbol{M}_x &= \varepsilon\left\{\boldsymbol{B}^{-1}\boldsymbol{A}^T\boldsymbol{M}_f^{-1}(\boldsymbol{F} - \boldsymbol{F}^0)(\boldsymbol{F} - \boldsymbol{F}^0)^T\boldsymbol{M}_f^{-1}\boldsymbol{A}\boldsymbol{B}^{-1}\right\}, \\ &= \boldsymbol{B}^{-1}\boldsymbol{A}^T\boldsymbol{M}_f^{-1}\varepsilon\{(\boldsymbol{F} - \boldsymbol{F}^0)(\boldsymbol{F} - \boldsymbol{F}^0)^T\}\boldsymbol{M}_f^{-1}\boldsymbol{A}\boldsymbol{B}^{-1} \end{aligned} \tag{4.72}$$

where $\varepsilon\{\boldsymbol{F} - \boldsymbol{F}^0)(\boldsymbol{F} - \boldsymbol{F}^0)^T\}$ is an estimation of the observed variance-covariance matrix $\boldsymbol{M}_f$. Therefore:

$$\boldsymbol{M}_x = \boldsymbol{B}^{-1}\boldsymbol{A}^T\boldsymbol{M}_f^{-1}\boldsymbol{M}_f\boldsymbol{M}_f^{-1}\boldsymbol{A}\boldsymbol{B}^{-1} = \boldsymbol{B}^{-1}\boldsymbol{A}^T\boldsymbol{M}_f^{-1}\boldsymbol{A}\boldsymbol{B}^{-1}. \tag{4.73}$$

Defining $\boldsymbol{B} = \boldsymbol{A}^T\boldsymbol{M}_f^{-1}\boldsymbol{A}$, then $\boldsymbol{M}_x = \boldsymbol{B}^{-1}$. That is, the variance-covariance matrix will be the inverse matrix of $\boldsymbol{B}$. Because only the relative values of variance and covariance can be estimated, the equation becomes:

$$\boldsymbol{M}_f = \sigma^2\boldsymbol{N}_f, \tag{4.74}$$

where σ^2 is the scaling factor and can scale $\boldsymbol{M}_f$ to the absolute value matrix $\boldsymbol{N}_f$. Therefore:

$$\boldsymbol{M}_f^{-1} = \frac{1}{\sigma^2}\boldsymbol{N}_f^{-1} \equiv \frac{1}{\sigma^2}\boldsymbol{P}_f, \tag{4.75}$$

$$B = A^T M_f^{-1} A = \frac{A^T P_f A}{\sigma^2}. \tag{4.76}$$

The variance and covariance matrix of the unknown values will follow:

$$M_x \equiv B^{-1} = \sigma^2 (A^T P_f A)^{-1}, \tag{4.77}$$

where P_f is often considered as the diagonal matrix elements formed by the weighted observed values. For a two-dimensional detector, several diffraction points will exist in one photo, where the linkage between different diffraction points cannot be ignored. The best estimation of σ^2, i.e., $\hat{\sigma}^2$, will be related to the amplitude of deviation V, where $\hat{\sigma}^2$ is the weighted average of the square deviation. The optimal $\hat{\sigma}^2$ is:

$$\hat{\sigma}^2 = \frac{V^T P_f V}{n - m}, \tag{4.78}$$

and when P_f is the dragonized matrix:

$$\hat{\sigma}^2 = \frac{\sum_{n}^{i=1} w_i \Delta_i^2}{n - m}. \tag{4.79}$$

The model becomes reasonable if the observations are given appropriate weights, i.e. $w_i \approx \Delta_i^{-2}$, and the goodness of fit, $\hat{\sigma}^2$, is close to 1. Substituting Eq. (4.79) into Eq. (4.77), the variance and covariance matrix of the unknown values follows:

$$M_x = \frac{\sum_{i=1}^{n} w_i \Delta_i^2}{n - m} (A^T P_f A)^{-1}. \tag{4.80}$$

The matrix element M_{ij} of the matrix M_x can often be written as $\sigma_i \sigma_j \gamma_{ij}$, where the relevant coefficient γ_{ij} is given by:

$$\gamma_{ij} = \frac{M_{ij}}{\sqrt{M_{ii} M_{jj}}}. \tag{4.81}$$

4.4.2 Maximum Entropy Method

The MEM is a technique based on information theory, which is based on equations in statistical thermodynamics. In its early stage, it was developed to enhance the signals of noisy data in the field of astronomy. The information entropy in the MEM will be the same as the statistical entropy in statistical thermodynamics, dealing with the maximum likelihood distribution. In statistical thermodynamics, this distribution is the distribution of particles in the position and momentum space. However, in information theory it refers to the distribution of digital quantities in a pixel ensemble.

When N particles are placed in m boxes, the probability that each box contains n_i particles P will be:

$$P = \frac{N!}{n_1! n_2! n_3! \dots n_m!}. \tag{4.82}$$

As in statistical thermodynamics, the entropy can be defined as *ln P*. Because the molecule in the above equation is:: a constant, the entropy can be written as:

$$S = -\sum_{i} n_i \ln n_i, \tag{4.83}$$

and we can use Stirling's formula:

$$\ln N! \approx N \ln N - N.$$

If the a prior probability q_i of box i containing n_i particles exist, according to the conditional probability formula, the probability P will become:

$$P = \frac{N!}{n_1!n_2!n_3! \dots n_m!} \times q_1^{n_1} q_2^{n_2} \dots q_m^{n_m}, \tag{4.84}$$

and the entropy becomes:

$$S = -\sum_{i} n_i \ln n_i + \sum_{i} n_i \ln q_i = -\sum_{i=1}^{m} n_i \ln \frac{n_i}{q_i}. \tag{4.85}$$

The MEM was introduced by Collins to the field of crystallography [8]. Based on the above equation, the electron density distribution information entropy can be written as the sum of M grid points in a unit cell:

$$S[\rho(\boldsymbol{r})] = -\sum_{j=1}^{M} p(\boldsymbol{r}) \ln \frac{p(\boldsymbol{r})}{m(\boldsymbol{r})}, \tag{4.86}$$

where $P(\boldsymbol{r})$ and $m(\boldsymbol{r})$ are defined respectively as:

$$\begin{aligned} p(\boldsymbol{r}_j) &= p_j = \frac{\rho(\boldsymbol{r}_j)}{\sum_{j=1}^{M} \rho(\boldsymbol{r}_j)} \\ m(\boldsymbol{r}_j) &= m_j = \frac{\rho_0(\boldsymbol{r}_j)}{\sum_{j=1}^{M} \rho_0(\boldsymbol{r}_j)}, \end{aligned} \tag{4.87}$$

where ρ_0 is the prior reference density, and this reference density can often be replaced using the electron density of a uniform distribution or independent atomic model. Without the restriction of the experimental structure factor, the maximum information entropy will be a uniform distribution, and any electron density information deviating from this distribution will be included in a data set. The $P(\boldsymbol{r})$ will be proportional to the probability of detecting an electron at position $\boldsymbol{r}$, and $m(\boldsymbol{r})$ will be proportional to the prior probability of detecting an electron at position $\boldsymbol{r}$.

Solving Eq. (4.86) is a self-consistent iteration process, and maximizing $S[\rho(\boldsymbol{r})]$ must be achieved under the following conditions:

$$C[\rho(\boldsymbol{r})] = \chi^2 = \sum_{k=1}^{N} \left| \frac{F_k^{\text{obs}}(\boldsymbol{H}) - F_k^{\text{calc}}(\boldsymbol{H})}{\sigma(F_k)} \right|^2 = N, \tag{4.88}$$

where N is the number of independent observations. This restriction condition will indicate that the structure factor from the model should be consistent with the measured value within the measurement error range and F^{calc} can be obtained from the sum of M grid points:

$$F_k^{\text{calc}}(\boldsymbol{H}) = \frac{V_{\text{unit cell}}}{M} \sum_{j=1}^{M} \rho(\boldsymbol{r}_j) \exp\{2\pi i \boldsymbol{H}_k \cdot \boldsymbol{r}_j\}. \tag{4.89}$$

Without the restrictions in Eq. (4.88), maximizing entropy will lead to uniformly distributed results.

Restriction conditions can be introduced by adding a Lagrange multiplier into the minimizing function $L(\lambda)$:

$$L(\lambda) = S(\rho(\boldsymbol{r})) - \lambda\chi^2, \tag{4.90}$$

and when convergence is achieved, the gradient of the minimized function will be zero:

$$\nabla_\rho(L) = \nabla_\rho(S) - \lambda\nabla_\rho(\chi^2) = 0. \tag{4.91}$$

Therefore, $\nabla_\rho(S) = \lambda\nabla_\rho(\chi^2)$.

For each grid point j:

$$\frac{\partial S}{\partial \rho_j} = \lambda \frac{\partial \chi^2}{\partial \rho_j}. \tag{4.92}$$

For the case where the prior density is evenly distributed, for every j there will be $\rho_0(\boldsymbol{r}_j) = \rho_{0j} = \rho_0$. According to the definition of Eq. (4.86), the S differential will be:

$$\frac{\partial S}{\partial \rho_j} = -\frac{1}{\sum_{j=1}^{M} \rho_j} \ln(\rho_j/A), \tag{4.93}$$

where A is a constant. Substituting this equation it (4.92) gives:

$$\rho_j = A \exp\left\{ -\lambda \left(\sum \rho_j \right) \frac{\partial C}{\partial \rho_j} \right\}, \tag{4.94}$$

where A is set as $A \approx \exp\{\sum p_j \ln \rho_j\}$. Because $\frac{\partial F_{\text{calc}}(\boldsymbol{H})}{\partial \rho_j} = \exp(2\pi i \boldsymbol{H} \cdot \boldsymbol{r}_j)$, the density at position $\boldsymbol{r}_j$ at the $n+1$ iteration step using Eq. (4.94) is given by:

$$\rho(\boldsymbol{r}_j, n+1) = \exp\left[\ln \sum_j p_j \ln \rho_j(n) + \lambda F(0) \sum_{\boldsymbol{H}} \frac{2}{\sigma(\boldsymbol{H})^2} \times |F_{\text{obs}}(\boldsymbol{H}) - F_{\text{calc}}(\boldsymbol{H})| \exp(2\pi i \boldsymbol{H} \cdot \boldsymbol{r}_j) \right]$$

$$F(0) = \frac{V_{\text{unitcell}}}{M} \sum_{j=1}^{M} \rho(\boldsymbol{r}_j). \tag{4.95}$$

Because maximizing entropy is a non-linear process, solving for it requires the use of the iteration algorithm. The $\lambda(n+1)$ and $\rho_j(n+1)$ at the $n+1$ step can be solved from $\lambda(n)$ and $\rho_j(n)$ at the n step. The starting value will be $\lambda(0) \approx 0$, where $\rho_j(0)$ is the prior density. The convergence can be achieved through two calculation steps, where the first step gets to the restriction condition $\chi^2 = N$, and the second step reaches maximum entropy S. In another method, λ is fixed at a small value to reach convergence.

The deviation of the diffraction point can be calculated by summation after the MEM iterative calculation. In the Fourier summation, the intensity of non-observed diffraction spots will be assumed as zero, while MEM provides the most possible value. The extinction correction must be conducted prior to MEM calculations, even if the abnormal scattering effect exists in the experimental data set. The independent atom model can often be used as the structure model for these corrections.

The MEM method has some significant features, namely, (i) there is no issue with parameter correlation, which is serious for some large systems and low-quality data sets, and (ii) it has similar processing methods for the ordered static disordered, or dynamic disordered parts. These features make MEM an appealing method for protein crystal analysis. Protein crystals usually contain many atoms, parameters, and disordered parts. Thus, the MEM analysis results can be used as a good starting point for multipole model refinement. The MEM method produces dynamic electron density, whereas the multipole model produces both dynamic and static electron density.

The MEM method reconstructs spin density differently from how it reconstructs electron density, where the electron density of each point in a unit cell will be positive, while the spin density can be positive or negative. Considering two channels with a total of $2M$ grid points for spin-up, $q_j = \rho_\uparrow(x_j, y_j, z_j)/\rho^\uparrow_{\text{cell}}$ $(j = 1 \sim M)$; while for spin-down, $q_j = \rho_\downarrow(x_j, y_j, z_j)/\rho^\downarrow_{\text{cell}}$ $(j = M \sim 2M)$. The entropy is:

$$S = -\sum_{2M}^{j=1} q_j \log(q_j). \tag{4.96}$$

The normalized spin density at position k will be $m_k : q_k - q_{k+M}$ $(k = 1, M)$. The m independent spin density can be used to calculate the magnetic structure factor $F_M^{\text{cal}}(\boldsymbol{k}_n)$, where $\boldsymbol{k}_n$ is the n scattering factor and $n = 1{\sim}N$. In addition, the MEM method can obtain the spin density by hmaximizing the entropy of Eq. (4.96) and minimizing χ^2 using the least square method:

$$\chi^2(m) = \frac{1}{N}\sum_{i=1}^{N}\frac{1}{\sigma_i^2}\left|F_M(\kappa_i) - F_M^{\text{cal}}(\kappa_i)\right|^2. \tag{4.97}$$

References

1 Gatti, C. and Macchi, P. (2012). A guided tour through modern charge density analysis. In: *Modern Charge-Density Analysis* (ed. C. Gatti and P. Macchi), 1–78. Springer: Berlin.

2 Stewart, R.F. and Feil, D. (1980). A theoretical study of elastic X-ray scattering. *Acta Crystallogr. A* 36: 503–509.

3 Madsen, A.Ø. (2012). Modeling and analysing thermal motion in experimental charge density studies. In: *Modern Charge-Density Analysis* (ed. C. Gatti and P. Macchi), 133–163. Berlin: Springer.

4 Schomaker, V. and Trueblood, K.N. (1968). On the rigid-body motion of molecules in crystals. *Acta Crystallogr. B* 24: 63–76.

5 Becker, P.J. and Coppens, P. (1975). Extinction within the limit of validity of the Darwin transfer equations. I. General formalisms for primary and secondary extinction and their application to spherical crystals. *Acta Crystallogr. A* 31: 129–147.

6 Gillon, B. and Becker, P. (2012). Magnetization densities in material science. In: *Modern Charge-Density Analysis* (ed. C. Gatti and P. Macchi), 277–302. Springer: Berlin.

7 Coppens, P. (1997). *X-ray Charge Densities and Chemical Bonding*. New York: Oxford University Press, Inc.

8 Collins, D.M. (1982). Electron density images from imperfect data by iterative entropy maximization. *Nature* 298: 49–51.

5

Pseudo-atom Models

5.1 Introduction

The establishment of a charge density model has been a fundamental breakthrough in the field of crystallography, confirming Debye's prediction (see the Preface of this book). In the 1970s, numerous electron density models were proposed, and among these models, the multipole expansion method of electron density has been considered the most operable and accurate model, also known as the so-called multipole model (MM). Among the various model frameworks, the one proposed by Hansen and Coppens is the most popular one. In the MM, the electron density of the crystal can be expanded as the contribution of atoms, called pseudo-atoms, where the atoms behave as a distribution of the charge density and are assumed to move strictly around the nucleus. The MM is a typical pseudo-atom model, and the electron density of a pseudo-atom can be simplified as the sum of the electron densities of the core and valence shells. In principle, each atomic shell should be refined individually. A pure orbital hypothesis cannot reconstruct molecular electron densities well, especially for elements in the second row of the periodic table or elements in the main group (only ns and np orbitals). According to the properties of spherical harmonics, p orbitals can only produce quadrupole moment densities; however, it is common to refine the octupole moments of C, N, and O atoms in real applications. In fact, the pseudo-atom expansion has its inherent limitations. The double-center density cannot be strictly explained based on the single-center density. Thus, to obtain a more convergent result, it is necessary to refine the high-order multipole moments.

For transition metals, the electron density will be highly monocentric due to the small overlap with the ligand orbital. Thus, multipole expansion can reconstruct the electron density of these metals; however, it requires a hexadecimal moment function. In fact, the $(n-1)$d orbital can generate the electron densities of the hexadecimal pole, quadrupole, and monopole moments, while the dipole and octupole moments will be generally negligible. For transition metal atoms or main group elements, multiple valence electron shells can also be refined simultaneously.

In the X-ray atomic orbital (XAO) model [1, 2], all atoms in a unit cell can be divided into electron subshells, where each subshell is regarded as an independent pseudo-atom. This allows atomic orbitals and their electron density populations to be expressed as a linear combination of each electron shell. When the working

Electronic Structure Crystallography and Functional Motifs of Materials, First Edition.
Guo-Cong Guo and Xiao-Ming Jiang.

conditions of a material are changed, the electron transfer between atomic orbitals in the crystal can be obtained by XAO analysis. The XAO model has been mainly used in studying the charge densities of ionic solids.

Molecular orbital information can be obtained through X-ray molecular orbital (XMO) model [3], where molecular orbitals can be represented as linear combinations of atomic orbitals. The initial molecular orbitals can be obtained from the ab initio self-consistent field calculations under the Hartree–Fock approximation and the basis function selected in this model should not produce sharp points in the atomic positions in the residual density map. The X-ray structure factor calculated from the molecular orbital can be used to simulate the measured structure factor, and the orthogonality of the molecular orbital must be maintained during the least-squares refinement process. Because the basis functions contain very similar Gaussian orbitals, there will be a high correlation between the molecular orbital parameters. Therefore, it is necessary to select certain parameters with less correlation for refinement. The molecular orbital model, along with the atomic orbital model, can in principle be used to analyze the molecular or atomic orbitals of most inorganic and organic compounds from the X-ray diffraction data.

As an alternative to the MM, molecular electron densities can be constructed from a set of defined molecular orbital bases and variable orbital occupancy numbers, according to the so-called molecular orbitals with variable occupation numbers (MOON) model [4, 5]. The advantages of this method include linear scaling, where the number of refinement parameters increases linearly with the system size and relatively easy computation of properties.

5.2 Independent Atom Model

Traditional reconstruction of crystal structures from diffraction data is based on the plausible assumption that the total scattering of the crystal mainly comes from the atoms, where the electrons are mainly localized around the nucleus; therefore, the total electron density can be regarded as the superposition of the electron density ρ^0 of the isolated atoms. This is known as the independent atom model (IAM), in which the real molecule is reduced to a promolecule, and the electron density of the promolecule can be defined as the sum of the spherical densities of isolated atoms k at the position $\boldsymbol{R}_k$, namely:

$$\rho_{\mathrm{IAM}}(\boldsymbol{r}) = \sum_k \rho_k^0(\boldsymbol{r} - \boldsymbol{R}_k). \tag{5.1}$$

IAM is a good approximation for heavy atoms, as valence electrons make up only a small fraction of the total density. However, there are issues with lighter atoms. For example, the lightest H atom has no inner electrons, and chemical bonding has a great influence on the electron density distribution of H atoms. Due to the influence of the covalent bonding of X–H (X = C, N, O), the electron density of H atoms can move toward the bonding direction. When using the IAM model to refine H atoms, as the center position of the electron density of the H atoms shifts toward the bond direction, the refined X—H bond length will be shorter than the real one.

The IAM model assumes that the atoms in a crystal are neutral, which contradicts the fact that atoms actually have a dipole moment and a higher electrostatic moment. The molecular dipole moment originates from the non-spherical distribution of electron density on atoms and occurs primarily from the charge transfer between atoms with different electronegativity. Atomic dipole moments can be attributed to specific electron populations on non-spherical atomic orbitals.

5.3 Kappa Model

A simple improvement to the IAM considers the charge transfer between atoms, known as the Kappa model (κ model). Separating the scattering of valence electrons from the inner shell electrons in this model allows for the adjustment of the population number (P_ν) and shrinkage coefficient (κ) of the valence shell. In practice, the two parameters P_ν and κ are added to the traditional IAM model refinement. These two parameters must be introduced into the model simultaneously, as the number of valence electrons can affect the exclusion effect between the electrons and affect the radial distribution of electrons.

In the Kappa model, the electron density of an atom can be expressed by:

$$\rho_{\text{atom}} = \rho_{\text{core}} + \rho'_{\text{valence}}(\kappa r) = \rho_{\text{core}} + P_\nu \kappa^3 \rho_{\text{valence}}(\kappa r), \tag{5.2}$$

where the parameter κ describes the radial scaling of the electron density. When $\kappa > 1$, the same electron density will appear at the position of a smaller r, and correspondingly, the valence electron shell will contract. Conversely, when $\kappa < 1$, the valence electron shell will expand. The coefficients κ^3 appear to satisfy the normalization requirement, as $N4\pi \int \rho_{\text{valence}}(\kappa r) r^2 \mathrm{d}r = 1$, yielding $N = \kappa^3$.

The scattering factor corresponding to the valence electron density is given by Eq. (5.2):

$$f'_{\text{valence}} = \int P_{\text{valence}} \kappa^3 \rho_{\text{valence}}(\kappa r) \exp(2\pi i \boldsymbol{S} \cdot \boldsymbol{r}) \mathrm{d}\boldsymbol{r}. \tag{5.3}$$

Substituting $\boldsymbol{r}$ and $\mathbf{S}$ in the exponent part with κr and S/κ, respectively, and rewriting $\kappa^3 \mathrm{d}\boldsymbol{r}$ with $4\pi\kappa^2 r^2 \mathrm{d}\kappa r$ yields:

$$f'_{\text{valence}}(S) = f_{\text{valence}}(S/\kappa). \tag{5.4}$$

The above equation illustrates the inverted relationship between the direct space and reciprocal space; namely, the contraction of electron density ($\kappa > 1$) corresponds to the expansion of the reciprocal space, and vice versa. Equation (5.4) also implies that the adjusted scattering factor can be calculated directly from the IAM scattering factor.

The total structure factor of the Kappa model density can be expressed by:

$$F(\boldsymbol{H}) = \sum_j [\{P_{j,c} f_{j,\text{core}}(H) + P_{j,\nu} f_{j,\text{valence}}(H/\kappa)\} \exp(2\pi i \boldsymbol{H} \cdot \boldsymbol{r}_j) T_j(\boldsymbol{H})], \tag{5.5}$$

where $f_{j,\text{core}}$ and $f_{j,\text{valence}}$ are both normalized to the scattering factor of one electron and $P_{j,c}$ and $P_{j,\nu}$ are the inner electrons and valence electron population, respectively.

The κ of the H atom cannot be determined by X-ray diffraction data refinement because H has no inner orbital electrons, and there is a strong correlation between κ and the temperature factor. An alternative method involves the temperature factor, which can be obtained using neutron diffraction.

5.4 Multipole Model

The Kappa model takes into account the charge transfer between atoms. However, an advanced electron density model must also account for non-spherical density functions. The MM has shown to be a more successful model, which is based on an atom-centered multipole function.

Atom-centric models cannot strictly account for the dual-center density term; however, this does not affect the usefulness of such models because bond charges can be projected onto atomic basis functions very efficiently, assuming bond charges are sufficiently diffused. In the non-atom-centered model, the bond electron density between atoms can be described by a cloud of electrons, and the lone pair electron region can also be described similarly. However, the refined results from such a model are difficult to interpret, as the thermal vibration effects cannot be considered; thus, they will not be suitable for a comparison of experimental and theoretical electrostatic properties.

5.4.1 Multipole Spherical Harmonics

All atomic density functions can be expressed as functions of three polar parameters, namely, r, θ, and ϕ. In the MM, the density function is the product of the radial distribution function and angular distribution function, while the radial distribution function is only related to r, and the angular distribution function is only related to θ and ϕ. The angular distribution functions are considered the product of the real spherical harmonics $y_{lm\pm}(\theta, \phi)$ and the normalization factor, and these functions can describe the angular distribution of s, p, d, and f orbits, which we are familiar with.

$y_{lm\pm}(\theta, \phi)$ (or y_{lmp}) is a linear combination of complex spherical harmonics Y_{lm}, and Y_{lm} can be defined as:

$$Y_{lm}(\theta, \phi) = (-1)^m \left[\frac{(2l+1)}{4\pi} \frac{(l-|m|)!}{(l+|m|)!} \right]^{1/2} P_l^{|m|}(\cos\theta) \exp i\, m\phi, \tag{5.6}$$

where $-1 \leq m \leq 1$, and $(-1)^m$ is the Condon–Shortley phase factor, and $P_l^{|m|}(\cos\theta)$ is the corresponding Legendre function, which can be defined as:

$$P_l^m(x) = (1-x^2)^{m/2} \left(\frac{\mathrm{d}}{\mathrm{d}x} \right)^{l+m} \frac{1}{2^l l!} (x^2-1)^l. \tag{5.7}$$

The real spherical harmonics can be represented by complex spherical harmonics, for $m = 0$: $y_{l0} = Y_{l0}$; where for $m > 0$:

$$\begin{aligned} y_{lm+} &= (-1)^m (Y_{lm} + Y_{l,-m})/2^{1/2} \\ y_{lm-} &= (-1)^m (Y_{lm} - Y_{l,-m})/(2^{1/2} i). \end{aligned} \tag{5.8}$$

Substituting Eq. (5.6) into the function yields:

$$
\begin{aligned}
y_{lm+}(\theta,\phi) &= \left[\frac{(2l+1)}{2\pi(1+\delta_{l0})}\frac{(l-\mid m\mid)!}{(l+\mid m\mid)!}\right]^{1/2} P_l^{|m|}(\cos\theta)\cos m\phi \\
&\equiv N_{lm}P_l^m(\cos\theta)\cos m\phi,
\end{aligned} \tag{5.9}
$$

where $0 \leq m \leq 1$, which is similar to:

$$
y_{lm-}(\theta,\psi) = N_{lm}P_l^m(\cos\theta)\sin m\phi. \tag{5.10}
$$

For $m = 0$, y_{l0+} is equal to Y_{l0}, and both are real numbers. For $l = 0$, because $0 \leq m \leq l$, m is equal to 0, and the function has a spherical symmetry, when $l \neq 0$ and $m = 0$, the function will be independent of ϕ, $Y_{lm} = y_{lm}$, and the function will have cylindrical symmetry around the z-axis.

The function for which l is an even number has atomic centrosymmetry, while the function for which l is an odd number is noncentrosymmetric. These functions allow the actual atomic positions to be imposed with certain symmetry conditions, i.e. a centrosymmetric atomic position has dipole moments and octupole moments of 0 when l is an odd number.

5.4.2 Real Spherical Harmonic Density Function

$y_{lmp}(p=\pm)$ represents atomic orbitals; thus, y_{lmp}^2 is a probability density distribution, and the integral should be 1. This normalization condition can be expressed by:

$$
\int y_{lmp}^2 \mathrm{d}\Omega = 1, \tag{5.11}
$$

where $\mathrm{d}\Omega$ is the volume element of the $\theta - \phi$ space, and the above equation is suitable for the normalization of the wave function. For the electron density function, a different normalization method should be used because the density is the integral of the first order of the function. The density function will usually be labeled as d_{lmp} and its normalization condition follows:

$$
\begin{aligned}
l>0: &\quad \int |d_{lmp}|\mathrm{d}\Omega = 2 \\
l=0: &\quad \int |d_{lmp}|\mathrm{d}\Omega = 1.
\end{aligned} \tag{5.12}
$$

This normalization condition implies that for the spherically symmetric $\theta - \phi$ function d_{00}, a population of 1 represents one electron. Aspheric functions of $l>0$ have positive and negative lobes, where each lobe region has an equal but opposite number of electrons, and the population number represents the number of electrons transferred from the negative region to the positive region.

Similar to Eqs. (5.9) and (5.10), d_{lmp} can be expressed by:

$$
\begin{aligned}
d_{lm+} &= N'_{lm}P_l^m(\cos\theta)\cos m\phi \\
d_{lm-} &= N'_{lm}P_l^m(\cos\theta)\sin m\phi,
\end{aligned} \tag{5.13}
$$

where the normalization factor N'_{lm} can be determined according to Eq. (5.12).

In addition, d_{lmp} can be expressed as the angular distribution function c_{lmp} in the Cartesian coordinate system, for example, $c_{21+} = xz$, and d_{lmp} will be equal to the product of c_{lmp} and coefficient L_{lm}, which depends on lm:

$$d_{lmp} = L_{lm} c_{lmp}. \tag{5.14}$$

Combined with the C_{lm} factor, c_{lmp} can be defined as:

$$C_{lm} c_{lmp} = P_l^m (\cos\theta)_{\sin m\varphi}^{\cos m\varphi}. \tag{5.15}$$

Combining Eqs. (5.13), (5.14), and (5.15) will yield:

$$N'_{lm} = \frac{L_{lm}}{C_{lm}}. \tag{5.16}$$

The spherical harmonic density function refers to the multipole moment, $l = 0$, 1, 2, 3, 4... corresponding to the monopole moment, dipole moment, quadrupole moment, octupole moment and hexadecimal pole moment, respectively. We can use c_{lmp} to express the electrostatic moment of electron density $\rho(\boldsymbol{r})$:

$$\mu_{lmp} = \int \rho(\boldsymbol{r}) c_{lmp} r^l \mathrm{d}\boldsymbol{r}, \tag{5.17}$$

where μ_{lmp} is the l-order static electrostatic moment.

5.4.3 Radial Distribution Functions

To preserve the structure of the angular distribution function of valence electron density, the radial distribution function can only be related to the atomic radial distance and adjusted by the κ scaling parameter.

The deformation function can be used to describe the bond charge density, and the atomic center term must be used in a single-center framework. It must also be sufficiently diffused and have different radial dependencies. Because the electron density is a superposition of atomic orbital products, a radially dependent parameter derived from the atomic orbital function can be used. Radial dependencies are based on hydrogen-like orbitals, which have Slater-type radial functions, including radial distribution functions for 2s and 2p hydrogen-like orbitals:

$$\begin{aligned} R_{2s} &= 2^{-1/2}(Z/a_0)^{3/2}(1 - Zr/2a_0)e^{-Zr/2a_0} \\ R_{2p} &= 24^{-1/2}(Z/a_0)^{5/2} r e^{-Zr/2a_0}, \end{aligned} \tag{5.18}$$

where Z is the nuclear charge, a_0 is the Bohr radius, and r is in Å. The polynomial in R_{2s} introduces a radial node, showing orthogonality with the 1s orbit, which is generally not required for density functions. A set of simple normalized nodeless density functions can be defined in terms of hydrogen-like atomic orbitals:

$$R_l(r) = \kappa'^3 \frac{\zeta^{n_l+3}}{(n_l+2)!} (\kappa' r)^{n(l)} \exp(-\kappa' \zeta_l r), \tag{5.19}$$

where κ' is used to describe the scaling parameter of the deformation function. These functions can better fit the non-spherical valence electron shells of first- and second-period element atoms. The κ' in this function will generally not be numerically equal to κ, which describes the spherical valence electron density.

Clementi and Raimondi calculated a more suitable isolated atom electron shell ζ value [6], which can serve as a starting point for least-squares refinement, adjusted by κ', and where the coefficient n_l must satisfy the condition of $n_l \geq l$. In addition, the radial dependence of the multipole density deformation function can be related to the orbital product. According to the multiplication rule of spherical harmonics, monopole, dipole, and quadrupole functions can be obtained from ss, sp, and pp-type orbital products.

According to $\mathrm{d}R/\mathrm{d}r = 0$, the radial maximum of the function in Eq. (5.19) will be $n_l/(\kappa\zeta)$.

For the second-period element atoms, the orbital product will lead to $n_l = 4$ for all deformation functions. However, this model gives the same radial function for all multipole moments with a radial maximum at 0.691 Å, which is significantly less than half the Si—Si bond length of 2.35 Å. This function cannot accurately describe the bond charge of Si—Si bonds and n_l can be corrected according to the actual situation to achieve the best-fitting results.

The difference in the treatment of spherical and aspherical electron densities can be attributed to the fact that for covalently bonded systems, the aspherical deformation density represents the bond charges between atoms. However, for transition metal atoms, the non-spherical distribution of electrons will be mainly affected by the orientational population of d orbitals under the crystal field, due to the low bond charge between the metal and the ligand. Other radial functions can also be used, including Gaussian and harmonic wave functions.

5.4.4 Multipole Model Framework

Combining the angular and radial distribution functions discussed above, a so-called MM can be obtained, in which the density of each atom can be expressed by:

$$\rho_{\text{at}}(\boldsymbol{r}) = P_c\rho_{\text{core}}(r) + P_v\kappa^3\rho_{\text{valence}}(\kappa r) + \sum_{l=0}^{l_{\max}}\kappa'^3 R_l(\kappa' r)\sum_{m=0}^{l} P_{lm\pm} d_{lm\pm}(\theta,\phi), \quad (5.20)$$

where the first two terms are the same as in Eq. (5.2) for the kappa model, and the aspherical density is described by adding a summation term to the kappa expression, which includes an additional monopole moment. The atoms of the first and second periods can be omitted; however, for transition metal atoms, the outer s electron shell must be increased, which is much more diffusive than the outermost d electron shell.

The MM proposed by Stewart [7] differs from the multipole model of Eq. (5.20) in several ways. In the Stewart model, the deformation function can be described by a Slater-type radially dependent multipole function, with no κ coefficient to adjust the scaling of the valence electron shell, whereas Eq. (5.20) often introduces chemical environmental confinement adjustments by using a local atomic coordinate system. In Hirshfeld's aspherical density model, the angular distribution function will be the sum of spherical harmonics.

5.4.5 Aspheric Atomic Scattering Factors

Taking the Fourier transform of the MM density and using Eq. (5.20), the form factor of the j atom can be obtained by using the Fourier transform:

$$\begin{aligned} f_j(\boldsymbol{S}) &= \int \rho_j(\boldsymbol{r})\exp(2\pi i\boldsymbol{S}\cdot\boldsymbol{r})\mathrm{d}\boldsymbol{r} \\ &= P_{j.c}f_{j.\mathrm{core}}(S) + P_{j.v}f_{j.\mathrm{valence}}(S/\kappa) + \sum_{l=0}^{l_{\mathrm{mas}}}\sum_{m=0}^{l}\sum_{p} P_{lmp}f_{lmp}(\boldsymbol{S}/\kappa'), \end{aligned} \tag{5.21}$$

where the multipole scattering factor $f_{lmp}(\boldsymbol{S})$ is the Fourier transform of the spherical harmonic function, $p = \pm$.

To obtain the Fourier transform of the spherical harmonics, the plane wave $\exp(2\pi i\boldsymbol{S}\cdot\boldsymbol{r})$ can be expanded into the product format of the spherical harmonics:

$$\exp(2\pi i\boldsymbol{S}\cdot\boldsymbol{r}) = 4\pi\sum_{l=0}^{\infty}\sum_{m=-l}^{l} i_l^l j_l(2\pi Sr)Y_{lm}(\theta,\phi)Y_{lm}^*(\beta,\gamma), \tag{5.22}$$

where j_l is a 1-order spherical Bessel function, θ and ϕ, β and γ are the angular coordinates of $\boldsymbol{r}$ and $\boldsymbol{S}$, respectively. Combining terms $m = -l$ and $m = l$ will yield:

$$\begin{aligned} \exp(2\pi i\boldsymbol{S}\cdot\boldsymbol{r}) &= \sum_{l=0}^{l} i^l j_l(2\pi Sr)(2l+1) \\ &\times \sum_{m=0}^{l}(2-\delta_{m0})\frac{(l-m)!}{(l+m)!}P_l^m(\cos\theta)P_l^m(\cos\beta)\cos\{m(\phi-\gamma)\}. \end{aligned} \tag{5.23}$$

We can calculate $f_{lm-}(\boldsymbol{S})$ and $f_{lm+}(\boldsymbol{S})$:

$$f_{lm+}(\boldsymbol{S}) = \int R_l(r)\mathrm{d}_{lm+}(\theta,\phi)\exp(2\pi i\boldsymbol{H}\cdot\boldsymbol{r})\mathrm{d}\boldsymbol{r}. \tag{5.24}$$

Substituting Eq. (5.22) and $d_{lm+}(\theta,\phi)$ into the above equation will yield:

$$\begin{aligned} f_{lm+}(\boldsymbol{S}) &= N'_{lm+}\iiint R_l(r)P_l^m(\cos\theta)\cos m\phi\sum_{l=0}^{\infty} i_l^l(2\pi Sr)(2l+1) \\ &\times \sum_{m'=0}^{l}(2-\delta_{m0})\frac{(l-m')!}{(l+m')!}P_l^{m'}(\cos\theta)P_l^{m'}(\cos\beta)\cos\{m'(\phi-\gamma)\}r^2\sin\theta\,\mathrm{d}\theta\,\mathrm{d}\phi\,\mathrm{d}r. \end{aligned} \tag{5.25}$$

Among these, the Legendre polynomial and spherical harmonics will form an orthogonal set, and only the terms of $l = l'$ and $m = m'$ will have non-zero integral values. In addition, for the θ integral

$$\int_0^{\pi} P_l^m(\cos\theta)^2\sin\theta\mathrm{d}\theta = \int_{-1}^{1} P_l^m(z)^2\mathrm{d}z = \frac{2}{(2l+1)}\frac{(l+|m|)!}{(l-|m|)!}, \tag{5.26}$$

and for ϕ:

$$\begin{aligned} m \neq 0: &\quad \int_0^{2\pi}\cos m\phi\cos m(\phi-\gamma)\mathrm{d}\phi = \pi\cos m\gamma \\ m = 0: &\quad \int_0^{2\pi}\cos m\phi\cos m(\phi-\gamma)\mathrm{d}\phi = 2\pi. \end{aligned} \tag{5.27}$$

Substituting these results into Eq. (5.24) gives:

$$f_{lm+}(\boldsymbol{S}) = 4\pi i^l \langle j_l \rangle N'_{lm+} P_l^m(\cos\beta)\cos m\gamma = 4\pi i^l \langle j_l \rangle d_{lm+}(\beta,\gamma), \tag{5.28}$$

which can also be expressed in the form:

$$f_{lmp}(\boldsymbol{S}) = 4\pi i^l \langle j_l \rangle d_{lmp}(\beta,\gamma), \tag{5.29}$$

where $\langle j_l \rangle$ is the Fourier–Bessel transform, a radial integral:

$$\langle j_l \rangle = \int j_l(2\pi Sr) R_l(r) r^2 \mathrm{d}r, \tag{5.30}$$

$$\langle j_0 \rangle = \int_0^\infty 4\pi r^2 \rho_j(r) \frac{\sin 2\pi Sr}{2\pi Sr} \mathrm{d}r \equiv \int_0^\infty 4\pi r^2 \rho_j(r) j_0 \mathrm{d}r. \tag{5.31}$$

Equations (5.28) and (5.29) show that the Fourier transform of the spherical harmonic function in the direct space will be the spherical harmonic function of the same order l, m in the reciprocal space, showing that the spherical harmonic function has Fourier transform invariance. This indicates that the dipole density along the c-axis of the unit cell is described by a d_{10} function will not contribute to the $(hk0)$ diffraction points, as these points are in the a^*b^* plane, which is the nodal plane of $d_{10}(\beta,\gamma)$.

In general, Eq. (5.21) can be summarized as:

$$F(\boldsymbol{H}) = \sum_j \left[\left\{ \begin{array}{c} P_{j,c} f_{j,\mathrm{core}}(H) + P_{j,v} f_{j,\mathrm{valence}}(H/\kappa) \\ +4\pi \sum\limits_{l=0}^{l_{\max}} \sum\limits_{m=0}^{l} P_{lm\pm} i^l \langle j_l(S/\kappa') \rangle d_{lm\pm}(\beta,\gamma) \end{array} \right\} \exp(2\pi i \boldsymbol{H}\cdot\boldsymbol{r}_j) T_j(\boldsymbol{H}) \right], \tag{5.32}$$

where $T_j(\boldsymbol{H})$ is the temperature factor.

5.4.6 Multipolar Model of Core Electron Expansion

In the standard MM refinement, the core electron density will be fixed and will not participate in refinement, making it suitable for most experimental charge analyses. However, in certain high-precision experimental charge analyses, it is necessary to consider the changes in the electron density of the core-shell when it is affected by chemical bonding. The core electron shell of carbon atoms in diamond undergo small but significant shrinkage under the formation of C—C covalent bonds. Fischer et al. noticed the effect of the isotropic core electron shell shrinkage of carbon atoms on the variation of scattering intensity with the diffraction angle. Similar to the effect of reducing the isotropic temperature factor, if the standard MM was used for the refinement, unreasonable results can be obtained, including a negative temperature factor [8]. To consider the influence of the core electron shell, Fischer et al. extended the Hansen–Coppens MM, where the expanded atomic electron density $\rho_{\mathrm{at}}(\boldsymbol{r})$ can be expressed as:

$$\rho_{\mathrm{at}}(\boldsymbol{r}) = P_c \kappa_c{}^3 \rho_{\mathrm{core}}(\kappa_c r) + P_v \kappa_v{}^3 \rho_{\mathrm{valence}}(\kappa_v r) + \sum_{l=0}^{l_{\max}} \kappa'^3 R_l(\kappa' r) \sum_{m=0}^{l} P_{lm\pm} d_{lm\pm}(\theta,\phi). \tag{5.33}$$

This extended model allows for the refinement of the core electron population P_c and shrinkage coefficient κ_c, in addition to the refinement of the parameters in the conventional multipole model.

Similar to Eq. (5.32), the structure factor can be expressed as:

$$F(\boldsymbol{H}) = \sum_j \left[\left\{ \begin{array}{l} P_{j,c} f_{j,\text{core}}(\boldsymbol{H}/\kappa_c) + P_{j,v} f_{j,\text{valence}}(\boldsymbol{H}/\kappa_v) \\ +4\pi \sum_{l=0}^{l_{\text{mar}}} \sum_{m=0}^{l} P_{lm\pm} i^l \langle j_l(S/\kappa') \rangle d_{lm\pm}(\beta, \gamma) \end{array} \right\} \exp(2\pi i \boldsymbol{H} \cdot \boldsymbol{r}_j) T_j(\boldsymbol{H}) \right]. \quad (5.34)$$

5.5 Spin Density Model

Assuming that the unit cell magnetization density is the sum of the magnetization densities $m_i(\boldsymbol{r}_i)$ of the pseudo-atoms in a unit cell, the symbol $\boldsymbol{r}_i = \boldsymbol{r} - \boldsymbol{R}_i$ can be defined and $\boldsymbol{R}_i$ is the equilibrium position of the i atom. The magnetic structure factor can be expressed as the sum of the contributions of n_a atoms in a unit cell:

$$\boldsymbol{F_M}(\boldsymbol{\kappa}) = \sum_{i=1}^{n_a} \mu_i F^i_{\text{mag}}(\boldsymbol{\kappa}) e^{i\boldsymbol{\kappa}\boldsymbol{R}_i} e^{-W_i}, \quad (5.35)$$

where $F^i_{\text{mag}}(\boldsymbol{\kappa})$ is the normalized magnetic form factor of the pseudo-atom i, the magnetic moment of atom i is μ_i, and W_i is the Debye–Waller factor of the pseudo-atom. In addition, the atomic magnetic form factor will be the Fourier transform of the magnetic moment, namely,

$$F^i_{\text{mag}}(\boldsymbol{\kappa}) = \frac{1}{\mu_i} \int m_i(\boldsymbol{r}) e^{i\boldsymbol{\kappa}\boldsymbol{r}} \mathrm{d}\boldsymbol{r}. \quad (5.36)$$

5.5.1 Pure Spin Contribution

If there is no orbital contribution, the magnetization density will be the spin density. This is similar to the case of organic radicals, where the magnetism is due to unpaired 2p electrons and the orbital magnetic moment is 0. Certain transition metal ions also have no angular momentum magnetic moment, including Mn^{2+} ($3d^5$, $e_g{}^2 t_{2g}{}^3$, $S = 5/2$) ions with half-filled 3d shells, Cr^{3+} ($3d^3$, $t_{2g}{}^3$, $S = 3/2$) ions in an octahedral field, and rare earth ion Gd^{3+} ($4f^7$, $S = 7/2$) ion, where the atomic magnetic moment is:

$$\mu_i = g_s S_i, \quad (5.37)$$

where g_s is the Lande factor, and S_i is the atomic spin operator.

5.5.1.1 Atomic Orbital Model of Spin Density

For an open-shell system with n_a atoms and n unpaired electrons, the unrestricted Hartree–Fock method must be used to account for the spin polarization, resulting in a negative spin density region, where $\phi_k^{\uparrow,\downarrow}(\boldsymbol{r})$ are the molecular orbitals describing

$n\uparrow$ spin-up and $n\downarrow$ spin-down, respectively. The spin density can be expressed as the difference between the spin-up and spin-down electron densities:

$$s(\boldsymbol{r}) = \rho^{\uparrow}(\boldsymbol{r}) - \rho^{\downarrow}(\boldsymbol{r}) = \sum_{k}^{n_\uparrow} \left|\phi_k^{\uparrow}(\boldsymbol{r})\right|^2 - \sum_{k}^{n_\downarrow} \left|\phi_k^{\downarrow}(\boldsymbol{r})\right|^2. \tag{5.38}$$

Molecular orbitals are linear combinations of n_a atomic orbitals $\psi_i(\boldsymbol{r}_i)$:

$$\phi_k^{\uparrow,\downarrow}(\boldsymbol{r}) = \sum_{i=1}^{n_a} c_i^{\uparrow,\downarrow} \psi_i(\boldsymbol{r}_i). \tag{5.39}$$

Under independent atom approximation (ignoring interatomic interactions), combining Eqs. (5.38) and (5.39) yield the spin density of atom i:

$$s_i(\boldsymbol{r}_i) = p_i |\psi_i(\boldsymbol{r}_i)|^2, \tag{5.40}$$

where p_i is the atomic spin population:

$$p_i = \sum_{k}^{n_\uparrow} c_i^{\uparrow 2} - \sum_{k}^{n_\downarrow} c_i^{\downarrow 2}. \tag{5.41}$$

The atomic spin population can be positive or negative, and satisfy:

$$\sum_{i}^{n_a} p_i = n_\uparrow - n_\downarrow = 2M_S. \tag{5.42}$$

Assuming that the magnetic moment of the i atom contains only one atomic shell, its quantum numbers will be N and L, where L ranges from 0 to $N-1$, and the corresponding single-electron atomic orbital is:

$$\psi_i(\boldsymbol{r}_i) = \mathcal{N} r_i^{N-1} e^{-\xi_L^i r_i} \sum_{M=-L}^{L} \alpha_{LM}^i \mathcal{N}_{\mathrm{ang}} Y_{LM}(\theta_i, \varphi_i), \tag{5.43}$$

where M is the magnetic quantum number and ranges from $-L$ to L, $Y_{LM}(\theta_i, \varphi_i)$ is the conventional spherical harmonic function, ξ_L^i is the Slater orbital index, $\mathcal{N}_{\mathrm{ang}}$ is the spherical harmonic function coefficient, and α_{LM}^i is the atomic orbital coefficient, which meets the normalization condition:

$$\sum_{M=-L}^{L} \left|\alpha_{LM}^i\right|^2 = 1. \tag{5.44}$$

The normalized magnetic structure factor can be expressed as:

$$F_{\mathrm{mag}}^i(\kappa) = p_i \sum_{\ell=0}^{\infty} \left\langle j_\ell^i(\kappa)\right\rangle \sum_{m=-\ell}^{\ell} C_{\ell m}^i Y_{\ell m}^*(\theta_\kappa, \varphi_\kappa), \tag{5.45}$$

where * denotes the complex conjugate.

$\left\langle j_\ell^i(\kappa)\right\rangle$ is the radial integral (Bessel–Fourier transform) of atom i, including the ℓ-order spherical Bessel function $\left\langle j_\ell^i(\kappa r)\right\rangle$, according to:

$$\left\langle j_\ell^i(\kappa)\right\rangle = \int_0^\infty r^2 \left(\mathcal{N} r_i^{N-1} e^{-\xi_L^i r_i}\right)^2 j_\ell^i(\kappa r) \mathrm{d}r, \tag{5.46}$$

where $C^i_{\ell m}$ is the linear combination of the product of orbital coefficients $\alpha^{i*}_{LM}\alpha^i_{LM'}$, and the values of ℓ and m satisfy $\ell \leq 2L$ and $m = M' - M$.

For the generalized magnetic structure factors in atomic orbital models:

$$F_M(\kappa) = \sum_{i=1}^{n_a} m_i \sum_{\ell=0}^{\infty} \left\langle j^i_\ell(\kappa) \right\rangle \sum_{m=-\ell}^{\ell} C^i_{\ell m} Y^*_{\ell m}(\theta_\kappa, \varphi_\kappa) e^{i\kappa R_i} e^{-W_i}, \tag{5.47}$$

where m_i is the atomic magnetic moment $m_i = \mu_i p_i$.

The atomic orbital $\psi_i(\boldsymbol{r}_i)$ can also be expressed in the form of the real spherical harmonic function $y^i_{\ell m\pm}$ and the atomic orbital coefficient α^i_{LM}, satisfying a normalization condition, yielding $F^{\text{cal}}_M(\kappa)$ similar to (5.47), where $y^i_{\ell m}$ is a real spherical harmonic function:

$$F_M(\boldsymbol{\kappa}) = \sum_{i=1}^{n_a} m_i \sum_{\ell=0}^{\infty} i^\ell \left\langle j^i_\ell(\kappa) \right\rangle \sum_{m=-\ell}^{\ell} A^i_{\ell m} y_{\ell m}(\theta_\kappa, \varphi_\kappa) e^{i\kappa R_i \ell} e^{-W_i}, \tag{5.48}$$

where $A^i_{\ell m}$ is the linear combination of orbital coefficients $\alpha^i_{LM}\alpha^i_{LM'}$.

Two refinement strategies can be utilized, as follows.

1. The magnetic form factor can be obtained from the literature to refine the atomic magnetic moments. The radial integrals of 3d and 4f ions $\langle j_\ell(\kappa) \rangle$ can be obtained from Ref. [9]. The magnetic form factors will usually be spherically approximated ($\ell = 0$):

$$F_M(\boldsymbol{\kappa}) = \sum_{i=1}^{n_a} m_i \left\langle j^i_0(\kappa) \right\rangle. \tag{5.49}$$

2. To obtain atomic orbital information, Eq. (5.48) can be used. The atomic orbital coefficients $A^i_{\ell m}$, the Slater radial exponent ξ^i_L, and the atomic magnetic moment m_i need to be refined. The initial Slater radial index can be calculated from the wave function in the literature [10].

The sum of the magnetic moments of n_a atoms will be the experimental value M_{exp} of the molecular magnetic moment $\sum_{i=1}^{n_a} m_i = M_{\text{exp}}$, and the normalized atomic spin population p_i can be calculated from the refined m_i:

$$p_i = m_i \frac{2M_S}{M_{\text{exp}}}. \tag{5.50}$$

5.5.1.2 Multipole Refinement of Spin Density

The MM of Hansen–Coppens can be used for the spin density refinement of centrosymmetric and noncentrosymmetric structures. In the multipole expansion of the electron density, the core term does not exist for the spin density, as only the valence electrons contribute to the magnetic moment. Multipole expansion of atomic spin density is given by:

$$s_i(\boldsymbol{r}_i) = P^i_0 \mathcal{N} r_i^{n_0} e^{-k'\varsigma^i_0 r_i} + \sum_{\ell=0}^{\ell_{\max}=4} \mathcal{N} r_i^{n_\ell} e^{-k''_i s^i_\ell r_i} \sum_{m=-\ell}^{\ell} P^i_{\ell m} y^i_{\ell m}(\theta_i, \varphi_i). \tag{5.51}$$

Most of the time, the mere second term in Eq. (5.51) can be used to describe the contribution of p- or d-type unpaired electron orbitals. The magnetic structure factor can be obtained after Fourier transform:

$$F_M(\boldsymbol{\kappa}) = \sum_{i=1}^{n_a} \sum_{\ell=0}^{\ell_{\max}} i^{\ell} \left\langle j_{\ell}^{i}(\kappa) \right\rangle \sum_{m=-\ell}^{\ell} P_{\ell m}^{i} y_{\ell m}(\theta_{\kappa}, \varphi_{\kappa}) e^{i\boldsymbol{\kappa}\boldsymbol{R}_i} e^{-W_i}, \tag{5.52}$$

where $\left\langle j_{\ell}^{i}(\kappa) \right\rangle$ is the ℓ-order Bessel–Fourier transform of the radial Slater function:

$$\left\langle j_{\ell}^{i}(\kappa) \right\rangle = \int_0^{\infty} r^2 \left(\mathcal{N}\, r_i^{n_{\ell}} e^{-k'' i s_{\ell}^{i} r_i} \right) j_{\ell}^{i}(\kappa r) \mathrm{d}r. \tag{5.53}$$

5.5.2 Spin and Orbital Contributions

In the presence of orbital contributions, it is necessary to fit the spin and orbital contributions to determine the ratio $\boldsymbol{\mu}_L/\boldsymbol{\mu}_S$, which is related to spin-orbit coupling and the crystal field. For transition metal and actinide metal ions, the magnetic moment can be expressed as:

$$\boldsymbol{\mu}_i = g_s S_i + g_L L_i, \tag{5.54}$$

where g_s and g_L are the spin and orbital components, respectively, and the sum of which is the Lande splitting factor g ($g = g_s + g_L$), where μ_i is the total atomic magnetic moment, which is equal to the sum of the spin and orbital contributions, namely, $\boldsymbol{\mu} = \boldsymbol{\mu}_s + \boldsymbol{\mu}_L$.

The magnetic form factor can be expressed as:

$$F_{\mathrm{mag}}^{i}(\boldsymbol{\kappa}) = \frac{\boldsymbol{\mu}_S^i}{\boldsymbol{\mu}_i} F_{\mathrm{mag}}^{iS}(\boldsymbol{\kappa}) + \frac{\boldsymbol{\mu}_L^i}{\boldsymbol{\mu}_i} F_{\mathrm{mag}}^{iL}(\boldsymbol{\kappa}). \tag{5.55}$$

Under dipole approximation, the spin part uses a spherical approximation, and the magnetic structure factor can be expressed as:

$$F_{\mathrm{mag}}^{i}(\boldsymbol{\kappa}) = \left\langle j_0^{i}(\kappa) \right\rangle + c_2 \left\langle j_2^{i}(\kappa) \right\rangle, \tag{5.56}$$

where $c_2 = \boldsymbol{\mu}_L/\boldsymbol{\mu}$ and $\boldsymbol{\mu}_L/\boldsymbol{\mu}_S = c_2/(1 - c_2)$.

For rare earth metals, the spin-orbit coupling will be much stronger than the crystal field, and the S_i operator must be replaced J_i, with eigenvalues:

$$J = |\, L \pm S \,|; \quad \boldsymbol{\mu}_i = g_j J_i. \tag{5.57}$$

If the crystal field splitting energy is smaller than the exchange energy and the magnetic field energy, the spin density will not depend on the angle between the magnetic field and the principal axis of the crystal field. Thus, the overall magnetic form factor can be described by a spherical approximation:

$$\begin{aligned} F_{\mathrm{mag}}^{i}(\boldsymbol{\kappa}) &= \langle j_0(\kappa) \rangle + c_2 \langle j_2(\kappa) \rangle \\ c_2 &= \frac{J(J+1) + L(L+1) - S(S+1)}{3J(J+1) - L(L+1) + S(S+1)}. \end{aligned} \tag{5.58}$$

5.5.3 Non-collinear Magnetism

Under weak fields, because the nonlinear effect of the external magnetic field is not important, the magnetic moment of atom i induced by the magnetic field H can be expressed as:

$$\mu_i = \chi_i \boldsymbol{H}, \tag{5.59}$$

where χ_i is the magnetic susceptibility tensor of the i atom. We can consider a case with only one magnetic atom in the asymmetric unit, where the magnetic structure factor can be expressed as:

$$\boldsymbol{F_M}(\boldsymbol{\kappa}) = F_{mag}(\boldsymbol{\kappa}) \sum_{\mathrm{i}=1}^{n_a} \chi_i \boldsymbol{H} \cdot e^{\mathrm{i}\boldsymbol{\kappa}\boldsymbol{R}_\mathrm{i}} e^{-W_i}, \tag{5.60}$$

where the summation has to be conducted at all equivalent points, and the symmetry operation on the tensor χ will be similar to the temperature factor $\boldsymbol{U}$. If the flipping ratios of the diffraction intensities under different magnetic fields are collected in the polarization neutron diffraction, the atomic magnetization parameter χ_{ij} can be determined in the refinement, as well as the temperature factor.

5.5.4 Combinatorial Refinement of Electron Density and Spin Density

One of the most challenging aspects in experimental electronic structure research is the simultaneous refinement of the electron density, spin density, and momentum density of the sample, which requires the simultaneous use of different scattering data, including X-ray diffraction, polarization neutron diffraction, and Compton scattering data. For details regarding the simultaneous refinement of electron and momentum density using X-ray diffraction and Compton scattering data, see Section 6.4.

X-ray diffraction and non-polarized neutron diffraction data include the integrated intensity of Bragg diffraction. Polarized neutron diffraction measures the ratio of diffraction intensities of the spin-up and spin-down incident neutrons to obtain the magnetization density of electrons, including pure spin and orbital contributions. The results have shown that X-ray diffraction and polarization neutron diffraction can form a good complementarity, where the former can obtain the total electron density $\rho(\boldsymbol{r})$, while the latter can obtain the spin density $\rho^\sigma(\boldsymbol{r})$, satisfying the following relationship:

$$\rho(\boldsymbol{r}) = \rho^\uparrow(\boldsymbol{r}) + \rho^\downarrow(\boldsymbol{r}), \tag{5.61}$$

$$\rho^\sigma(\boldsymbol{r}) = \rho^\uparrow(\boldsymbol{r}) - \rho^\downarrow(\boldsymbol{r}). \tag{5.62}$$

Obviously, the spin-resolved electron density information in magnetic crystals can be obtained by the joint refinement of high-resolution X-ray diffraction and polarization neutron diffraction data. To achieve such joint refinement, a spin-resolved pseudoatomic model is required [11]. Based on the Hansen–Coppens multipole model, $\rho(\boldsymbol{r})$ can be expressed by:

$$\rho_i(\mathbf{r}) = \rho_{i,\text{core}}(r) + P^{\uparrow}_{i,\text{valence}}\kappa^{\uparrow 3}_i \rho_{i,\text{valence}}\left(\kappa^{\uparrow}_i \mathbf{r}\right) + P^{\downarrow}_{i,\text{valence}}\kappa^{\downarrow 3}_i \rho_{i,\text{valence}}\left(\kappa^{\downarrow}_i \mathbf{r}\right)$$
$$+ \sum_{l=0,l_{\max}} \kappa'^{\uparrow 3}_i R_{i,l}\left(\kappa'^{\uparrow}_{i,l}\mathbf{r}\right) \sum_{m=0,l} P^{\uparrow}_{i,lm\pm} y_{lm\pm}(\mathbf{r}/r)$$
$$+ \sum_{l=0,l_{\max}} \kappa'^{\downarrow 3}_i R_{i,l}\left(\kappa'^{\downarrow}_{i,l}\mathbf{r}\right) \sum_{m=0,l} P^{\downarrow}_{i,lm\pm} y_{lm\pm}(\mathbf{r}/r), \quad (5.63)$$

where $P^{\uparrow}$ and $P^{\downarrow}$ are the spin-up and spin-down parameters, respectively, and the core and valence in the subscript represent the core and valence electrons, respectively. The model has to use the X-ray diffraction and polarized neutron diffraction data, and the parameters that must be refined include $P^{\uparrow}$ and $P^{\downarrow}$, as well as $\kappa^{\uparrow}$ and $\kappa^{\downarrow}$.

5.6 Other Electron Density Models

5.6.1 The X-ray Atomic Orbital (XAO) Model

The XAO model [1, 2] is a method based on atomic orbitals that uses a traditional least squares algorithm, which minimizes the difference between the observed and model-calculated values of the structure factor:

$$S = \sum_{\mathbf{h}} w_{\mathbf{h}}[|F_{\text{obs}}(\mathbf{h})| - k|F_{\text{calc}}(\mathbf{h})|]^2, \quad (5.64)$$

where $w_{\mathbf{h}}$ is the weight of each diffraction, F_{obs} and F_{calc} are the observed and model-calculated values of the structure factor, respectively, and k is the scale factor. In XAO analysis, each atom is broken into subshell electrons (p, d, and f electrons), which combine to form a pseudo-atom. During the refinement process, individual atomic orbitals have to maintain orthogonality, which reduces the parameter correlation in refinement. XAO does not consider electron correlation, and the refined non-integer occupancy number may reflect this issue from the side. The XAO model can determine a single central atomic orbital and its electron population and is more suitable for systems with a high degree of electron localizations, including transition metal, rare earth metal, and ionic compounds. In addition, XAO does not consider the double-center term in the molecular orbital; thus, XAO can be regarded as an electron population analysis in which the atomic orbitals keep strictly orthogonal conditions, and the contribution of the double-center term will be generally very small.

5.6.1.1 Atomic Single-electron Orbitals in a Crystal Field

Subshell atomic orbitals ψ_k will degenerate if not affected by a crystal field. Under the action of a crystal field, the ith atomic orbita $\Psi_{\alpha,i}(\mathbf{r})$ of the αth atom can be assumed to be expressed as a linear combination of ψ_k:

$$\Psi_{\alpha,i}(\mathbf{r}) = \pm\sum_{k=1}^{k_{\max}} a_{ik}\psi_k(\mathbf{r}), \quad (5.65)$$

where a_{ik} is the coefficient to be determined in the least squares refinement, and for the angle quantum number l, $k_{\max}$. When considering spin-orbit coupling, $k_{\max}$, $j = l+1/2$ or $|l-1/2|$. The $\pm$ in Eq. (5.65) cannot be determined from the X-ray diffraction data because the wave function is not observable, nor is it for molecular orbitals. The orbits of the degenerate atomic orbitals through unitary transformation will also be the solution to the Schrödinger equation. The unitary transformation will not change the value of electron density, nor will it change the orthogonalization condition. Thus, we only have to determine one of the infinite groups of atomic orbitals. For example, the d orbital wave function Ψ_k is one of $d_{x^2-y^2}$, d_{z^2}, d_{yz}, d_{zx} and d_{xy}, which can be expressed as a linear combination of a hydrogen-like orbital $\phi_{nlm_l}(\boldsymbol{r})$:

$$\psi_k(\boldsymbol{r}) = \sum_{m_l=-l}^{l} d_{km_l}\phi_{nlm_l}(\boldsymbol{r}), \tag{5.66}$$

where d_{km_l} is a known constant, and $\phi_{nlm_l}(\boldsymbol{r})$ is the product of the radial function $R_{nl}(r)$ and spherical harmonic function $Y_{lm_l}(\theta,\phi)$, namely:

$$\phi_{nlm_l}(\boldsymbol{r}) = R_{nl}(r)Y_{lm_l}(\theta,\phi). \tag{5.67}$$

Nonrelativistic radial functions calculated by Mann [12] and the relativistic radial functions calculated by the HEX program [13] can be used in XAO analysis. In refinement, the specific form of $R_{nl}(r)$ can be changed to introduce the contraction factor κ, where the p, d, and f orbital basis functions $\psi_k(\boldsymbol{r})$ for the half-integral j values are the spin orbitals, defined as:

$$\psi_k(\boldsymbol{r}) = \sum_{m_l}\sum_{m_s}\phi_{nlm_l}s(m_s)\left\langle l\frac{1}{2}m_lm_s\middle|l\frac{1}{2}jm_j\right\rangle$$
$$m_j = j, j-1, \ldots, -j+1, -j \tag{5.68}$$

where $s(m_s)$ and m_s are the spin-orbit and spin quantum numbers, respectively. The angle brackets on the right side of Eq. (5.68) refer to the values of Condon and Shortley [14]. In addition, $m_j = m_l + m_s$, where the subscript κ is an integer from 1 to $2j+1$, corresponding to the values of $m_j = j, j-1, \cdots, -j+1, -j$, respectively. For the specific form of the basis functions, see Table 5.1. When j is an integer, a real basis function can be selected. Table 5.1 does not list the real basis function of the f orbital because the spin-orbit coupling effect of heavy atoms cannot be ignored;

Table 5.1, the basis functions ψ_k for the p, d, and f orbitals.

$k = 1, 2, \cdots, 2j+1$ corresponds to the states of (j, j), $(j, j-1)$, $\cdots$, $(j, -j)$, respectively, and α and β corresponds to the spin functions with eigenvalues $\hbar/2$ and $-\hbar/2$, respectively.

In nonlinear least squares refinement, the initial values of the parameters should be as close as possible to the true values. Thus, $\Psi_{\alpha,i}(r)$ in Eq. (5.65) and energy level splitting can be calculated using the first-order perturbation theory; that is, the crystal field is assumed to be a perturbation. The secular equation is given by:

$$\det\left(H_{kk'} - E_n^{(1)}\delta_{kk'}\right) = 0, \tag{5.69}$$

Table 5.1 The basis functions ψ_k for the p, d, and f orbitals.

p orbital basis function	$j = 1/2$	$\psi_1 = \sqrt{2/3}\phi_{1,1}\beta - \sqrt{1/3}\phi_{1,0}\alpha$
		$\psi_2 = \sqrt{1/3}\phi_{1,0}\beta - \sqrt{2/3}\phi_{1,-1}\alpha$
	$j = 1$ (real basis function)	$p_x = \sqrt{1/2}(-\phi_{1,1} + \phi_{1,-1})$
		$p_y = i\sqrt{1/2}(\phi_{1,1} + \phi_{1,-1})$
		$p_z = \phi_{1,0}$
	$j = 3/2$	$\psi_1 = \phi_{1,1}\alpha$
		$\psi_2 = \sqrt{1/3}\phi_{1,1}\beta + \sqrt{2/3}\phi_{1,0}\alpha$
		$\psi_3 = \sqrt{3/2}\phi_{1,0}\beta+, \psi_4 = \phi_{1,-1}\beta$
d orbital basis function	$j = 3/2$	$\psi_1 = \sqrt{4/5}\phi_{2,2}\beta - \sqrt{1/5}\phi_{2,1}\alpha$
		$\psi_2 = \sqrt{3/5}\phi_{2,1}\beta - \sqrt{2/5}\phi_{2,0}\alpha$
		$\psi_3 = \sqrt{2/5}\phi_{2,0}\beta - \sqrt{3/5}\phi_{2,-1}\alpha$
		$\psi_4 = \sqrt{1/5}\phi_{2,-1}\beta - \sqrt{4/5}\phi_{2,-2}\alpha$
	j=2 (real basis function)	$d_{x^2-y^2} = \sqrt{1/2}(\phi_{2,2} + \phi_{2,-2})$
		$d_{z^2} = \phi_{2,0}$
		$d_{yz} = i\sqrt{1/2}(\phi_{2,1} + \phi_{2,-1})$
		$d_{zx} = \sqrt{1/2}(-\phi_{2,1} + \phi_{2,-1})$
		$d_{xy} = i\sqrt{1/2}(-\phi_{2,2} + \phi_{2,-2})$
	$j = 5/2$	$\psi_1 = \phi_{2,2}\alpha$
		$\psi_2 = \sqrt{1/5}\phi_{2,2}\beta + \sqrt{4/5}\phi_{2,1}\alpha$
		$\psi_3 = \sqrt{2/5}\phi_{2,1}\beta + \sqrt{3/5}\phi_{2,0}\alpha$
		$\psi_4 = \sqrt{3/5}\phi_{2,0}\beta + \sqrt{2/5}\phi_{2,-1}\alpha$
		$\psi_5 = \sqrt{4/5}\phi_{2,-1}\beta + \sqrt{1/5}\phi_{2,-2}\alpha$
		$\psi_6 = \phi_{2,-2}\beta$
f orbital basis function	$j = 5/2$	$\psi_1 = \sqrt{6/7}\phi_{3,3}\beta - \sqrt{1/7}\phi_{3,2}\alpha$
		$\psi_2 = \sqrt{5/7}\phi_{3,2}\beta - \sqrt{2/7}\phi_{3,1}\alpha$
		$\psi_3 = \sqrt{4/7}\phi_{3,1}\beta - \sqrt{3/7}\phi_{3,0}\alpha$
		$\psi_4 = \sqrt{3/7}\phi_{3,0}\beta - \sqrt{4/7}\phi_{3,-1}\alpha$
		$\psi_5 = \sqrt{2/7}\phi_{3,-1}\beta - \sqrt{5/7}\phi_{3,-2}\alpha$
		$\psi_6 = \sqrt{1/7}\phi_{3,-2}\beta - \sqrt{6/7}\phi_{3,-3}\alpha$
	$j = 7/2$	$\psi_1 = \phi_{3,3}\alpha$
		$\psi_2 = \sqrt{1/7}\phi_{3,3}\beta + \sqrt{6/7}\phi_{3,2}\alpha$
		$\psi_3 = \sqrt{2/7}\phi_{3,2}\beta + \sqrt{5/7}\phi_{3,1}\alpha$
		$\psi_4 = \sqrt{3/7}\phi_{3,1}\beta + \sqrt{4/7}\phi_{3,0}\alpha$
		$\psi_5 = \sqrt{4/7}\phi_{3,0}\beta + \sqrt{3/7}\phi_{3,-1}\alpha$
		$\psi_6 = \sqrt{5/7}\phi_{3,-1}\beta + \sqrt{2/7}\phi_{3,-2}\alpha$
		$\psi_7 = \sqrt{6/7}\phi_{3,-2}\beta + \sqrt{1/7}\phi_{3,-3}\alpha$
		$\psi_8 = \phi_{3,-3}\beta$

where $E_n^{(1)}$ and $\delta_{k,k'}$ are the first-order energy correction and Kronecker δ, respectively, and

$$H_{k,k'} = \int \psi_k^* \hat{H}' \psi_{k'} \mathrm{d}\mathbf{r}, \tag{5.70}$$

where$H'$is the crystal field perturbation:

$$\hat{H}' = v_{\text{crystal}}(r, \theta, \phi). \tag{5.71}$$

If each atom is compacted to a point charge, then $v_{\text{crystal}}(r, \theta, \phi)$ can be expressed in a spherical harmonic format. By placing the point charge of -Ze on the λth atom (coordinate $\boldsymbol{R}_\lambda$), whose polar coordinates are $(R_\lambda, \theta_\lambda, \phi_\lambda)$, the potential of the electron at $\boldsymbol{r}$ can be obtained by:

$$v_{\text{crys}}(\boldsymbol{r}) = \sum_\lambda Z_\lambda e^2/(\boldsymbol{R}_\lambda - \boldsymbol{r}). \tag{5.72}$$

Because the atomic orbital region is smaller than the bond length, $\frac{1}{|\boldsymbol{R}_\lambda - \boldsymbol{r}|}$ can be expanded by spherical harmonics:

$$\frac{1}{|\boldsymbol{R}_\lambda - \boldsymbol{r}|} = \frac{1}{R_\lambda} \sum_{k=0}^{\infty} \left(\frac{4\pi}{2k+1}\right) \left(\frac{r}{R_\lambda}\right)^k \sum_{m=-k}^{k} Y_{km}(\theta, \phi) Y_{km}^*(\theta_\lambda, \phi_\lambda), \tag{5.73}$$

to yield:

$$v_{\text{crys}}(r, \theta, \phi) = \sum_{k=0}^{\infty} \sum_{m=-k}^{k} r^k q_{km} C_m^{(k)}(\theta, \phi), \tag{5.74}$$

where q_{km} is a function of the point charge coordinates:

$$q_{km} = \sqrt{\frac{4\pi}{2k+1}} \sum_\lambda \left(\frac{Z_\lambda e^2}{R_\lambda^{k+1}}\right) Y_{km}^*(\theta_\lambda, \phi_\lambda), \tag{5.75}$$

$$C_m^{(k)}(\theta, \phi) = \sqrt{\frac{4\pi}{2k+1}} Y_{km}(\theta, \phi). \tag{5.76}$$

The nonzero q_{km} values of the 32-point groups are shown in Table 5.1, where the quantized z-axis is set to the principal axis direction of the point group. For a point group with a secondary axis perpendicular to the principal axis or the mirror face containing the principal axis, the quantized x-axis or y-axis will be set to the secondary axis or perpendicular to the mirror face. When the point group has no secondary axes or mirrors, the x and y axes can be set to any direction.

$H_{k,k'}$ in Eq. (5.70) contains the following integrals:

$$c^K\left(lm_l, l'm_{l'}\right) = \int Y_{lm_l}^*(\theta, \phi) C_M^{(K)}(\theta, \phi) Y_{l'm_{l'}}(\theta, \phi) \mathrm{d}\tau. \tag{5.77}$$

For the specific values of $c^K\left(lm_l, l'm_{l'}\right)$, please refer to Condon and Shortley [14]. For $c^K\left(lm_l, l'm_{l'}\right) = (-1)^{m_l - m_{l'}} c^K\left(l'm_{l'}, lm_l\right)$, the equation will not be equal to 0 only when the following equation is satisfied:

$$\begin{aligned} &K + l + l' = \text{even} \\ &|l - l'| \leq K \leq l + l' \\ &M = m_l - m_{l'}. \end{aligned} \tag{5.78}$$

q_{km} and ψ_k jointly determine non-zero $H_{kk'}$, and only when q_{km} is a complex number, $H_{kk'}$ is a complex number. When j is a half-integer and m is arranged from $|(j \pm 1)/2|$ to $-(j \pm 1)/2$, the formed Hermitian matrix $\boldsymbol{H} = \{H_{kk'}\}$ satisfy the relation:

$$
\begin{aligned}
H_{k,k} &= H_{2j+2-k,2j+2-k} \\
H_{2j+2-k,k} &= 0 \\
H_{k,k'} &= (-1)^{k+k'} H_{2j+2-k',2j+2-k} \quad (k > k').
\end{aligned}
\tag{5.79}
$$

When the valence shell has more than one electron, XAO analysis requires that an antisymmetric product wave function is used in the Slater form.

5.6.1.2 Electron Density and Structure Factor

The electron density $\rho_\alpha(r)$ at the α atom r in the asymmetric unit can be decomposed into two contributions from the core orbital $\rho_{\alpha,\,\text{core}}(r)$ and valence orbital $\rho_{\alpha,\,\text{valence}}(r)$, according to:

$$\rho_\alpha(\boldsymbol{r}) = \rho_{\alpha,\text{core}}(r) + \rho_{\alpha,\text{valence}}(\boldsymbol{r}), \tag{5.80}$$

where the $\rho_{\alpha,\,\text{valence}}(r)$ can be expressed as the sum of the electron densities of all atomic orbitals i of the α atom at $\boldsymbol{r}_\alpha^{\text{atom}}$, namely:

$$\rho_{\alpha,\text{valence}}(\boldsymbol{r}) = \sum_i n_{\alpha,i} \Psi^*_{\alpha,i}\left(\kappa_i\left(\boldsymbol{r} - \boldsymbol{r}_\alpha^{\text{atom}}\right)\right) \Psi_{\alpha,i}\left(\kappa_i\left(\boldsymbol{r} - \boldsymbol{r}_\alpha^{\text{atom}}\right)\right), \tag{5.81}$$

where $n_{\alpha,\,i}$ is the occupied number of $\Psi_{\alpha,\,i}$ orbitals. In the crystal field, if the i orbital expands, then $\kappa_i < 1$, and if the i orbital contracts, then $\kappa_i > 1$.

We can define an electron coordinate vector $\boldsymbol{r}_i$ from the nucleus of an atom to the ith orbital:

$$\boldsymbol{r}_i = \boldsymbol{r} - \boldsymbol{r}_\alpha^{\text{atom}}. \tag{5.82}$$

The α atom at $\boldsymbol{r}_\alpha^{\text{atom}}$ in the asymmetric unit can be transformed to the equivalent atom $\boldsymbol{r}_{\alpha\sigma}^{\text{atom}}$ under the symmetry operation (R_σ, t_σ):

$$\boldsymbol{r}_{\alpha\sigma}^{\text{atom}} = R_\sigma \boldsymbol{r}_\alpha^{\text{atom}} + \boldsymbol{t}_\sigma, \tag{5.83}$$

where R_σ and $\boldsymbol{t}_\sigma$ are the rotation and translation matrices of the σ-symmetric operation, respectively. In addition, $\boldsymbol{r}$ in Eq. (5.82) can be equivalently symmetric to $\boldsymbol{r}_\sigma$:

$$\boldsymbol{r}_\sigma = R_\sigma \boldsymbol{r} + \boldsymbol{t}_\sigma = \left(R_\sigma \boldsymbol{r}_\alpha^{\text{atom}} + \boldsymbol{t}_\sigma\right) + R_\sigma \boldsymbol{r}_i = \boldsymbol{r}_{\alpha\sigma}^{\text{atom}} + \boldsymbol{r}_{i\sigma}, \tag{5.84}$$

where $\boldsymbol{r}_{i\sigma} = R_\sigma \boldsymbol{r}_i$.

Based on these relations, the structure factor can be written as:

$$F(\boldsymbol{k}) = \sum_\alpha \omega_\alpha \sum_\sigma F_{\alpha\sigma}(\boldsymbol{k}) T_{\alpha\sigma}(\boldsymbol{k}), \tag{5.85}$$

where ω_α is the multiplicity of the α atom, and $T_{\alpha\sigma}(\boldsymbol{k})$ is the temperature factor of the α atom after the σ symmetry operation, including anharmonic vibrations. In addition, $F_{\alpha\sigma}(\boldsymbol{k})$ can be divided into the contribution of the core-shell and valence shell:

$$F_{\alpha\sigma}(\boldsymbol{k}) = F_{\alpha\sigma}^{\text{core}}(k) + F_{\alpha\sigma}^{\text{valence}}(\boldsymbol{k}), \tag{5.86}$$

where $F_{\alpha\sigma}^{\text{core}}(k)$ is the sum of the $\rho_{\alpha,\text{core}}(r_\sigma)$ Fourier transform and anomalous scattering contribution of α atom. $F_{\alpha\sigma}^{\text{valence}}(\boldsymbol{k})$ can be expressed as the Fourier transform sum of all atomic orbital electron densities of all atoms in a unit cell:

$$F_{\alpha\sigma}^{\text{valence}}(\boldsymbol{k}) = \sum_i n_{\alpha,i} F_{\alpha\sigma,i}(\boldsymbol{k}), \tag{5.87}$$

where

$$F_{\alpha\sigma,i}(\boldsymbol{k}) = \int \Psi_{\alpha,i}^*\left(\kappa_i\left(\boldsymbol{r}_\sigma - \boldsymbol{r}_{\alpha\sigma}^{\text{atom}}\right)\right)\exp(i\boldsymbol{k}\cdot\boldsymbol{r}_\sigma) \times \Psi_{\alpha,i}\left(\kappa_i\left(\boldsymbol{r}_\sigma - \boldsymbol{r}_{\alpha\sigma}^{\text{atom}}\right)\right) \mathrm{d}\boldsymbol{r}_\sigma. \tag{5.88}$$

Because $\boldsymbol{r}_{\alpha\sigma}^{\text{atom}}$ is a constant, Eq. (5.88) can be expressed as:

$$F_{\alpha\sigma,i}(\boldsymbol{k}) = \exp\left(i\boldsymbol{k}\cdot\boldsymbol{r}_{\alpha\sigma}^{\text{atom}}\right)\int \Psi_{\alpha,i}^*(\kappa_i\boldsymbol{r}_{i\sigma})\exp(i\boldsymbol{k}\cdot\boldsymbol{r}_{i\sigma})\Psi_{\alpha,i}(\kappa_i\boldsymbol{r}_{i\sigma})\mathrm{d}\boldsymbol{r}_{i\sigma}, \tag{5.89}$$

and according to the symmetry rule, the above equation can be further expressed as:

$$F_{\alpha\sigma,i}(\boldsymbol{k}) = \exp\left(i\boldsymbol{k}\cdot\boldsymbol{r}_{\alpha\sigma}^{\text{atom}}\right)\int \Psi_{\alpha,i}^*(\kappa_i\boldsymbol{r}_i)\exp\left(iR_\sigma^{-1}\mathbf{k}\cdot\boldsymbol{r}_i\right)\Psi_{\alpha,i}(\kappa_i\boldsymbol{r}_i)\mathrm{d}\boldsymbol{r}_i. \tag{5.90}$$

Therefore, the scattering factor of the electron on the ith atomic orbital of the α atom after σ symmetry operation is the product of the phase factor of the α atom at $\boldsymbol{r}_{\alpha\sigma}^{\text{atom}}$ and the Fourier transform of the ith orbital in the asymmetric unit at the diffraction $R_\sigma^{-1}\mathbf{k}$. We can use Eqs. (5.65), (5.66), and (5.67) to express $F_{\alpha\sigma,i}(\boldsymbol{k})$ in the form ϕ_{nlm_l}:

$$F_{\alpha\sigma,i}(\boldsymbol{k}) = \exp\left(i\boldsymbol{k}\cdot\boldsymbol{r}_{\alpha\sigma}^{\text{atom}}\right)\sum_k\sum_{k'} a_{i,k}^* a_{i,k'}\sum_{m_l}\sum_{m_{l'}} d_{km_l}^* d_{k'm_{l'}} g_{\sigma,m_l,m_{l'}}(\boldsymbol{k}), \tag{5.91}$$

where

$$g_{\sigma,m_l,m_{l'}}(\boldsymbol{k}) = \int \phi_{nlm_l}^*(\kappa_i\boldsymbol{r}_i)\exp\left(iR_\sigma^{-1}\mathbf{k}\cdot\boldsymbol{r}_i\right)\phi_{n'l'm_{l'}}(\kappa_i\boldsymbol{r}_i)\mathrm{d}\boldsymbol{r}_i. \tag{5.92}$$

Eq. (5.92) can be further expanded by $\exp\left(iR_\sigma^{-1}\mathbf{k}\cdot\boldsymbol{r}_i\right)$:

$$\exp\left(iR_\sigma^{-1}\mathbf{k}\cdot\boldsymbol{r}_i\right) = 4\pi\sum_{K=0}^{\infty}\sum_{m=-K}^{K} i^K j_K(kr)Y_{Km}^*(\theta,\phi)Y_{Km}(\beta_\sigma,\gamma_\sigma), \tag{5.93}$$

where (r,θ,ϕ) and $(k,\beta_\sigma,\gamma_\sigma)$ are the polar coordinates of the electron and the scattering vector $R_\sigma^{-1}\mathbf{k}$, respectively, which are defined by the quantized axis of the electron in the atomic orbital in the asymmetric unit and $j_K(kr)$ is the k -order Bessel function. When the quantum numbers of atoms i and i' are (n,l,m_l) and $(n',l',m_{l'})$, the specific form of $g_{\sigma,m_l,m_{l'}}(\boldsymbol{k})$ will be:

$$g_{\sigma,m_l,m_{l'}}(\boldsymbol{k}) = \sum_{K=0}^{\infty}\sum_{M=-K}^{K} i^K\langle j_K\rangle\sqrt{2(2K+1)}c^K\left(lm_l;l'm_{l'}\right) \times \Theta_K^M(\beta_\sigma)\exp(iM\gamma_\sigma), \tag{5.94}$$

where $\langle j_K \rangle$ is the Fourier–Bessel transform of $R_{n,l}(r)R_{n'l'}(r)$:

$$\langle j_K \rangle = \int R_{n,l}(r)R_{n',l'}(r)j_K(kr)r^2\mathrm{d}r. \tag{5.95}$$

Because the phase factor $\exp\left(i\boldsymbol{k} \cdot \boldsymbol{r}_{\alpha,\sigma}^{\text{atom}}\right)$ of all the atomic orbitals of the α atom is the same, the scattering factor $f_{\alpha,\sigma,i}(k)$ of the i orbital of the α atom after the σ symmetry operation will be:

$$f_{\alpha,\sigma,i}(\boldsymbol{k}) = \sum_{k}\sum_{k'} a_{i,k}^{*}a_{i,k'} \sum_{K=|l-l'|}^{l+l'} \langle j_K \rangle \sum_{M=-K}^{K} i^K \sqrt{2(2K+1)}\Theta_K^M(\beta_\sigma)$$
$$\times \exp(iM\gamma_\sigma) \sum_{m_l=-l}^{l} \sum_{m_{l'}=-l'}^{l'} d_{k,m_l}^{*} d_{k',m_{l'}} c^K\left(lm_l, l'm_{l'}\right). \tag{5.96}$$

5.6.2 X-ray Molecular Orbital Model (XMO)

XMO model [3] analysis provides a method for obtaining molecular orbitals under the Hartree–Fock approximation based solely on the measured X-ray diffraction data.

5.6.2.1 Molecular Orbital and Electron Density

A molecular orbital $\Psi_{\text{mo}}(\boldsymbol{r})$ can be expressed as a linear combination of symmetric orbitals $\boldsymbol{\Phi}_{\text{so}}(\boldsymbol{r})$, where $\boldsymbol{\Phi}_{\text{so}}(\boldsymbol{r})$ has the same symmetry as $\Psi_{\text{mo}}(\boldsymbol{r})$:

$$\Psi_{\text{mo}}(\boldsymbol{r}) = \sum_{\text{so}} a_{\text{mo,so}}\Phi_{\text{so}}(\boldsymbol{r}), \tag{5.97}$$

where mo and so have values in $1{\sim}M$ and $1{\sim}N$, respectively ($M \leq N$). In addition, $a_{\text{mo, so}}$ is assumed to be a real number, which must be determined from the X-ray diffraction data in XMO analysis. Because the wave function is not an observable quantity, the phase of the molecule cannot be determined experimentally; thus, the purpose of introducing the symmetric orbital $\boldsymbol{\Phi}_{\text{so}}(\boldsymbol{r})$ herein is to maintain the constraints between the molecular symmetry on the $a_{\text{mo, so}}$ parameters. $\boldsymbol{\Phi}_{\text{so}}(\boldsymbol{r})$ can be expressed as a linear combination of the suitable basis functions $\varphi_{\text{bo}(A)}(\boldsymbol{r} - \boldsymbol{A})$, according to:

$$\Phi_{\text{so}}(\boldsymbol{r}) = \sum_{\text{bo}=1}^{n_{\text{bo}}} b_{\text{so,bo}}\varphi_{\text{bo}(A)}(\boldsymbol{r} - \boldsymbol{A}), \tag{5.98}$$

where $b_{\text{so, bo}}$ is a constant that depends on the symmetry of the crystal field, including the normalization constant, and n_{bo} is the number of basis functions used to construct symmetrical orbitals. Both $\boldsymbol{\Phi}_{\text{so}}(\boldsymbol{r})$ and $\varphi_{\text{bo}(A)}$ are normalized, where the subscript bo(A) represents the bo-th basis function of the A atom (coordinate $\boldsymbol{A}$).

Because $\Psi_{\mathrm{mo}}(\boldsymbol{r})$ in Eq. (5.97) is normalized and orthogonal to each other, we can obtain:

$$\sum_{\mathrm{so}=1}^{N}\sum_{\mathrm{so}'=1}^{N} a_{\mathrm{mo,so}}^{*} a_{\mathrm{mo',so'}} \int \Phi_{\mathrm{so}}^{*}(\boldsymbol{r})\Phi_{\mathrm{so}'}(\boldsymbol{r})\mathrm{d}\boldsymbol{r}$$
$$\equiv \sum_{\mathrm{so}=1}^{N}\sum_{\mathrm{so}'=1}^{N} a_{\mathrm{mo,so}}^{*} a_{\mathrm{mo',so'}} s_{\mathrm{so,so'}} = \delta_{\mathrm{mo,mo'}}, \tag{5.99}$$

where $s_{\mathrm{so,so'}}$ is the overlapping integral of symmetric orbitals $\Phi_{\mathrm{so}}(\boldsymbol{r})$ and $\Phi_{\mathrm{so}'}(\boldsymbol{r})$, only when $\mathrm{mo} = \mathrm{mo}'$, $\delta_{\mathrm{mo,mo'}} = 1$, otherwise $\delta_{\mathrm{mo,mo'}} = 0$. Equation (5.99) can be expressed in the matrix form:

$$\boldsymbol{CS}(\boldsymbol{C}^{*})^{t} = \boldsymbol{I}, \tag{5.100}$$

where $\boldsymbol{C} = \{a_{\mathrm{mo,\,so}}\}$, $\boldsymbol{S} = \{s_{\mathrm{so,so'}}\}$, $\boldsymbol{I}$ are the matrices of M×N, N×N, and M×M, respectively, and I is the identity matrix, where superscript t represents the transpose. Assuming that both $a_{\mathrm{mo,\,so}}$ and the molecular orbitals are real numbers, the electron density $\rho_{\mathrm{mo}}(\boldsymbol{r})$ of the mo-th molecular orbital will be:

$$\rho_{\mathrm{mo}}(\boldsymbol{r}) = \Psi_{\mathrm{mo}}^{*}(\mathbf{r})\Psi_{\mathrm{mo}}(\boldsymbol{r}), \tag{5.101}$$

where the total electron density $\rho(\boldsymbol{r})$ of a molecule can be expressed as:

$$\rho(\boldsymbol{r}) = \sum_{\mathrm{mo}=1}^{M} g_{\mathrm{mo}} p_{\mathrm{mo}} \rho_{\mathrm{mo}}(\boldsymbol{r}), \tag{5.102}$$

where p_{mo} is the electron population of the molecular orbital $\Psi_{\mathrm{mo}}(\boldsymbol{r})$, and g_{mo} is usually 1; thus, it is necessary to adjust g_{mo} when the molecule and asymmetric unit do not coincide. We can decompose $\rho_{\mathrm{mo}}(\boldsymbol{r})$ into single-center and double-center electron densities according to:

$$\rho_{\mathrm{mo}}(\boldsymbol{r}) = \rho_{\mathrm{mo},1}(\boldsymbol{r}) + \rho_{\mathrm{mo},2}(\boldsymbol{r}). \tag{5.103}$$

Let $\boldsymbol{r} - \boldsymbol{A} \equiv \boldsymbol{r}_A$, then these two terms can be expressed as:

$$\rho_{\mathrm{mo},n}(\boldsymbol{r}) = \sum_{\mathrm{so}}\sum_{\mathrm{so}'}\sum_{\mathrm{bo}}\sum_{\mathrm{bo}'} a_{\mathrm{mo,so}}^{*} a_{\mathrm{mo,so'}} b_{\mathrm{so,bo}}^{*} b_{\mathrm{so',bo'}} \varphi_{\mathrm{bo}(A)}^{*}$$
$$\times (\boldsymbol{r}_A)\varphi_{\mathrm{bo}'(A')}\left(\boldsymbol{r}_{A'}\right), \tag{5.104}$$

which will be $\rho_{\mathrm{mo},\,1}(\boldsymbol{r})$ when $A = A'$ and will be $\rho_{\mathrm{mo},\,2}(\boldsymbol{r})$ when $A \neq A'$.

5.6.2.2 Structure Factors for Monocentric and Bicentric Terms

The sth crystal symmetry operation $\widehat{S}_s(R_s, \boldsymbol{t}_s)$ can be defined as the linear combination of the rotation matrix R_s and the translation vector $\boldsymbol{t}_s$, according to:

$$\widehat{S}_s\boldsymbol{r} = R_s\boldsymbol{r} + \boldsymbol{t}_s \quad (s = 1,2,\dots,N_s). \tag{5.105}$$

The structure factor can be calculated from the atoms in the asymmetric unit. Because the asymmetric unit and molecule may not be consistent, it is first necessary to use the molecular symmetry operation $\widehat{S}_m(R_m, \boldsymbol{t}_m)$ on the asymmetric unit to construct the molecule and then calculate the molecular orbital. The molecule can then be manipulated by crystal symmetry to obtain the entire unit cell.

The structure factor $F(\boldsymbol{k})$ is a function of the scattering vector k, which can be obtained by the Fourier transform of the electron density $\rho(\boldsymbol{r})$ in the unit cell:

$$F(\boldsymbol{k}) = \sum_{s=1}^{N_s}\sum_{\text{mo}} g_{\text{mo}} p_{\text{mo}} \int \rho^s_{\text{mo},1}(\widehat{S}_s\boldsymbol{r}) + \rho^s_{\text{mo},2}(\widehat{S}_s\boldsymbol{r}) \times \exp[i\boldsymbol{k}\cdot(\widehat{S}_s\boldsymbol{r})]\mathrm{d}(\widehat{S}_s\boldsymbol{r}). \tag{5.106}$$

To calculate the double-center electron density integral, the basis function $\varphi^s_{\text{bo}(A)}$ of the A atom after the sth symmetry operation in Eq. (5.98) can be expressed as a linear combination of primitive Gaussian-type orbitals (GTO) $\chi_{\text{bo}(A),\,\text{go}}(\boldsymbol{r}_A)$:

$$\varphi^s_{\text{bo}(A)}\left(\boldsymbol{r}^s_A\right) = \sum_{\text{go}=1}^{N_g} c_{\text{bo,go}}\chi^s_{\text{bo}(A),\text{go}}\left(\boldsymbol{r}^s_A\right)$$

$$\chi^s_{\text{bo}(A),\text{go}}\left(\boldsymbol{r}^s_A\right) = R_{\text{go}}\left(x^s_A\right)^{l_{\text{go}}}\left(y^s_A\right)^{m_{\text{go}}}\left(z^s_A\right)^{n_{\text{go}}}\exp\left[-\alpha_{\text{bo,go}}\left(\boldsymbol{r}^s_A\right)^2\right], \tag{5.107}$$

where N_g is the number of original GTOs required to construct the basis function, R_{go} is the normalization factor for the original GTO, $c_{\text{bo, go}}$ and $\alpha_{\text{bo, go}}$ are constants, and $c_{\text{bo, go}}$ is introduced here to normalize each basis function. In addition, l_{go}, m_{go} and n_{go} are integers, and when $l_{\text{go}} + m_{\text{go}} + n_{\text{go}}$ is 0, 1, and 2, this will correspond to s, p, and d type GTO, respectively. $\boldsymbol{r}^s_A$ in Eq. (5.107) is the coordinate of the A atom after the sth symmetry operation:

$$\boldsymbol{r}^s_A \equiv \begin{pmatrix} x^s_A \\ y^s_A \\ z^s_A \end{pmatrix} = \widehat{S}_s\boldsymbol{r} - \widehat{S}_s\boldsymbol{A} = R_s(\boldsymbol{r} - \boldsymbol{A}) \equiv R_s\boldsymbol{r}_A, \tag{5.108}$$

where $\boldsymbol{r}^s_A$ and $\boldsymbol{k}^t = (k_x, k_v, k_z)$ are defined on the Cartesian coordinate system of the quantized axis.

The structure factor $F(\boldsymbol{k})$ can be divided into two parts: the monocentric structure factor $F^s_{\text{mo},1}$ and the bicentric structure factor $F^s_{\text{mo},2}$. A strict expression for $F(\boldsymbol{k})$ can be obtained by treating the two parts separately:

$$F(\boldsymbol{k}) = \sum_{s=1}^{N_s}\sum_{\text{mo}=1}^{M} g_{\text{mo}} p_{\text{mo}}\left[F^s_{\text{mo},1}(\boldsymbol{k}) + F^s_{\text{mo},2}(\boldsymbol{k})\right] \equiv F_1(\boldsymbol{k}) + F_2(\boldsymbol{k}), \tag{5.109}$$

where

$$F^s_{\text{mo},n} = \sum_{\text{so}}\sum_{\text{so}'} a^*_{\text{mo,so}} a_{\text{mo}',\text{so}'} \sum_{\text{bo}(A)}\sum_{\text{bo}'(A')} b^*_{\text{so,bo}(A)} b_{\text{so}',\text{bo}'(A')} \times \sum_{\text{go}}\sum_{\text{go}} c^*_{\text{bo}(A),\text{go}} c_{\text{bo}'(A'),\text{go}'} f^s_{\text{bo}(A),\text{go},\text{bo}'(A'),\text{go}',n}(\boldsymbol{k}),$$

$$n = 1,2. \tag{5.110}$$

$$f^s_{\text{bo}(A),\text{go},\text{bo}'(A'),\text{go}',n}(\boldsymbol{k}) = \int \chi_{\text{bo,go}}\left(\boldsymbol{r}^s_A\right)\exp(i\boldsymbol{k}\cdot\widehat{S}_s\boldsymbol{r})\chi_{\text{bo}',\text{go}'}\left(\boldsymbol{r}^s_{A'}\right)\mathrm{d}(\widehat{S}_s\boldsymbol{r}),$$

$$n = 1,2. \tag{5.111}$$

$f^s_{\mathrm{bo}(A),\mathrm{go},\mathrm{bo}'(A'),\mathrm{go}',n}(\boldsymbol{k})$ can be abbreviated as $f^s_n(\boldsymbol{k})$. To calculate $f^s_n(\boldsymbol{k})$, we can define a point as $\boldsymbol{P}$ on the bond formed by atoms $\boldsymbol{A}$ and $\boldsymbol{A}'$:

$$\boldsymbol{P} = \frac{\alpha_{\mathrm{bo}(A),\mathrm{go}}}{\gamma}\boldsymbol{A} + \frac{\alpha_{\mathrm{bo}'(A'),\mathrm{go}'}}{\gamma}\boldsymbol{A}'$$
$$\gamma = \alpha_{\mathrm{bo}(A),\mathrm{go}} + \alpha_{\mathrm{bo}'(A'),\mathrm{go}'}. \tag{5.112}$$

When $A = A'$, $\boldsymbol{P} = \boldsymbol{A}$ and since $\boldsymbol{r} = \boldsymbol{r}_P + \boldsymbol{P}$,

$$\widehat{S}_s\boldsymbol{r} = \boldsymbol{R}_s(\boldsymbol{r}_P + \boldsymbol{P}) + \boldsymbol{t}_s = \boldsymbol{R}_s\boldsymbol{P} + \boldsymbol{t}_s + \boldsymbol{R}_s\boldsymbol{r}_P. \tag{5.113}$$

According to Eqs. (5.111), (5.112) and (5.113) we can obtain:

$$f^s_n(\boldsymbol{k}) = \exp[i\boldsymbol{k}\cdot(\boldsymbol{R}_s\boldsymbol{P} + \boldsymbol{t}_s)]\int \chi_{\mathrm{bo},\mathrm{go}}(\boldsymbol{R}_s\boldsymbol{r}_A)$$
$$\times \exp(i\boldsymbol{k}\cdot\boldsymbol{R}_s\boldsymbol{r}_P)\chi_{\mathrm{bo}',\mathrm{go}'}\left(\boldsymbol{R}_s\boldsymbol{r}_{A'}\right)\mathrm{d}(\boldsymbol{R}_s\boldsymbol{r}_P). \tag{5.114}$$

According to the symmetry rule, Eq. (5.114) can be expressed as:

$$f^s_n(\boldsymbol{k}) = \exp\left[i\left(R_s^{-1}\boldsymbol{k}\cdot\boldsymbol{P} + \boldsymbol{k}\cdot\boldsymbol{t}_s\right)\right]\int \chi_{\mathrm{bo},\mathrm{go}}(\boldsymbol{r}_A)$$
$$\times \exp\left[i\left(R_s^{-1}\boldsymbol{k}\cdot\boldsymbol{r}_P\right)\right]\chi_{\mathrm{bo}',\mathrm{go}'}\left(\boldsymbol{r}_{A'}\right)\mathrm{d}\boldsymbol{r}_P, \tag{5.115}$$

where $\boldsymbol{R}_s^{-1}$ is the inverse matrix of $\boldsymbol{R}_s$. Because the GTO in the integrand can be defined on the quantized axis coordinate system, the reciprocal lattice vector $\boldsymbol{R}_s^{-1}\boldsymbol{k} \equiv \boldsymbol{k}_s$ can be expressed in the form of a quantized axis coordinate:

$$R_s^{-1}\boldsymbol{k} \equiv \boldsymbol{k}^{\mathrm{q}}_s = k^{\mathrm{q}}_{s,x}\boldsymbol{I}^{\mathrm{q}} + k^{\mathrm{q}}_{s,y}J^{\mathrm{q}} + k^{\mathrm{q}}_{s,z}\boldsymbol{K}^{\mathrm{q}}, \tag{5.116}$$

where $(\boldsymbol{I}^{\mathrm{q}}, \boldsymbol{J}^{\mathrm{q}}, \boldsymbol{K}^{\mathrm{q}})$ is the unit vector along the quantization axis, and the superscript q represents the vector defined on the quantized axis as $\left(\boldsymbol{r}^{\mathrm{q}}_P\right)^t = \left(x^{\mathrm{q}}_P, y^{\mathrm{q}}_P, z^{\mathrm{q}}_P\right)$. The integrand in Eq. (5.115) can be reduced to a $\boldsymbol{r}_P$ function:

$$f^s_n(\boldsymbol{k}) = R_{\mathrm{go}}R_{\mathrm{go}'}\exp[i(\boldsymbol{k}_s\cdot\boldsymbol{P} + \boldsymbol{k}\cdot\boldsymbol{t}_s)]$$
$$\times \exp\left[-\alpha_{\mathrm{bo},\mathrm{go}}\alpha_{\mathrm{bo}',\mathrm{go}'}(\boldsymbol{A}'^{\mathrm{q}} - \boldsymbol{A}^{\mathrm{q}})^2/\gamma\right] I_x I_y I_z, \tag{5.117}$$

$$I_x = \int_{-\infty}^{\infty}\sum_{l=0}^{l_{\mathrm{go}}+l_{\mathrm{go}'}} f_l\left[l_{\mathrm{go}}, l_{\mathrm{go}'}, (\vec{AP})^{\mathrm{q}}_x, (\vec{A'P})^{\mathrm{q}}_x\right]\left(x^{\mathrm{q}}_P\right)^l$$
$$\times \exp\left[-\gamma\left(x^{\mathrm{q}}_P\right)^2 + ik^{\mathrm{q}}_{s,x}x^{\mathrm{q}}_P\right]\mathrm{d}x^{\mathrm{q}}_P, \tag{5.118}$$

where $(\vec{AP})^{\mathrm{q}}_x$ and $(\vec{A'P})^{\mathrm{q}}_x$ are the x-components of the $\boldsymbol{P}^{\mathrm{q}} - \boldsymbol{A}^{\mathrm{q}}$ and $\boldsymbol{P}^{\mathrm{q}} - \boldsymbol{A}'^{\mathrm{q}}$ vectors, respectively, which can be defined on the quantized axis, and $\boldsymbol{P}$ and $\boldsymbol{k}$ in Eq. (5.117) can be defined on the lattice and reciprocal lattice, respectively. Furthermore, f_l can be defined as:

$$\left(x^{\mathrm{q}}_A\right)^{l_{\mathrm{go}}}\left(x^{\mathrm{q}}_{A'}\right)^{l_{\mathrm{go}'}} = \left\{x^{\mathrm{q}}_P + (\vec{AP})^{\mathrm{q}}_x\right\}^{l_{\mathrm{go}}}\left\{x^{\mathrm{q}}_P + (\vec{A'P})^{\mathrm{q}}_x\right\}^{l_{\mathrm{go}'}}$$
$$\equiv \sum_{l=0}^{l_{\mathrm{go}}+l_{\mathrm{go}'}} f_l\left[l_{\mathrm{go}}, l_{\mathrm{go}'}, (\vec{AP})^{\mathrm{q}}_x, (\vec{A'P})^{\mathrm{q}}_x\right]\left(x^{\mathrm{q}}_P\right)^l, \tag{5.119}$$

and I_y, I_z, f_m and f_n can be defined in a similar way.

I_x can be expressed as:

$$I_x = \left(\frac{\pi}{\gamma}\right)^{1/2} \exp\left[\frac{-\left(k_{s,x}^{q}\right)^2}{4\gamma}\right]$$
$$\times \sum_{l=0}^{l_{go}+l_{go'}} \left[\frac{i}{2(\gamma)^{1/2}}\right]^l f_l\left(x_P^q\right)^l H_l\left[\frac{k_{s,x}^q}{2(\gamma)^{1/2}}\right], \tag{5.120}$$

$$H_n(z) = n! \sum_{i=0}^{[n/2]} \frac{(-1)^i (2z)^{n-2i}}{i!(n-2i)!}, \tag{5.121}$$

where $[x]$ is the floor function, and $l_{go} + l_{go'}$, $m_{go} + m_{go'}$ and $n_{go} + n_{go'}$ are set to l_1, m_1 and n_1, respectively. In the case of $\boldsymbol{A} = \boldsymbol{A}'$, when l, m and n are equal to l_1, m_1 and n_1, respectively, then f_l, f_m and f_n will be equal to 1, respectively, and otherwise equal to 0. The bicentric scattering factor can finally be expressed as:

$$f_2^s(\boldsymbol{k}) = R_{go}R_{go'} \exp[i(\boldsymbol{k}_s \cdot \boldsymbol{P} + \boldsymbol{k} \cdot \mathbf{t}_s)]$$
$$\times \exp\left[\frac{-\alpha_{bo,go}\alpha_{bo',go'}}{\gamma}(\boldsymbol{A}'^q - \boldsymbol{A}^q)^2\right]\left(\frac{\pi}{\gamma}\right)^{3/2}$$
$$\times \exp\left[-\frac{\left(k_s^q\right)^2}{4\gamma}\right] \sum_{l=0}^{l_1}\sum_{m=0}^{m_1}\sum_{n=0}^{n_1}\left[\frac{i}{2(\gamma)^{1/2}}\right]^{l+m+n}$$
$$\times f_l f_m f_n H_l\left[\frac{k_{s,x}^q}{2(\gamma)^{1/2}}\right] H_m\left[\frac{k_{s,y}^q}{2(\gamma)^{1/2}}\right] H_n\left[\frac{k_{s,z}^q}{2(\gamma)^{1/2}}\right]$$
$$\equiv \exp[i(\boldsymbol{k}_s \cdot \boldsymbol{P} + \boldsymbol{k} \cdot \mathbf{t}_s)] g_2^s(\boldsymbol{k}). \tag{5.122}$$

The monocentric scatter factor can be written in a similar form:

$$f_1^s(\boldsymbol{k}) = R_{go}R_{go'} \exp[i(\boldsymbol{k}_s \cdot \boldsymbol{A} + \boldsymbol{k} \cdot \boldsymbol{t}_s)]\left(\frac{\pi}{\gamma}\right)^{3/2}$$
$$\times \exp\left[-\frac{\left(k_s^q\right)^2}{4\gamma}\right]\left[\frac{i}{2(\gamma)^{1/2}}\right]^L H_{l_1}\left[\frac{k_{s,x}^q}{2(\gamma)^{1/2}}\right] H_{m_1}\left[\frac{k_{s,y}^q}{2(\gamma)^{1/2}}\right]$$
$$\times H_{n_1}\left[\frac{k_{s,z}^q}{2(\gamma)^{1/2}}\right]$$
$$\equiv \exp[i(\boldsymbol{k}_s \cdot \boldsymbol{A} + \boldsymbol{k} \cdot \boldsymbol{t}_s)] g_1^s(\boldsymbol{k}), \tag{5.123}$$

where $L = l_1 + m_1 + n_1$. When $A = A'$, $f_2^s(\boldsymbol{k})$ can be simplified as $f_1^s(\boldsymbol{k})$.

5.6.2.3 Processing of Temperature Factors

Due to atomic vibrations, the vector P in Eq. (5.112) will also vibrate. Assuming that the vibrations of atoms A and A' are independent, then when their relative equilibrium positions $\boldsymbol{A}_0$ and $\boldsymbol{A}_0'$ deviate from $\boldsymbol{u}_A$ and $\boldsymbol{u}_{A'}$, respectively, the time average of the phase factor $\langle \exp(i\boldsymbol{k}_s \cdot \boldsymbol{P})\rangle$ in Eq. (5.122) can be expressed by:

$$\langle \exp(i\boldsymbol{k}_s \cdot \boldsymbol{P})\rangle = \exp(i\boldsymbol{k}_s \cdot \boldsymbol{P}_0)$$
$$\times \left\langle \exp\left[i\boldsymbol{k}_s \cdot \left(\alpha_{bo,go}\boldsymbol{u}_A + \alpha_{bo',go'}\boldsymbol{u}_{A'}\right)/\gamma\right]\right\rangle, \tag{5.124}$$

where $\boldsymbol{P}_0$ is the value obtained by replacing $\boldsymbol{A}$ and $\boldsymbol{A}'$ in Eq. (5.112) with $\boldsymbol{A}_0$ and $\boldsymbol{A}'_0$, respectively, and $\langle\rangle$ indicates that $\boldsymbol{u}_A$ and $\boldsymbol{u}_{A'}$ are averaged. The quantities on the right-hand side of $\langle\rangle$ can be defined as the temperature factors of the bicentric scattering factors $T_{A,A',\mathrm{go},\mathrm{go}'}(\boldsymbol{k})$:

$$T_{A,A',\mathrm{go},\mathrm{go}'}(\boldsymbol{k}) = [\boldsymbol{T}_A(\boldsymbol{k})]^{\frac{\alpha_{\mathrm{bo,go}}}{\gamma}} \left[T_{A'}(\boldsymbol{k})\right]^{\frac{\alpha_{\mathrm{bo',go'}}}{\gamma}}, \tag{5.125}$$

where $T_A(\boldsymbol{k}) = \langle\exp(i\boldsymbol{k}_s \cdot \boldsymbol{u}_A)\rangle$ and $T_{A'}(\boldsymbol{k}) = \left\langle \exp\left(i\boldsymbol{k}_s \cdot \boldsymbol{u}_{A'}\right)\right\rangle$ are the temperature factors of A and A', respectively.

When considering anharmonic vibrations, the monocentric temperature factor $T_A(\boldsymbol{k})$ can be divided into two parts, namely, the centric and acentric parts, which can be denoted by superscripts c and a, respectively:

$$\begin{aligned} T_A(\boldsymbol{k}) &= T_H(\boldsymbol{k}) \left[T_A^{U,\mathrm{c}}(\boldsymbol{k}) + iT_A^{U,\mathrm{a}}(\boldsymbol{k})\right] = |T_A(\boldsymbol{k})| \exp\{i[\delta_A(\boldsymbol{k})]\} \\ &= T_A^{\mathrm{c}}(\boldsymbol{k}) + iT_A^{\mathrm{a}}(\boldsymbol{k}), \end{aligned} \tag{5.126}$$

where $T_A^{U,\mathrm{c}}(\boldsymbol{k})$ and $T_A^{U,\mathrm{a}}(\boldsymbol{k})$ are the centric and acentric terms of the anharmonic temperature factor of the A atom, respectively, and $T_H(\boldsymbol{k})$ is the harmonic part of the temperature factor, yielding:

$$\begin{aligned} |T_A(\boldsymbol{k})| &= T_H(\boldsymbol{k})\left\{\left[T_A^{U,\mathrm{c}}(\boldsymbol{k})\right]^2 + \left[T_A^{U,\mathrm{a}}(\boldsymbol{k})\right]^2\right\}^{1/2} \\ \delta_A &= \tan^{-1}\left[\frac{T_A^{U,\mathrm{a}}(\boldsymbol{k})}{T_A^{U,\mathrm{c}}(\boldsymbol{k})}\right], \end{aligned} \tag{5.127}$$

According to Eqs. (5.125), (5.126), and (5.127), the bicentric temperature factor can also be divided into two parts, namely, the centric and acentric parts:

$$\begin{aligned} T_{A,A',\mathrm{go},\mathrm{go}'}(\boldsymbol{k}) &= |T_A(\boldsymbol{k})|^{\frac{\alpha_{\mathrm{bo,go}}}{\gamma}} |T_{A'}(\boldsymbol{k})|^{\frac{\alpha_{\mathrm{bo',go'}}}{\gamma}} \\ &\quad \times \exp\left[\frac{i\left(\alpha_{\mathrm{bo,go}}\delta_A + \alpha_{\mathrm{bo',go'}}\delta_{A'}\right)}{\gamma}\right] \\ &\equiv T^{\mathrm{c}}_{A,A',\mathrm{go},\mathrm{go}'} + iT^{\mathrm{a}}_{A,A',\mathrm{go},\mathrm{go}'}(\boldsymbol{k}). \end{aligned} \tag{5.128}$$

We can divide the structure factor $F(\boldsymbol{k})$ in Eq. (5.109) multiplied by the temperature factor into two parts, real and imaginary:

$$F(\boldsymbol{k}) = F_1(\boldsymbol{k}) + F_2(\boldsymbol{k}) \equiv A_1(\boldsymbol{k}) + A_2(\boldsymbol{k}) + i[B_1(\boldsymbol{k}) + B_2(\boldsymbol{k})], \tag{5.129}$$

where subscripts 1 and 2 denote monocentric and bicentric terms, respectively.

The real and imaginary parts of the phase factor of the monocentric scattering factor in Eq. (5.123) are:

$$\begin{aligned} A^s_{\mathrm{bo}(A)}(\boldsymbol{k}) &= \cos(\boldsymbol{k}_s \cdot \boldsymbol{A}_0 + \boldsymbol{k} \cdot \boldsymbol{t}) \\ B^s_{\mathrm{bo}(A)}(\boldsymbol{k}) &= \sin(\boldsymbol{k}_s \cdot \boldsymbol{A}_0 + \boldsymbol{k} \cdot \boldsymbol{t}). \end{aligned} \tag{5.130}$$

The real and imaginary parts of $f_1^s(\boldsymbol{k})$ can be expressed as $A^s_{\mathrm{bo}(A)}(\boldsymbol{k})$ and $B^s_{\mathrm{bo}(A)}(\boldsymbol{k})$ separately multiplied by $g_1^s(\boldsymbol{k})$ from Eq. (5.123):

$$A^{s,f}_{\mathrm{bo}(A)}(\boldsymbol{k}) = A^{s}_{\mathrm{bo}(A)}(\boldsymbol{k})g_1^{s,\mathrm{c}}(\boldsymbol{k}) - B^{s}_{\mathrm{bo}(A)}(\boldsymbol{k})g_1^{s,\mathrm{a}}(\boldsymbol{k})$$
$$B^{s,f}_{\mathrm{bo}(A)}(\boldsymbol{k}) = A^{s}_{\mathrm{bo}(A)}(\boldsymbol{k})g_1^{s,\mathrm{c}}(\boldsymbol{k}) + B^{s}_{\mathrm{bo}(A)}(\boldsymbol{k})g_1^{s,\mathrm{a}}(\boldsymbol{k}). \tag{5.131}$$

We can multiply these terms by the temperature factor to yield:

$$A_1^{s,ft}(\boldsymbol{k}) = A^{s,f}_{\mathrm{bo}(A)}(\boldsymbol{k})T_A^{\mathrm{c}}(\boldsymbol{k}) - B^{s,f}_{\mathrm{bo}(A)}(\boldsymbol{k})T_A^{\mathrm{a}}(\boldsymbol{k})$$
$$B_1^{s,ft}(\boldsymbol{k}) = A^{s,f}_{\mathrm{bo}(A)}(\boldsymbol{k})T_A^{\mathrm{c}}(\boldsymbol{k}) + B^{s,f}_{\mathrm{bo}(A)}(\boldsymbol{k})T_A^{\mathrm{a}}(\boldsymbol{k}), \tag{5.132}$$

where the subscript "1" of $A_1^{s,ft}(\boldsymbol{k})$, $B_1^{s,ft}(\boldsymbol{k})$, $g_1^{c}(\boldsymbol{k})$ and $g_1^{a}(\boldsymbol{k})$ presents bo(A), go, bo$'$(A)go$'$, 1.

Similarly, the phase factor of the bicentric scattering factor of Eq. (5.122) can be expressed as:

$$A^{s}_{\mathrm{go,go'}}(\boldsymbol{k}) = \cos(\boldsymbol{k}_s \cdot \boldsymbol{P}_0 + \boldsymbol{k} \cdot \boldsymbol{t})$$
$$B^{s}_{\mathrm{go,go'}}(\boldsymbol{k}) = \sin(\boldsymbol{k}_s \cdot \boldsymbol{P}_0 + \boldsymbol{k} \cdot \boldsymbol{t}). \tag{5.133}$$

According to Eq. (5.122), we can obtain:

$$A^{s,f}_{\mathrm{go,go'}}(\boldsymbol{k}) = A^{s}_{\mathrm{go,go'}}(\boldsymbol{k})g_2^{s,\mathrm{c}}(\boldsymbol{k}) - B^{s}_{\mathrm{go,go'}}(\boldsymbol{k})g_2^{s,\mathrm{a}}(\boldsymbol{k})$$
$$B^{s,f}_{\mathrm{go,go'}}(\boldsymbol{k}) = A^{s}_{\mathrm{go,go'}}(\boldsymbol{k})g_2^{s,\mathrm{a}}(\boldsymbol{k}) + B^{s}_{\mathrm{go,go'}}(\boldsymbol{k})g_2^{s,\mathrm{c}}(\boldsymbol{k}). \tag{5.134}$$

According to Eq. (5.128), we can obtain:

$$A_2^{s,ft}(\boldsymbol{k}) = A^{s,f}_{\mathrm{go,go'}}(\boldsymbol{k})T^{\mathrm{c}}_{AA',\mathrm{go,go'}} - B^{s,f}_{\mathrm{go,go'}}(\boldsymbol{k})T^{\mathrm{a}}_{AA',\mathrm{go,go'}}$$
$$B_2^{s,ft}(\boldsymbol{k}) = A^{s,f}_{\mathrm{go,go'}}(\boldsymbol{k})T^{\mathrm{a}}_{AA',\mathrm{go,go'}} + B^{s,f}_{\mathrm{go,go'}}(\boldsymbol{k})T^{\mathrm{c}}_{AA',\mathrm{go,go'}}. \tag{5.135}$$

When $A \neq A'$, the subscript 2 of $A_2^{s,ft}(\boldsymbol{k})$, $B_2^{s,ft}(\boldsymbol{k})$, $g_2^{c}(\boldsymbol{k})$ and $g_2^{a}(\boldsymbol{k})$ denotes *bo*(A), go, bo$'$(A'), go$'$, 2. Therefore, the final form of $F_n(\boldsymbol{k})$ in Eq. (5.109) will be:

$$F_n(\boldsymbol{k}) = \sum_s \sum_{\mathrm{mo}} g_{\mathrm{mo}} p_{\mathrm{mo}} \sum_{\mathrm{so}} \sum_{\mathrm{so'}} a^*_{\mathrm{mo,so}} a_{\mathrm{mo',so'}}$$
$$\times \sum_{\mathrm{bo}(A)} \sum_{\mathrm{bo'}(A')} b^*_{\mathrm{so,ao}(A)} b_{\mathrm{so',ao'}(A')} \sum_{\mathrm{go}} \sum_{\mathrm{go'}} c^*_{\mathrm{bo}(A),\mathrm{go}} c_{\mathrm{bo'}(A'),\mathrm{go'}}$$
$$\times \left(A_n^{s,ft} + iB_n^{s,ft}\right), \tag{5.136}$$

where $n = 1, 2$ correspond to the monocentric and bicentric structure factors, respectively.

5.6.3 Molecular Orbitals with Variable Occupation Numbers Model (MOON)

The main idea of the molecular orbitals with variable occupation numbers model (MOON) [4, 5] is to maintain the molecular orbital form obtained by the variational (Hartree–Fock or DFT) method and refine the occupied number $\{n_j\}$ of the occupied and unoccupied molecular orbitals $\{\phi_j(\boldsymbol{r})\}$. The number of occupations must satisfy the constraints $0 < n_j < n_{\max}$ and $\sum n_j = F(000)/Z$. In this method, molecular orbitals can be expanded on the basis of compressed Gaussian atomic orbitals. Under this framework, the electron density $\rho(\boldsymbol{r})$ can be expressed as:

$$\rho(\boldsymbol{r}) = \sum_{J=1}^{K} n_j \phi_j(\boldsymbol{r})^* \phi_j(\boldsymbol{r}). \tag{5.137}$$

The structural factor of this model can be expressed as:

$$F_c(\boldsymbol{s}) = \sum_{j=1}^{K} n_j \sum_{l=1}^{K} \sum_{m=1}^{K} c_{jl} c_{jm} \langle \psi_l | e^{2\pi i \boldsymbol{s} \cdot \boldsymbol{r}} | \psi_m \rangle \exp(2\pi i \boldsymbol{s} \cdot \boldsymbol{r}_{lm}) T_{lm}(\boldsymbol{s}), \tag{5.138}$$

where c_{jl} is the fixed molecular orbital expansion coefficient, $\boldsymbol{r}_{lm}$ is the position vector of the product of the lmth basis function, and T_{lm} is the anisotropic temperature factor, which can be obtained by averaging the temperature factors of the two atoms where the basis function is located. In addition, the Fourier transform product $\langle \psi_l | e^{2\pi i s \cdot \boldsymbol{r}} | \psi_m \rangle$ of the basis function can be calculated using Stewart's algorithm [15]. These integrals can be calculated in the initial step and will remain unchanged in subsequent refinements; thus, they will need to be recalculated if the atomic coordinates change.

Two refinement methods exist for MOON. (1) Refinement using the constraint $n_j \leq 2$ can obtain the electron transition information from the occupied orbital to the virtual orbital, and the electron density will be similar to the full configuration correlation (full-CI) – single particle density obtained by the variational method. (2) No restriction is imposed on the occupancy number, i.e. $n_{\max} \gg 2$, and although this method lacks the rigor of quantum mechanics, it has good flexibility and good fit with the experimental data.

Equation (5.138) does not contain strictly anomalous (inelastic) scattering contributions, and anomalous scattering mainly originates from core electron orbitals. There are several ways to deal with this problem, namely, (1) for compounds containing only light atoms, we can ignore these contributions, (2) we can correct the $\langle \psi_l | e^{2\pi i s \cdot \boldsymbol{r}} | \psi_m \rangle$ integral of the core orbital using the correction factor of the atomic scattering factor (including phase and amplitude), and (3) we can use the diffractions from which the anomalous scattering contribution has been eliminated to refine.

References

1 Tanaka, K. (1988). X-ray analysis of wavefunctions by the least-squares method incorporating orthonormality. i. general formalism. *Acta Cryst. A* 44: 1002–1008.

2 Tanaka, K., Makita, R., Funahashi, S. et al. (2008). X-ray atomic orbital analysis. I. Quantum mechanical and crystallographic framework of the method. *Acta Cryst. A* 64: 437–449.

3 Tanaka, K. (2018). X-ray molecular orbital analysis. I. Quantum mechanical and crystallographic framework. *Acta Cryst. A* 74: 345–356.

4 Hibbs, D.E., Howard, S.T., Huke, J.P., and Waller, M.P. (2005). A new orbital-based model for the analysis of experimental molecular charge densities: an application to (Z)-n-methyl-c-phenylnitrone. *Phys. Chem. Chem. Phys.* 7: 1772–1778.

5 Waller, M.P., Howard, S.T., Platts, J.A. et al. (2006). Novel properties from experimental charge densities: an application to the zwitterionic neurotransmitter taurine. *Chem. Eur. J.* 12: 7603–7614.

6 Clementi, E. and Raimondi, D.L. (1963). Atomic screening constants from scf functions. *J. Chem. Phys.* 38: 2686–2689.

7 Stewart, R.F. (1976). Electron population analysis with rigid pseudoatoms. *Acta Cryst. A* 32: 565–574.

8 Fischer, A., Tiana, D., Scherer, W. et al. (2011). Experimental and theoretical charge density studies at subatomic resolution. *J. Phys. Chem. A* 115: 13061–13071.

9 Brown, P.J. (1992). Magnetic form factors. In: *International Tables for Crystallography*, vol. C (ed. A.J.C. Wilson), 391–399. Dordrecht: Kluwer.

10 Clementi, E. and Roetti, C. (1974, Tab 14). Roothaan-Hartree-Fock atomic wavefunctions. Basis functions and their coefficients for ground and certain excited states of neutral and ionized atoms, $Z \leq 54$. *Atom Data Nucl. Data* 177.

11 Macchi, P., Gillet, J.-M., Taulelle, F. et al. (2015). Modelling the experimental electron density: only the synergy of various approaches can tackle the new challenges. *IUCrJ* 2: 441–451.

12 Mann, J.B. (1968). *Report LA3691*. New Mexico, USA: Los Alamos National Laboratory.

13 Liberman, D.A., Cromer, D.T., and Waber, J.T. (1971). Relativistic self-consistent field program for atoms and ions. *Comput. Phys. Commun.* 2: 107–113.

14 Condon, E.U. and Shortley, G.H. (1967). *Theory of Atomic Spectra*. Cambridge University Press.

15 Stewart, R.F. (1969). Generalized X-ray scattering factors. *J. Chem. Phys.* 51: 4569–4577.

6

Density Matrix Model

6.1 Introduction

A system can be described by the N-electron wavefunction Ψ. The N-electron density matrix, which can be obtained by the product of the wavefunction and its conjugate wavefunction Ψ^*, will be equivalent to the N-electron wavefunction in a pure state. The two-electron reduced-density matrix (2-RDM) can be obtained by integrating the variables of all other electrons except for two electrons in the N-electron density matrix. Similarly, the one-electron 1-RDM can be obtained by integrating the variables of all other electrons except for one electron. Because electrons interact with each other through Coulomb exclusion, the energies of atoms and molecules, as well as other electron structural properties, can be directly calculated based on 2-RDM. The 2-RDM contains all critical physical and chemical information of the multi-electron wavefunction, which implies that for studying a given molecular system, it suffices to only obtain the 2-RDM without the need to obtain the multi-electron wavefunction. In fact, if an electron structure is calculated by 2-RDM, the 2-RDM must also be calculated based on the wavefunction, meaning the 2-RDM must satisfy the so-called N-representability conditions [1, 2].

In the field of electron structure crystallography, and with the assistance of experimental data, we could simplify the theoretical model, namely the 1-RDM within the HF framework. In most cases, 1-RDM will be good enough to describe the experimental results. Similarly, 1-RDM must also satisfy the N-representability conditions.

According to the RDM functional theory, the expectation of any observed variables of the ground state of the system will be the function of the 1-RDM of the ground state. Therefore, 1-RDM is also a basic physical variable for the theoretical and experimental electron structure studies of materials.

This chapter mainly introduces the concept and properties of the 1-RDM and its connection with experimental observations. Among the experimentally observed variables, the electron density and momentum density are the most important, and together they determine the 1-RDM. This chapter also introduces the reconstruction and refinement theory of experimental 1-RDM.

Electronic Structure Crystallography and Functional Motifs of Materials, First Edition.
Guo-Cong Guo and Xiao-Ming Jiang.

6.2 Density Matrix Model

6.2.1 Definition of the Density Matrix

N-electron density matrix in the pure state can be expressed by the N-electron wavefunction:

$$\Gamma^{(N)}\left(\boldsymbol{x}_1 \ldots \boldsymbol{x}_N; \boldsymbol{x}_1' \ldots \boldsymbol{x}_N'\right) = \langle \boldsymbol{x}_1 \ldots \boldsymbol{x}_N | \Psi\rangle \left\langle \Psi | \boldsymbol{x}_1' \ldots \boldsymbol{x}_N' \right\rangle, \tag{6.1}$$

where x_i is the generalized spin position coordinates of electron i. Without considering the spin and assuming $\sigma_i = \sigma_i'$ for electrons, the summation of all spin coordinates will yield:

$$\gamma^{(N)}\left(\boldsymbol{r}_1 \ldots \boldsymbol{r}_N; \boldsymbol{r}_1' \ldots \boldsymbol{r}_N'\right) = \sum_{\sigma_1,\ldots\sigma_N} \Gamma^{(N)}\left(\boldsymbol{x}_1 \ldots \boldsymbol{x}_N; \boldsymbol{x}_1' \ldots \boldsymbol{x}_N'\right). \tag{6.2}$$

We defined the N-electron RDM as:

$$\begin{aligned} &\gamma^{(N)}\left(\boldsymbol{r}_1 \ldots \boldsymbol{r}_n; \boldsymbol{r}_1' \ldots \boldsymbol{r}_n'\right) \\ &= \binom{N}{n} \int \gamma^{(N)}\left(\boldsymbol{r}_1 \ldots \boldsymbol{r}_N; \boldsymbol{r}_1' \ldots \boldsymbol{r}_N'\right) \mathrm{d}\boldsymbol{r}_{n+1} \ldots \mathrm{d}\boldsymbol{r}_N \mathrm{d}\boldsymbol{r}_{n+1}' \ldots \mathrm{d}\boldsymbol{r}_N'. \end{aligned} \tag{6.3}$$

1-RDM and 2-RDM can be related to the electron density $\rho(\boldsymbol{r}_1)$ and pair distribution function $P(\boldsymbol{r}_1, \boldsymbol{r}_2)$:

$$\begin{aligned} \rho(\boldsymbol{r}_1) &= \gamma^{(1)}\left(\boldsymbol{r}_1; \boldsymbol{r}_1' = \boldsymbol{r}_1\right) \\ P(\boldsymbol{r}_1, \boldsymbol{r}_2) &= \gamma^{(2)}\left(\boldsymbol{r}_1, \boldsymbol{r}_2; \boldsymbol{r}_1' = \boldsymbol{r}_1, \boldsymbol{r}_2' = \boldsymbol{r}_2\right) \end{aligned}. \tag{6.4}$$

6.2.2 Localized Model of the Density Matrix

The N-electron wavefunction can be written in single determinant form:

$$\psi(\boldsymbol{r}_1, \boldsymbol{r}_2, \ldots, \boldsymbol{r}_N) = \frac{1}{\sqrt{N!}} \sum_P (-1)^p \hat{P}[\varphi_1(\boldsymbol{r}_1)\varphi_2(\boldsymbol{r}_2) \ldots \varphi_n(\boldsymbol{r}_n)]. \tag{6.5}$$

Each molecular orbital could be written as a linear combination of the atomic orbitals centered on $\boldsymbol{R}_A$:

$$\varphi_n(\boldsymbol{r}_1) = \sum_A \sum_{j \in A} c_{jn}^A \varphi_{j\in A}(\boldsymbol{r}_1 - \boldsymbol{R}_A). \tag{6.6}$$

Thus, the population matrix $\boldsymbol{P}$ related to the atomic basis sets could be defined as:

$$P_{ij}^{AB} = \sum_n c_{in}^A c_{jn}^B. \tag{6.7}$$

It is generally not required to determine the N-electron wavefunction, as only one- or two-electron properties can be experimentally observed. Therefore, the established density matrix model requires a 2-RDM at most, which can be further simplified into a 1-RDM under the HF framework. In most cases, 1-RDM will be sufficient to describe the experimental results. 1-RDM on one-electron atomic orbitals can be expanded as:

$$\gamma^1(\boldsymbol{r}_1';\boldsymbol{r}_2'') = \sum_{A,B}\sum_{\substack{i\in A\\ j\in B}} P_{ij}^{AB}\phi_{i\in A}(\boldsymbol{r}_1' - \boldsymbol{R}_A)\,\phi_{j\in B}(\boldsymbol{r}_1'' - \boldsymbol{R}_B)\,, \tag{6.8}$$

where the first summation includes all atom pairs *(A, B)*, and for a given atom pair, the second summation includes all atomic orbitals of each atom $\varphi_i(\boldsymbol{r}_1 - \boldsymbol{R}_A)$. In addition, the orbitals are real numbers, for convenience, and $\boldsymbol{P}_{ij}^{AB}$ is the population matrix.

For convenience, the 1-RDM can be decomposed into the form of atomic contributions. One feasible decomposition method is given by:

$$\begin{aligned}\gamma^{(1)}(\boldsymbol{r}_1';\boldsymbol{r}_1'') &= \sum_A \gamma_A^{(1)}(\boldsymbol{r}_1';\boldsymbol{r}_1'')\\ &= \sum_A\left[\gamma_{AA}^{(1)}(\boldsymbol{r}_1';\boldsymbol{r}_1'') + \frac{1}{2}\sum_{B\neq A}\left(\gamma_{AB}^{(1)}(\boldsymbol{r}_1';\boldsymbol{r}_1'') + \gamma_{BA}^{(1)}(\boldsymbol{r}_1';\boldsymbol{r}_1'')\right)\right],\end{aligned} \tag{6.9}$$

where $\gamma_A^{(1)}$ can be interpreted as the contribution of the atoms to the overall density matrix and can be constructed by the monocentric term (containing only the product of the orbitals of this atom) and the bicentric term (containing the product of the orbitals of two different atoms):

$$\gamma_{AB}^{(1)}(\boldsymbol{r}_1';\boldsymbol{r}_1'') = \sum_{\substack{i\in A\\ j\in B}} P_{ij}^{AB}\phi_{i\in A}(\boldsymbol{r}_1' - \boldsymbol{R}_A)\,\phi_{j\in B}(\boldsymbol{r}_1'' - \boldsymbol{R}_B). \tag{6.10}$$

It is worth mentioning that for a certain pair of atoms, $\gamma_{AB}^{(1)}$ and $\gamma_{BA}^{(1)}$ should be symmetric to maintain the permutation invariance of the electron coordinates.

In infinite perfect crystal systems, translation invariance must be considered, which can be achieved by considering the so-called inner coordinate $\boldsymbol{s}_1 = \boldsymbol{r}_1' - \boldsymbol{r}_1''$ and outer coordinate $\boldsymbol{r}_1 = (\boldsymbol{r}_1' + \boldsymbol{r}_2'')/2$, according to:

$$\begin{aligned}\gamma_{\mathbf{L},A}^{(1)}(\boldsymbol{r}_1 - \boldsymbol{R}_A - \boldsymbol{L};\boldsymbol{s}_1) = \sum_{\mathbf{L}',B} P_{ij}^{AB}(\boldsymbol{L}')\phi_{i\in A}(\boldsymbol{r}_1 - \boldsymbol{L} - (\boldsymbol{s}_1 + \boldsymbol{L}')/2 - \boldsymbol{R}_A)\\ (\boldsymbol{r}_1 - \boldsymbol{L} + (\boldsymbol{s}_1 + \boldsymbol{L}')/2 - \boldsymbol{R}_B).\end{aligned} \tag{6.11}$$

Thus, 1-DRM can be written as:

$$\gamma^{(1)}(\mathbf{r}_1;\mathbf{s}_1) = \sum_{\mathbf{L}}\sum_{A\in\mathbf{L}} \gamma_{\mathbf{L},A}^{(1)}(\boldsymbol{r}_1 - \boldsymbol{R}_A - \boldsymbol{L};\boldsymbol{s}_1), \tag{6.12}$$

where $\boldsymbol{L}$ represents the lattice translation. This decomposition method emphasized the position of the atomic center of the 1-RDM, and the bicentric contribution was included in the contribution of each atom.

The pseudo-atomic model of Hansen and Coppens can be observed as an extreme case of this representation, in which the contribution of the bicentric term can be ignored. Still, the pseudo-atomic model has been remarkably successful. In addition, it can be directly checked that the bicentric contribution to the structure factor will be negligible, and the bicentric structure factor can be significantly reduced with orbital overlaps.

Monocentric structure factor can be written as:

$$
\begin{aligned}
F(\mathbf{Q}) &= \int \gamma^{(1)}(\boldsymbol{r}_1; 0) e^{i\mathbf{Q}.\boldsymbol{r}_1} \mathrm{d}\boldsymbol{r}_1 \\
&= \sum_A \left[\int \gamma_{AA}^{(1)}(\boldsymbol{r}_1; 0) e^{i\mathbf{Q}.\boldsymbol{r}_1} \mathrm{d}\boldsymbol{r}_1 + \frac{1}{2} \sum_{B \neq A} \right. \\
&\quad \left. \times \left(\int \gamma_{AB}^{(1)}(\boldsymbol{r}_1; 0) e^{i\mathbf{Q}.\boldsymbol{r}_1} \mathrm{d}\boldsymbol{r}_1 + \int \gamma_{BA}^{(1)}(\boldsymbol{r}_1; 0) e^{i\mathbf{Q}.\boldsymbol{r}_1} \mathrm{d}\boldsymbol{r}_1 \right) \right]
\end{aligned}
\tag{6.13}
$$

and bicentric structure factor is:

$$
\begin{aligned}
F_{AB}(Q) &= \int \gamma_{AB}^{(1)}(\mathbf{r}_1; 0) e^{i\mathbf{Q}.\mathbf{r}_1} d\mathbf{r}_1 \\
&= e^{i\mathbf{Q}.\mathbf{R}_{AB}} \sum_{\substack{i \in A \\ j \in B}} P_{ij}^{AB} \int \phi_{i\in A}(\mathbf{r} - \mathbf{S}_{AB}) \phi_{j\in B}(\mathbf{r} + \mathbf{S}_{AB}) e^{i\mathbf{Q}.\mathbf{r}} d\mathbf{r}
\end{aligned}
\tag{6.14}
$$

where $\boldsymbol{R}_{AB} = (\boldsymbol{R}_A + \boldsymbol{R}_B)/2$, $\boldsymbol{S}_{AB} = (\boldsymbol{R}_A - \boldsymbol{R}_B)/2$.

Therefore, to approximate the density matrix based on X-ray diffraction experiments, the bicentric contribution does not have to be considered. Instead, it can be revealed that the bicentric term is essential to describe the momentum space density. We can consider the electron density in the momentum space:

$$
\begin{aligned}
\pi(\boldsymbol{p}) &= \int \gamma^{(1)}(\boldsymbol{r}_1; \boldsymbol{s}_1) e^{i\boldsymbol{p}.\boldsymbol{s}_1} \mathrm{d}\boldsymbol{r}_1 \mathrm{d}\boldsymbol{s}_1 \\
&= \sum_A \left[\int \gamma_{AA}^{(1)}(\boldsymbol{r}_1; \boldsymbol{s}_1) e^{i\boldsymbol{p}.\boldsymbol{s}_1} \mathrm{d}\boldsymbol{r}_1 \mathrm{d}\boldsymbol{s}_1 \right. \\
&\quad \left. + \frac{1}{2} \sum_{B \neq A} \left(\int \gamma_{AB}^{(1)}(\boldsymbol{r}_1; \boldsymbol{s}_1) e^{i\boldsymbol{p}.\boldsymbol{s}_1} \mathrm{d}\boldsymbol{r}_1 \mathrm{d}\boldsymbol{s}_1 + \int \gamma_{BA}^{(1)}(\boldsymbol{r}_1; \boldsymbol{s}_1) e^{i\boldsymbol{p}.\boldsymbol{s}_1} \mathrm{d}\boldsymbol{r}_1 \mathrm{d}\boldsymbol{s}_1 \right) \right]
\end{aligned}
, \tag{6.15}
$$

where the bicentric term can be written as:

$$
\pi_{AB}(\boldsymbol{p}) = \int \gamma_{AB}^{(1)}(\boldsymbol{r}_1; \boldsymbol{s}_1) e^{i\boldsymbol{p}.\boldsymbol{s}_1} \mathrm{d}\boldsymbol{r}_1 \mathrm{d}\boldsymbol{s}_1 = \sum_{\substack{i \in A \\ j \in B}} P_{ij}^{AB} \tilde{\phi}_{i\in A}(\mathbf{p}) \tilde{\phi}_{j\in B}(\mathbf{p}) e^{2i\boldsymbol{p}.\boldsymbol{S}_{AB}}. \tag{6.16}
$$

In the momentum space, the bicentric term will be mainly determined by the product of atomic orbitals. Of course, two atoms with a large distance from each other will produce a rapidly oscillating contribution, which is difficult to be detect using instruments with a momentum resolution of $\sigma(\boldsymbol{p}) > \pi/S_{AB}$.

6.3 Correlation of Density Matrix to Scattering Experiments

6.3.1 Dynamic Scattering Factor

The information provided by X-ray scattering experiments will be contained in the dynamic scattering factor $S(\boldsymbol{k}, \omega)$ which is the spatiotemporal Fourier transform of the correlation function $G(\boldsymbol{r}, t)$:

$$S(\mathbf{k}, \omega) = \frac{1}{2\pi} \iint G(\mathbf{r}, t) e^{-\mathrm{i}\kappa.\mathbf{r}} e^{i\omega t} \mathrm{d}\boldsymbol{r}\mathrm{d}t, \tag{6.17}$$

where $\boldsymbol{k}$ is the scattering vector and ω is the angular frequency related to the energy transferred into the system, and $\hbar\omega = \hbar(\omega_1 - \omega_2)$.

The correlation function $G(\boldsymbol{r}, t)$ can be written as:

$$G(\boldsymbol{r}, t) = \int \langle \hat{\rho}(\boldsymbol{r}', t) \hat{\rho} \left(\boldsymbol{r}' - \boldsymbol{r}, 0\right) \rangle \mathrm{d}\boldsymbol{r}', \tag{6.18}$$

where the symbol $\langle \dots \rangle$ represents the canonical ensemble average of all possible quantum states of the scattering system, and each state is multiplied by the Boltzmann probability factor p_n:

$$\langle \dots \rangle = \sum p_n \langle \Psi_n \dots | \Psi_n \rangle. \tag{6.19}$$

$\hat{\rho}(\mathbf{r}, t)$ is the electron density operator:

$$\hat{\rho}(\boldsymbol{r}, t) = \sum_i \delta(\boldsymbol{r} - \boldsymbol{r}_i(t)). \tag{6.20}$$

The information contained in the observations can be related to the mode of scattering (the form of energy transfer during the interaction of photons with electrons).

6.3.2 Static Structure Factor

Without analyzing the energy, if all photons within a scattering angle are collected, and all energy transfer processes are integrated, the static structure factor can be obtained as:

$$S(\mathbf{k}) = \int S(\mathbf{k}, \omega) \mathrm{d}\omega = \int G(\boldsymbol{r}, t = 0) e^{-\mathrm{i}k.\boldsymbol{r}} \mathrm{d}\mathbf{r}. \tag{6.21}$$

This form of scattering will be useful for probing the electron pair distribution function, because

$$\begin{aligned} G(\boldsymbol{r}, t = 0) &= \int \sum_{i,j} \langle \delta(\boldsymbol{r}' - \boldsymbol{r}_i(0)) \delta(\boldsymbol{r}' - \boldsymbol{r} - \boldsymbol{r}_j(0)) \rangle \mathrm{d}\boldsymbol{r}' \\ &= \int \langle \gamma^{(1)}(\boldsymbol{r}', \boldsymbol{r}' - \boldsymbol{r}) \rangle \delta(\boldsymbol{r}) \mathrm{d}\boldsymbol{r}' + \int \langle P(\boldsymbol{r}', \boldsymbol{r}' - \boldsymbol{r}) \rangle \mathrm{d}\boldsymbol{r}' \end{aligned} . \tag{6.22}$$

6.3.3 Elastic Scattering

Elastic scattering, which occurs when there is no energy transfer between incident photons and electrons, is a process dominated by long time scales:

$$S(\mathbf{k}, \omega) \approx \delta(\omega) \int G(\boldsymbol{r}, t = \infty) e^{-\mathrm{i}\mathbf{k}.\mathbf{r}} \mathrm{d}^3\boldsymbol{r}. \tag{6.23}$$

In this case, the movement correlation of electrons can be disregarded:

$$\begin{aligned} G(\mathbf{r}, t = \infty) &= \int \langle \hat{\rho}(\boldsymbol{r}', t = \infty) \hat{\rho}(\boldsymbol{r}' - \mathbf{r}, 0) \rangle \mathrm{d}\boldsymbol{r}' \\ &= \int \langle \hat{\rho}(\boldsymbol{r}', t = \infty) \rangle \langle \hat{\rho}(\boldsymbol{r}' - \mathbf{r}, 0) \rangle \mathrm{d}\boldsymbol{r}' \end{aligned} . \tag{6.24}$$

Therefore,

$$\begin{aligned} S(\boldsymbol{k}, \omega = 0) &\approx \left| \int \langle \gamma^{(1)}(\boldsymbol{r}, \boldsymbol{r}) \rangle e^{-i\boldsymbol{k}.\boldsymbol{r}} \mathrm{d}\boldsymbol{r} \right|^2 \\ &= \left| \int \langle \hat{\rho}(\boldsymbol{r}) \rangle e^{-i\mathbf{k}.\mathbf{r}} \mathrm{d}\boldsymbol{r} \right|^2 = |F(\boldsymbol{k})|^2 \end{aligned} . \tag{6.25}$$

When the Bragg condition is satisfied (i.e. $\boldsymbol{k}$ is a reciprocal lattice vector ($\boldsymbol{H}$)), X-ray scattering, under elastic scattering, will produce a signal proportional to the square of the modulus of the coherent elastic Bragg structure factor ($F(\boldsymbol{h})$), which can be related to the ensemble average of the electron density distribution in the coordinate space; hence, to the diagonal part of the 1-RDM of the position space. It should be noted that it is not possible to obtain pure-state electron density, as phonons will also contribute to scattering. A major problem in elastic scattering experiments involves obtaining the electron density by the deconvolution of the lattice vibrations and the separation of static disorders.

6.3.4 Inelastic Scattering

As indicated in the previous discussion, most inelastic scattering will originate from the correlation of a short time scale, namely:

$$S(\boldsymbol{k}, \omega) = \frac{1}{2\pi} \iint \langle \hat{\rho}(\boldsymbol{r}', t) \hat{\rho}(\boldsymbol{r}' - \boldsymbol{r}, 0) \rangle e^{-i\boldsymbol{k}.\boldsymbol{r}} e^{i\omega t} \mathrm{d}\boldsymbol{r}' \mathrm{d}\boldsymbol{r} \mathrm{d}t. \tag{6.26}$$

Using the electron density operator in (6.20), we can obtain:

$$\begin{aligned} S(\boldsymbol{k}, \omega) &= \frac{1}{2\pi} \iint \sum_{l,j} \langle \delta(\boldsymbol{r}' - \boldsymbol{r}_j(t)) e^{-i\boldsymbol{k}\cdot\boldsymbol{r}'} \delta(\boldsymbol{r} - \boldsymbol{r}_l(0)) e^{-i\boldsymbol{k}\cdot\boldsymbol{r}} \rangle e^{i\omega t} \mathrm{d}\boldsymbol{r}' \mathrm{d}\boldsymbol{r} \mathrm{d}t \\ &= \frac{1}{2\pi} \int \sum_{l,j} \left\langle e^{-i\boldsymbol{k}\cdot\boldsymbol{r}_j(t)} e^{-i\boldsymbol{k}\cdot\boldsymbol{r}_l(0)} \right\rangle e^{i\omega t} \mathrm{d}t \end{aligned} . \tag{6.27}$$

The dynamical behavior of electrons is not straightforward to illustrate; however, if $\hat{T}$ and $\hat{V}$ are the kinetic energy and potential energy operators, respectively, we can rewrite $e^{-i\mathbf{k}.\mathbf{r}_j(t)}$ as:

$$e^{-i\mathbf{k}.\boldsymbol{r}_j(t)} = e^{i(\hat{T}+\hat{V})t/\hbar} e^{-i\mathbf{k}.\boldsymbol{r}_j} e^{-i(\hat{T}+\hat{V})t/\hbar}. \tag{6.28}$$

We can also use the Zassenhaus expansion formula of noncommutative operators:

$$e^{i(\hat{T}+\hat{V})t/\hbar} = e^{i\hat{T}t/\hbar} e^{i\hat{V}t/\hbar} e^{\hat{C}_2(it/\hbar)^2} e^{\hat{C}_3(it/\hbar)^3} \dots, \tag{6.29}$$

where

$$\begin{aligned} \hat{C}_2 &= -\frac{1}{2}[\hat{T}, \hat{V}] \\ \hat{C}_3 &= \frac{1}{3}[\hat{T}, [\hat{T}, \hat{V}]] \end{aligned} . \tag{6.30}$$

If the potential of the electron system does not change significantly during the interaction time (i.e. all $e^{\hat{C}_n(it/\hbar)^n}$ can be regarded as static compared to the phase factor $e^{i\omega t}$), the following approximations can be applied:

$$\begin{aligned} e^{-i\mathbf{k}.\mathbf{r}_j(t)} &\approx e^{i\hat{T}\frac{t}{\hbar}} e^{i\hat{V}\frac{t}{\hbar}} e^{-i\mathbf{k}\cdot\mathbf{r}_j} e^{-i\hat{V}\frac{t}{\hbar}} e^{-i\hat{T}\frac{t}{\hbar}} \\ &= e^{i\frac{p_j^2}{2m}\frac{t}{\hbar}} e^{-i\mathbf{k}\cdot\mathbf{r}_j} e^{-i\frac{p_j^2}{2m}\frac{t}{\hbar}} = e^{-i\mathbf{k}\cdot\mathbf{r}_j} e^{i(2\mathbf{p}_j\mathbf{k}-\hbar\mathbf{k}^2)\frac{t}{2m}} \end{aligned} . \tag{6.31}$$

This is known as the so-called sudden or impulse approximation (IA). Assuming large energy transfer between the electrons and X-ray photons, the dynamic structure factor will become:

$$S_{IA}(\mathbf{k},\omega) = \frac{1}{2\pi}\int \sum_{l,j} \left\langle e^{i\mathbf{k}.(\mathbf{r}_j-\mathbf{r}_l)} e^{i(2\mathbf{p}_j\mathbf{k}-\hbar\mathbf{k}^2)\frac{t}{2m}} \right\rangle e^{i\omega t}\mathrm{d}t. \tag{6.32}$$

Dynamic structure factors can be categorized into single-electron contributions $S_{IA}^{(1)}$ and electron-pair contributions $S_{IA}^{(2)}$, where the former does not include a coordinate operator for the electron, and the magnitude of the latter will strongly depend on the transferred momentum. If the transferred momentum is large, the position-dependent phase factor will exhibit fast oscillations, providing a non-negligible contribution to the signal. The incoherent process with large energy and momentum transfer is called X-ray Compton scattering. Under the IA framework,

$$\begin{aligned} S_{IA}^{(1)}(\mathbf{k},\omega) &= \frac{1}{2\pi}\int \sum_{j} \left\langle e^{i(2\mathbf{p}_j\mathbf{k}-\hbar\mathbf{k}^2)\frac{t}{2m}} \right\rangle e^{i\omega t}\mathrm{d}t \\ &= \left\langle \sum_j \delta\left(\frac{2\mathbf{p}_j\mathbf{k}-\hbar\mathbf{k}^2}{2m}+\omega\right)\right\rangle \end{aligned} \tag{6.33}$$

Introducing the momentum symbol $q = \frac{\hbar k}{2} - m\frac{\omega}{k}$, whose unit vector is situated along the scattering vector $\boldsymbol{u}(\boldsymbol{q} = q\boldsymbol{u})$, we can obtain:

$$S_{IA}^{(1)}(\mathbf{k},\omega) = \frac{m}{k}\left\langle \sum_j \delta(\mathbf{p}_j.\boldsymbol{u} - q)\right\rangle = \frac{m}{k}J(\boldsymbol{q}), \tag{6.34}$$

where $J(\boldsymbol{q})$ is the so-called directional Compton profile (DCP), which is associated with the behavior of the electron cloud in the momentum space. Actually, $J(\boldsymbol{q})$ is the projection of the momentum density $\pi(\mathbf{p})$ on the scattering vector:

$$J(q\boldsymbol{e}_z) = \iint \langle \pi(\mathbf{p})\rangle \mathrm{d}p_x \mathrm{d}p_y. \tag{6.35}$$

If the z-axis is regarded as the scattering vector, the Fourier transform of DCP (i.e. the reciprocal space form factor) $B(\boldsymbol{s})$ will be:

$$B(\boldsymbol{s}) = \int \left\langle \sum_j \delta(\mathbf{p}_j - \mathbf{p})\right\rangle e^{-i\mathbf{p}.\boldsymbol{s}}\mathrm{d}\mathbf{p} = \int \langle \pi(\mathbf{p})\rangle e^{-i\mathbf{p}.\boldsymbol{s}}\mathrm{d}\mathbf{p}, \tag{6.36}$$

or

$$B(s\boldsymbol{u}) = \int J(q\boldsymbol{u})e^{-iqs}\mathrm{d}q = \int \left\langle \sum_j \delta(\mathbf{p}_j.\boldsymbol{u} - q)\right\rangle e^{-iqs}\mathrm{d}q. \tag{6.37}$$

This has been widely used during the reconstruction of 3D momentum density with limited DCP data. Finally, we can obtain:

$$B(\mathbf{s}) = \int \langle \gamma^{(1)}(\mathbf{r},\mathbf{r}')\rangle e^{-i\mathbf{p}.(\mathbf{r}-\mathbf{r}')}e^{-i\mathbf{p}.\mathbf{s}}\mathrm{d}\mathbf{r}'\mathrm{d}\mathbf{r}\mathrm{d}\mathbf{p} = \int \langle \gamma^{(1)}(\mathbf{r},\mathbf{r}+\mathbf{s})\rangle \mathrm{d}\mathbf{r}. \tag{6.38}$$

Therefore, coherent elastic X-ray scattering provides information regarding the diagonal part of the 1-RDM (i.e. the electron density in the coordinate space), whereas incoherent inelastic scattering (under the framework of impulse approximation) provides abundant information related to the off-diagonal 1-RDM electron structure information.

6.4 Reconstruction and Refinement of the Density Matrix

6.4.1 Bayesian Method

Reconstructing 1-RDM from a limited data set requires a mathematical model. Here, we introduce the Bayesian method, which considers the N data points ($D_{i=1,N}$) and arranges them into an N-vector $\boldsymbol{D}$. The confidence can be estimated by the experimental variance-covariance matrix $\boldsymbol{\Sigma}$. For convenience, $\boldsymbol{\Sigma}$ can usually be approximated using its diagonal form $\boldsymbol{\Sigma}_{ii} = \sigma_i^2$, where σ_i is the standard deviation. According to a specific physics theory, a model ($\boldsymbol{M}$) containing a set of vectors $\boldsymbol{\alpha}$ arranged into n-dimensional parameters (i.e. $\{\alpha_n\}$) can be constructed. The goal of refinement is to find the most suitable parameter estimates so that the calculated $\boldsymbol{M}(\overline{\boldsymbol{\alpha}})$ will be closest to the experimentally observed values ($\boldsymbol{D}$). Hence, the goal is to find the most probabilistic parameter values for given $\boldsymbol{D}$ and $\boldsymbol{\Sigma}$; thus, requiring an expression to describe the probability associated with the parameter $\boldsymbol{\alpha}$. According to Bayes' theorem, the posterior probability distribution function (PDF) $p(\{\alpha\}|\boldsymbol{D}, \boldsymbol{\Sigma}, \boldsymbol{I})$, which depends on the model parameters, experimental data, covariance matrix, and other information ($\boldsymbol{I}$), can be written as:

$$p(\{\alpha\}|\boldsymbol{D}, \boldsymbol{\Sigma}, \boldsymbol{I}) = p(\boldsymbol{D}|\{\alpha\}, \boldsymbol{\Sigma}, \boldsymbol{I}) \times \frac{p(\{\alpha\}|\boldsymbol{\Sigma}, \boldsymbol{I})}{p(\boldsymbol{D}|\boldsymbol{\Sigma}, \boldsymbol{I})}, \tag{6.39}$$

where $p(\boldsymbol{D}|\{\alpha\}, \boldsymbol{\Sigma}, \boldsymbol{I})$ is the likelihood function, $p(\{\alpha\}|\boldsymbol{\Sigma}, \boldsymbol{I})$ and $p(\boldsymbol{D}|\boldsymbol{\Sigma}, \boldsymbol{I})$ are the prior and evidential PDF, respectively. Typically, we have fair prior knowledge regarding the values of these parameters, and the prior probability distribution will be relatively smooth. In fact, the prior probability distribution can be considered a uniform distribution, while the likelihood function will be a very sharp peak. The evidential PDF can be regarded as a normalization factor for the posterior PDF. To find the most suitable set of parameters for a given experimental result, it is necessary to maximize Eq. (6.39).

The exact form of the likelihood function is very critical. We can assume that a suitable PDF will be the result of maximizing the total information entropy, and the best parameters should be able to obtain the expectation of each data:

$$\langle \boldsymbol{D} \rangle = \int \boldsymbol{D} p(\boldsymbol{D}\,|\boldsymbol{\alpha}, \boldsymbol{\Sigma}, \boldsymbol{I}) d^N D, \tag{6.40}$$

and the variance:

$$\langle D_i D_j - \langle D_i \rangle \langle D_j \rangle \rangle = \sum_{ij}. \tag{6.41}$$

This means that the model can take into account all of the physical effects associated with the measurement, and the average of multiple measurements of a data point should approximate the estimated value of the model $\langle \boldsymbol{D} \rangle = \boldsymbol{M}(\overline{\boldsymbol{\alpha}})$. If D_{ij} is an output with a probability p_{ij} of D_i, the entropy to be maximized will be:

$$S_i = -\sum_j p_{ij} \log(p_{ij}), \tag{6.42}$$

and meet the constraints:

$$\begin{aligned} \langle D_i \rangle &= M_i \\ \left\langle D_i^2 - \langle D_i \rangle^2 \right\rangle &= \sigma_i^2, \end{aligned} \tag{6.43}$$

thus, maximizing the following:

$$-\sum_j p_{ij} \log(p_{ij}) + \lambda \left[1 - \sum_j p_{ij} \right] + \mu \left[\sigma_i^2 - \sum_j p_{ij} (D_{ij} - M_i)^2 \right], \tag{6.44}$$

where λ and μ are the Lagrange multiplier. Eliminating its partial derivatives for p_{ij}, $\lambda$and μ can yield a Gaussian PDF:

$$p_{ij} = \left(\pi\sigma_i^2\right)^{-\frac{1}{2}} \exp\left(-\frac{(D_{ij} - M_i(\boldsymbol{\alpha}))^2}{\sigma_i^2} \right). \tag{6.45}$$

Under the approximation that the possible output will be a continuous distribution, we can obtain:

$$p_i(D_i) = \left(\pi\sigma_i^2\right)^{-\frac{1}{2}} \exp\left(-\frac{(D_i - M_i(\boldsymbol{\alpha}))^2}{\sigma_i^2} \right). \tag{6.46}$$

For data not related to each other, the total probability of $\boldsymbol{D}$ will be:

$$p(\boldsymbol{D} \,|\, \boldsymbol{\alpha}, \boldsymbol{\Sigma}, \boldsymbol{I}) = \prod_i p_i \left(D_i \middle| \boldsymbol{\alpha}, \sigma_i^2, \boldsymbol{I} \right) = \prod_i \left(\pi\sigma_i^2\right)^{-\frac{1}{2}} \exp\left(-\frac{(D_i - M_i(\boldsymbol{\alpha}))^2}{\sigma_i^2} \right), \tag{6.47}$$

which is a PDF with $N - n$ degrees of freedom. Using this equation as a likelihood function, the Bayesian theory suggests that the most suitable parameters for model $\boldsymbol{M}$ will be the maximized value in Eq. (6.47), or the minimized mean squared residual:

$$\chi^2 = \sum_i \frac{(D_i - M_i(\boldsymbol{\alpha}))^2}{\sigma_i^2}. \tag{6.48}$$

In the extreme case of numerous data points, χ^2 will be $N - n$ and the average broadening will be $\sqrt{N}$, where the sharper the minimum value, the more accurate the estimate of the parameter.

6.4.2 Combined Refinement of Different Types of Data

Different scattering experiments performed on the same system will be essentially different representations of the same wavefunction; therefore, verifying the correctness of a model does not require restriction to a single dataset. Simultaneously, using two or more sets of data refinement can improve the data parameter ratio.

A typical example is the simultaneous treatment of elastic X-ray diffraction and Compton profile data in different crystallographic directions. Both experiments can provide information regarding electron density in crystals, except that the former corresponds to the position space and the latter is the momentum space. This complementarity will be equally applicable to the study of the spin density distribution of magnetic systems by combining polarized neutron diffraction and magnetic Compton scattering.

We can consider the minimization case of two datasets $\boldsymbol{D}_A$ and $\boldsymbol{D}_B$ that correspond to two different experiments, $\boldsymbol{A}$ and $\boldsymbol{B}$, respectively. A feasible approach involves minimizing the combined residuals:

$$\chi^2 = \chi_A^2 + \chi_B^2 = \sum_{E=A,B} \left[\sum_{i\in E}^{N_E} \frac{(D_{E,i} - M_{E,i}(\boldsymbol{\alpha}))^2}{\sigma_{E,i}^2} \right]. \tag{6.49}$$

We must consider two important aspects: ① if the data volume of $\boldsymbol{A}$ is much larger than that of data $\boldsymbol{B}$, $\boldsymbol{A}$ will have a greater contribution to χ^2; thus, a suitable weighting scheme will be required. ② The change of the parameter vector will depend on the gradient vector, which is inversely proportional to the variance:

$$\frac{\partial \chi^2}{\partial \alpha} = -2 \sum_{E=A,B} \sum_{i\in E}^{N_E} \left[\frac{D_{E,i} - M_{E,i}(\boldsymbol{\alpha})}{\sigma_{E,i}^2} \right] \frac{\partial M_{E,i}(\boldsymbol{\alpha})}{\partial \alpha}. \tag{6.50}$$

In combinatorial refinement, it is very important to assign a credible weight to all datasets, as is the weight of each data point in each dataset. We can overcome the above problems by using a simple approach. We can assume that the value of individual $\sigma_{E,i}^2$ will not be credible for some unknown reasons, but the ratio $\sigma_{E,i}^2/\sigma_{E,j}^2$ will be credible. Therefore, $\sigma_{A,i}^2/\sigma_{B,j}^2$ cannot fairly reflect the credibility of each dataset itself. This situation can be resolved by applying an overall scale factor correction to the variance of each dataset: $\sum_E \to \eta_E^2 \sum_E$ or $\left\{\sigma_{E,i}^2\right\} \to \left\{\eta_E^2 \times \sigma_{E,i}^2\right\}$.

Therefore, the posterior PDF for each set of observations can be obtained by integrating over all possible scale factors $p(\eta_E)$, i.e.

$$\begin{aligned} p\left(\boldsymbol{D}_E | \boldsymbol{\alpha}, \sum_E, I\right) &= \int p\left(\boldsymbol{D}_E \middle| \boldsymbol{\alpha}, \eta_E \sum_E, I\right) p(\eta_E) \mathrm{d}\eta_E \\ &= \int \prod_i \left(\pi \eta_E^2 \sigma_{E,i}^2\right)^{-\frac{1}{2}} \exp\left(-\frac{(D_{E,i} - M_{E,i}(\boldsymbol{\alpha}))^2}{\eta_E^2 \sigma_{E,i}^2} p(\eta_E) \mathrm{d}\eta_E \right) \end{aligned} \tag{6.51}$$

Jeffreys [3] indicated that $p(\eta_E) \propto \eta_E^{-1}$; therefore,

$$\begin{aligned} p\left(\boldsymbol{D}_E | \alpha, \sum_E, I\right) &\propto \left[\prod_{i=1} \left(\pi \sigma_{E,i}^2\right)^{-\frac{1}{2}} \right] \int \left(\frac{1}{\eta_E}\right)^{N_E+1} e^{-\frac{\chi_E^2}{\eta_E^2}} \mathrm{d}\eta_E \\ &= \left[\prod_{i=1} \left(\pi \sigma_{E,i}^2\right)^{-\frac{1}{2}} \right] \times \Gamma\left(\frac{N_E}{2}\right) \times \left(\chi_E^2\right)^{-\frac{N_E}{2}} \end{aligned}. \tag{6.52}$$

The global probability of parameters in combined data is given by:

$$p\left(\boldsymbol{\alpha}\,\middle|\,\boldsymbol{D}_A, \sum_A, \boldsymbol{D}_B, \sum_B, I\right) \propto \left(\chi_A^2\right)^{-\frac{N_A}{2}} \left(\chi_B^2\right)^{-\frac{N_B}{2}}. \tag{6.53}$$

Therefore, the most probable parameters can be obtained by minimizing the following:

$$L\left(\boldsymbol{\alpha}\,\middle|\,\boldsymbol{D}_A, \sum_A, \boldsymbol{D}_B, \sum_B, I\right) \propto N_A \log\left(\chi_A^2\right) + N_B \log\left(\chi_B^2\right). \tag{6.54}$$

It should be noted that only in the fully convergent neighborhood, when $N_E \gg n$, $\chi_A^2 \approx N_A - n$ and $\chi_B^2 \approx N_B - n$, will its gradient be:

$$\frac{\partial}{\partial \alpha} L\left(\boldsymbol{\alpha}\,\middle|\,\boldsymbol{D}_A, \sum_A, \boldsymbol{D}_B, \sum_B, I\right) \propto \frac{N_A}{\chi_A^2}\frac{\partial \chi_A^2}{\partial \alpha} + \frac{N_B}{\chi_B^2}\frac{\partial \chi_B^2}{\partial \alpha}, \tag{6.55}$$

which is equivalent to the traditional gradient (minimizing $\chi_A^2 + \chi_B^2$).

6.4.3 Refinement of the One-electron Reduced Density Matrix (1-RDM)

Extracting 1-RDM from experimental data is obviously not as simple as obtaining electron density. This is about as complex as putting together datasets from different experiments. However, the nature of 1-RDM allows the union of coordinate space and momentum space, providing an opportunity to establish a stable model of the electron structure that can be tested by different experiments. Compared to the electron density, 1-RDM exhibits a more prominent characteristic of chemical bonds, and also allows for the calculation of the total energy under the approximation of a mean field. Thanks to the development of computing power and experimental techniques in recent years, the quality and quantity of experimentally reconstructed 1-RDM have significantly improved.

Currently, all related research has focused on some simple systems and parametric models, such as independent electron gas. In this framework, 1-RDM can be expressed in the form of natural spin orbitals (NSO) occupied by integers (n_i):

$$\gamma^{(1)}(\boldsymbol{r}', \boldsymbol{r}) = \sum n_i \Phi_i^*(\boldsymbol{r}')\Phi_i(\boldsymbol{r}). \tag{6.56}$$

Expanding each NSO on the atomic basis set, we can obtain:

$$\Phi_i(\boldsymbol{r}) = \sum_j c_{ij}\chi_j(\boldsymbol{r}). \tag{6.57}$$

Then 1-RDM can be written in the following form:

$$\begin{aligned}\gamma^{(1)}(\boldsymbol{r}', \boldsymbol{r}) &= \langle \boldsymbol{r}' | \left[\sum_{i,j,k} n_i c_{ij} c_{ik}^* |\chi_j\rangle\langle\chi_k|\right] |\boldsymbol{r}\rangle \\ &= \langle \boldsymbol{r}' | \left[\sum_{j,k} P_{jk} |\chi_j\rangle\langle\chi_k|\right] |\boldsymbol{r}\rangle\end{aligned} \tag{6.58}$$

To reduce the computation time, almost all research has focused on determining the matrix elements P_{jk} of the population matrix $\boldsymbol{P}$. Recently, introducing a masking parameter to allow the expansion and contraction of atomic orbitals has been proposed.

The 1-RDM needs to satisfy the N-representability conditions, which is the statistical mixture associated with the N-electron wavefunction (pure state) or N-representative pure state 1-RDM. Although it is generally almost impossible to express the N-representability constraint, it is sufficient in terms of the isodensity constraint in the HF framework. Therefore, most 1-RDM reconstructions will satisfy the population matrix:

$$\boldsymbol{PSP} = 2\boldsymbol{P}, \tag{6.59}$$

where $\boldsymbol{S}$ is the overlapping matrix of the basis functions and factor 2 only applies to closed-shell systems. This condition, which is critical to the 1-RDM model, can be obtained by a purification approach, as proposed by McWeeny, or by constraining the matrix to a specific analytical form. The former uses an iterative approach, which assumes that the initial population matrix will be very close to the isodensity condition and operate recursively:

$$\frac{3}{2}\boldsymbol{P}_n\boldsymbol{S}\boldsymbol{P}_n - \frac{1}{2}\boldsymbol{P}_n\boldsymbol{S}\boldsymbol{P}_n\boldsymbol{S}\boldsymbol{P}_n = \boldsymbol{P}_{n+1}. \tag{6.60}$$

This refinement starts with an initial theoretical prediction for $\boldsymbol{P}$, and the purification process operates within the constraints that the expectations match the experimental values.

Schmider et al. proposed an alternative by writing the NSO coefficient matrix $\boldsymbol{C}$ as the product of $\boldsymbol{C}_0$ (e.g. obtained from a first-principle computation) and a unit matrix $\boldsymbol{C}_R(\boldsymbol{\alpha})$ containing the parameters, with $\boldsymbol{C}_R(\boldsymbol{\alpha})$ in the form of the product of planar rotation matrices

$$\boldsymbol{C}_R(\{\alpha\}) = \prod_{ij} \boldsymbol{C}_{ij}(\alpha_{ij}), \tag{6.61}$$

where the subscript covers all pairs of basis functions, and the matrix $\boldsymbol{C}_{ij}(\alpha_{ij})$ corresponds to the transformation of the basis function invariant:

$$\begin{aligned} \chi_i(\boldsymbol{r}) &\rightarrow \cos(\alpha_{ij})\chi_i(\boldsymbol{r}) + \sin(\alpha_{ij})\chi_j(\boldsymbol{r}) \\ \chi_j(\boldsymbol{r}) &\rightarrow -\sin(\alpha_{ij})\chi_i(\boldsymbol{r}) + \cos(\alpha_{ij})\chi_j(\boldsymbol{r}) \end{aligned}. \tag{6.62}$$

Although this strategy is computationally intensive, it does work, and in principle, allows for electron correlations to be considered. It is worth mentioning that the combinatorial refinement of 1-RDM in the position space and momentum space will not be as successful even for small systems in the single determinant limit when different types of data are used simultaneously. Therefore, to obtain reliable 1-RDM, it is not sufficient to rely only on isodensity conditions. However, key information regarding the position space and momentum space can be obtained. Studies have revealed that 1-RDM matrix elements related to the bond properties cannot be obtained by a single data set. Without sufficient Compton profile data, even if the high-resolution Bragg diffraction and DCP data are simultaneously used for combinatorial refinement, it will still be impossible to reconstruct all the off-diagonal matrix elements of the 1-RDM [4].

6.4.4 Combinatorial Refinement of Structure Factor and Compton Profile Data

We can consider a pure-state 1-RDM [5] defined by an N-electron wavefunction $\Gamma\left(\boldsymbol{x}_1', \boldsymbol{x}_1\right)$:

$$\Gamma\left(\boldsymbol{x}_1', \boldsymbol{x}_1\right) = N \int \psi^*\left(\boldsymbol{x}_1', \boldsymbol{x}_2, \ldots, \boldsymbol{x}_N\right) \psi\left(\boldsymbol{x}_1', \boldsymbol{x}_2, \ldots, \boldsymbol{x}_N\right) \mathrm{d}\boldsymbol{x}_2 \ldots \mathrm{d}\boldsymbol{x}_N, \tag{6.63}$$

where $\boldsymbol{X}_i$ is the spin and position variables of an electron i. Here, we do not consider the spin-resolved case, which is the density matrix $\Gamma(\boldsymbol{r}', \boldsymbol{r})$ after spin integration:

$$\Gamma(\boldsymbol{r}', \boldsymbol{r}) = \int \Gamma(\boldsymbol{x}', \boldsymbol{x})_{\sigma'=\sigma} \mathrm{d}\sigma, \tag{6.64}$$

where σ is the spin variable.

The structure factor is the Fourier transform of the thermally averaged electron density; therefore, it will be related to the diagonal part of the 1-RDM of the coordinate representation, i.e.

$$F(\boldsymbol{Q}) = \int \langle \Gamma(\boldsymbol{r}', \boldsymbol{r})_{\mathbf{r}'=\mathbf{r}} \rangle_T \exp(i\boldsymbol{Q} \cdot \boldsymbol{r}) \mathrm{d}\boldsymbol{r}. \tag{6.65}$$

Based on this equation, if only the X-ray diffraction data is used for refinement, it will be difficult to obtain the meaningful off-diagonal matrix elements of the 1-RDM without strong constraints. Inelastic scattering experiments provide projection information on the electron momentum density, which can be used to probe the off-diagonal matrix elements of the 1-RDM. Under the impulse approximation, the DCP is given by:

$$J(\boldsymbol{u}, q) = \frac{1}{2\pi\hbar} \int \Gamma(\boldsymbol{r}, \boldsymbol{r} + (\boldsymbol{u} \cdot \boldsymbol{s})\boldsymbol{u}) \exp(iq\boldsymbol{u} \cdot \boldsymbol{s}) \mathrm{d}\boldsymbol{r} \mathrm{d}\boldsymbol{s}, \tag{6.66}$$

where $\boldsymbol{u}$ is the unit vector of the inelastic scattering vector. By collecting a large amount of DCP data, it is possible to reconstruct credible information for 2D and 3D electron momentum density.

Equations (6.65) and (6.66) show the experimental approaches in position space and momentum space, which probe different parts of the 1-RDM and are strongly complementary to each other. Bragg scattering is known for its ability to obtain the localized density distribution of electrons, but will be severely affected by disorder, temperature, and multiple scattering. By contrast, Compton scattering, which is based on an incoherent process, is not particularly sensitive to crystal quality. Thus, DCP will be more conducive to the study of delocalized electrons. In addition, electrons with small momentum will be barely affected by thermal vibrations.

6.4.5 Spin-resolved One-order Reduced Density Matrix (1-SRDM) Refinement

6.4.5.1 Basic Framework

In the position space, the spin density corresponding to the diagonal elements of 1-SRDM [6] is given by:

$$\rho(\boldsymbol{r}_1) = \int [\delta(s_1 - s^{\uparrow}) - \delta(s_1 - s^{\downarrow})] \Gamma^{(1)}\left(\boldsymbol{x}_1; \boldsymbol{x}_1'\right)_{\boldsymbol{x}_1=\boldsymbol{x}_1'} \mathrm{d}s_1. \tag{6.67}$$

The Fourier transform of the spin density gives the Magnetic structure factor (MSF):

$$F_{\mathrm{M}}(\boldsymbol{Q}) = \int [\delta(s_1 - s^{\uparrow}) - \delta(s_1 - s^{\downarrow})] \times \Gamma^{(1)}(\boldsymbol{x}_1; \boldsymbol{x}_1')_{\boldsymbol{x}_1 = \boldsymbol{x}_1'} \exp(i\boldsymbol{Q} \cdot \mathbf{r}) \mathrm{d}\boldsymbol{r}, \tag{6.68}$$

where $\boldsymbol{Q}$ is the scattering vector.

Compton scattering can provide information regarding the electron momentum density, where the spin-resolved electron momentum density is the off-diagonal matrix elements of 1-SRDM. The correlation of the magnetic Compton profile (MCP) to 1-SRDM is given by:

$$J(\boldsymbol{u}, q) = \frac{1}{2\pi\hbar} \int [\delta(s_1 - s^{\uparrow}) - \delta(s_1 - s^{\downarrow})] \times \Gamma^{(1)}(\boldsymbol{x}_1; \boldsymbol{x}_1')_{\boldsymbol{x}_1 = \boldsymbol{x}_1'} \exp(iq\boldsymbol{u} \cdot \boldsymbol{r}') \mathrm{d}\boldsymbol{r} \mathrm{d}\boldsymbol{r}', \tag{6.69}$$

where $\boldsymbol{u}$ is the unit vector of the inelastic scattering vector. The projection of the momentum space spin density $n(\boldsymbol{p})$ in the $\boldsymbol{u}$ direction can be obtained by the MCP with momentum $\boldsymbol{q}$:

$$J(\boldsymbol{u}, q) = \int n(\boldsymbol{p}) \delta(q - \boldsymbol{p} \cdot \boldsymbol{u}) \mathrm{d}\boldsymbol{p}. \tag{6.70}$$

Therefore, to reconstruct the entire 1-SRDM, we need both the MSF to describe the position space and the MCP experimental data to describe the momentum space.

6.4.5.2 Molecular Modeling

Molecular orbitals are formed by linear combinations of basis functions:

$$\Phi_i(\boldsymbol{r}) = \sum_{j=1} C_{i,j} \chi_j(\boldsymbol{r}), \tag{6.71}$$

where $C_{i,j}$ is the molecular orbital coefficient, and $\chi_j(\boldsymbol{r})$ is the basis function. Under the Gaussian framework, the basis function can be defined as a linear combination of Gaussian functions:

$$\chi_j(\mathbf{r}) = \sum_{k=1} d_{j,k} N(\alpha_k)(x - A_{x_k})^{a_k}(y - A_{y_k})^{b_k}(z - A_{z_k})^{c_k} \times \exp\left[-\alpha_k(\boldsymbol{r} - \boldsymbol{R}_k)^2\right], \tag{6.72}$$

where a_k, b_k, and c_k determine the angular momentum quantum number $\boldsymbol{L}$, satisfying $L = a_k + b_k + c_k$, and $d_{j,k}$ is the contraction factor of the jth orbital and corresponding kth Gaussian function. When the expanded equation of the molecular orbital is obtained, 1-SRDM can be expressed using the natural molecular orbital and corresponding occupancy number n_i:

$$\Gamma^{(1)}(\boldsymbol{r}, \boldsymbol{r}') = \sum_i n_i \Phi_i(\boldsymbol{r}) \Phi_i(\boldsymbol{r}'). \tag{6.73}$$

According to Eqs. (6.71), (6.72), and (6.73), 1-SRDM can be expressed in the form of the product of the spin population matrix and the wavefunction:

$$\Gamma^{(1)}(\boldsymbol{r}, \boldsymbol{r}') = \Gamma_{\uparrow}^{(1)}(\boldsymbol{r}, \boldsymbol{r}') - \Gamma_{\downarrow}^{(1)}(\boldsymbol{r}, \boldsymbol{r}') = \sum_{j,k} P_{j,k} \chi_j(\boldsymbol{r}) \chi_k(\boldsymbol{r}'), \tag{6.74}$$

where the spin population matrix $\boldsymbol{P}$ can be defined by the occupancy number of each spin state and the molecular orbital coefficient:

$$P_{j,k} = \sum_{i=1} n_i^{\uparrow} C_{i,j}^{\uparrow} C_{i,k}^{\uparrow} - \sum_{i=1} n_i^{\downarrow} C_{i,j}^{\downarrow} C_{i,k}^{\downarrow}. \tag{6.75}$$

6.4.5.3 Magnetic Structure Factor and Magnetic Compton Profile

When the 1-SRDM is obtained, all one-electron observations of the system can be expressed as a function of the product of the spin population matrix and the basis function. The MSF $F_{\mathbf{M}}(\boldsymbol{Q})$ can be defined as the Fourier transform of the 1-SRDM diagonal. According to Eqs. (6.68) and (6.74), the MSF can be written as the summation of the products of the spin population matrix and the Fourier transform of the product of the basis functions:

$$F_{\mathrm{M}}(\boldsymbol{Q}) = \sum P_{j,k} \int \chi_j(\boldsymbol{r})\chi_k(\boldsymbol{r}) \exp(i\boldsymbol{Q} \cdot \boldsymbol{r})\mathrm{d}\boldsymbol{r}. \tag{6.76}$$

Substituting Eq. (6.74) into Eqs. (6.69) and (6.70), the MCP can be expressed by:

$$J(\boldsymbol{u}, q) = \frac{1}{2\pi} \sum_{j,k} P_{j,k} \iint \left[\int S_{j,k}(\boldsymbol{r}) \exp(-i\boldsymbol{p} \cdot \boldsymbol{r})\mathrm{d}\boldsymbol{r} \right] \times \exp(i\boldsymbol{p} \cdot \boldsymbol{u}t)\mathrm{d}\boldsymbol{p} \exp(-iqt)\mathrm{d}t, \tag{6.77}$$

where

$$S_{j,k}(\boldsymbol{r}) = \int \chi_j(\boldsymbol{r}')\chi_k(\boldsymbol{r}' + \boldsymbol{r})\mathrm{d}\boldsymbol{r}'. \tag{6.78}$$

Therefore, the MSF, similar to the MCP, will be a function of the product of the spin population matrix and the wavefunction.

6.4.5.4 Variation of the Basis Functions

To obtain an accurate 1-SRDM, first, we need to change the radial expansion of each atomic orbital to make it possible to consider dispersion effects. To achieve this, the basis functions need to be rescaled by changing the index of the atomic orbitals $j(\{\alpha\} \Rightarrow \{\alpha_0\} \times \zeta_j)$:

$$\chi_j(\zeta_j, \boldsymbol{r}) = \sum_{k=1} d_{j,k} N(\zeta_j \alpha_{0,k})(x - A_{x_k})^{a_k}(y - \boldsymbol{A}_{y_k})^{b_k}(z - \boldsymbol{A}_{z_k})^{c_k} \times \exp\left[-\zeta_k \alpha_{0,k}(\boldsymbol{r} - \boldsymbol{R}_k)^2\right]. \tag{6.79}$$

In the minimization calculations, the variational operation on the atomic orbitals should be carried out under the following conditions:

$$0.8 \lesssim \zeta_j \lesssim 1.2, \quad \int \mathrm{d}\boldsymbol{r}\Phi_j(\boldsymbol{r})\Phi_k(\boldsymbol{r}) = \delta_{j,k} \tag{6.80}$$

where the first condition controls the extent of the expansion of atomic orbitals, and the second condition involves orthogonality between molecular orbitals. The variation of the basis functions will be clearly reflected in the MSF and the MCP:

$$F_{\mathrm{M}}(\boldsymbol{Q}) = \sum_{j,k} P_{j,k} \int \chi_j(\zeta_j, \boldsymbol{r})\chi_k(\zeta_k, \boldsymbol{r}) \exp(i\boldsymbol{Q} \cdot \boldsymbol{r})\mathrm{d}\boldsymbol{r}, \tag{6.81}$$

$$J(\boldsymbol{u}, q) = \frac{1}{2\pi} \sum_{j,k} P_{j,k} \iint \left[\int S_{j,k}(\{\zeta\}, \boldsymbol{r}) \exp(-i\boldsymbol{p} \cdot \boldsymbol{r}) \mathrm{d}\boldsymbol{r} \right], \\ \times \exp(i\boldsymbol{p} \cdot \boldsymbol{u}t) \mathrm{d}\boldsymbol{p} \exp(-iqt) \mathrm{d}t \tag{6.82}$$

where

$$S_{j,k}(\{\zeta\}, \boldsymbol{r}) = \int \chi_j(\zeta_j, \boldsymbol{r}')\chi_j(\zeta_k, \boldsymbol{r}' + \boldsymbol{r})\mathrm{d}\boldsymbol{r}'. \tag{6.83}$$

6.4.5.5 Variation of Spin Population Matrices

After finding the best basis function, a variation of the spin population matrix must be performed:

$$n_i^\sigma = n_{i,0}^\sigma + \delta n_i^\sigma, \tag{6.84}$$

where $n_{i,0}^\sigma$ is the initial occupancy number of the ith state and $\sigma = [\uparrow, \downarrow]$ is the spin state. The variation of the occupancy number only has to be performed on a few states near the Fermi level. Variational operations need to be performed under the N-representability condition of 1-SRDM:

$$\begin{aligned} &0 \leq n_i^\sigma \leq 1 \\ &\sum_{i=1} \left(n_i^\uparrow + n_i^\downarrow\right) = N_{\text{electron}} \\ &\sum_{i=1} \left(n_i^\uparrow - n_i^\downarrow\right) = N_{\text{spin}} \end{aligned}, \tag{6.85}$$

where the first condition reflects the Pauli exclusion principle, and the second and third conditions indicate the consistency with the experimental electron number and the spin value, respectively. The second condition can be applied by minimizing the following:

$$C[\chi^2(\{n\})] = \sum_j \log\left[\chi_j^2(\{n\})\right] \\ -\mu_1\left(\sum_{i=1} n_i^\uparrow - N_{\text{electron}}^\uparrow\right) - \mu_2\left(\sum_{i=1} n_i^\downarrow - N_{\text{electron}}^\downarrow\right), \tag{6.86}$$

where j represents a definitive experiment and μ_1, μ_2 are the Lagrange multipliers. The first term in Eq. (6.86) is the difference between the model and the experiment, while the latter two terms limit the number of electrons in each spin state, actually satisfying the third condition in Eq. (6.85). The variation of the occupancy number results in the change of the spin population matrix:

$$P_{j,k} = P_{j,k}^0 + \delta P_{j,k}, \tag{6.87}$$

where

$$\begin{aligned} P_{j,k}^0 &= \sum_{i=1} n_{i,0}^\uparrow C_{i,j}^\uparrow C_{i,k}^\uparrow - \sum_{i=1} n_{i,0}^\downarrow C_{i,j}^\downarrow C_{i,k}^\downarrow \\ \delta P_{j,k} &= \sum_{i=1} \delta n_i^\uparrow C_{i,j}^\uparrow C_{i,k}^\uparrow - \sum_{i=1} \delta n_i^\downarrow C_{i,j}^\downarrow C_{i,k}^\downarrow \end{aligned}. \tag{6.88}$$

The first term represents the initial spin population matrix, expressed by using the initial occupancy number and molecular orbital coefficients. The second term, which represents the variation of the spin population matrix, is a function of the variation of the occupancy number and the molecular orbital coefficient.

The density matrix model includes a parameter set $\{x\}$, of which the optimum $\{x\}$ is obtained by minimizing $C(\chi^2) = \sum_Y \log\left(\chi_Y^2\right)$, and χ^2 can be defined as:

$$\chi_Y^2(\{x\}) = \sum_i \frac{\left|Y_i^{\mathrm{m}}(\{x\}) - Y_i^{\mathrm{c}}(\{x\})\right|^2}{\sigma_i^2}, \tag{6.89}$$

where it is necessary for i to traverse all of the measured/calculated MSFs ($Y^{\mathrm{m}}/Y^{\mathrm{c}} = F^{\mathrm{m}}/F^{\mathrm{c}}$) and MCPs ($Y^{\mathrm{m}}/Y^{\mathrm{c}} = J^{\mathrm{m}}/J^{\mathrm{c}}$), and $\sigma_{(i)}$ is the standard deviation of $Y_i^{\mathbf{m}}$. For the variation of the basis functions, the minimization function $\boldsymbol{C}$ can be defined as:

$$C[\chi^2(\{\zeta\})] = \log\left[\chi_J^2(\{\zeta\})\right] + \log\left[\chi_F^2(\{\zeta\})\right], \tag{6.90}$$

where $\boldsymbol{F}$ and $\boldsymbol{J}$ denote the MSF and the MCP, respectively.

For the variation of the spin population matrix, the minimization function $\boldsymbol{C}$ can be defined as:

$$C[\chi^2(\{n\})] = \log\left[\chi_J^2(\{n\})\right] + \log\left[\chi_F^2(\{n\})\right]. \tag{6.91}$$

References

1 Mazziotti, D.A. (2007). *Reduced-Density-Matrix Mechanics: With Application to Many-Electron Atoms and Molecules*, vol. 134. Wiley.

2 Weyrich W. *One-Electron Density Matrices and Related Observables*, 1996.

3 Jeffreys, H. (1939). *Theory of Probability*. Oxford: Clarendon.

4 Gillet, J.-M., Becker, P.J., and Cortona, P. (2001). Joint refinement of a local wave-function model from Compton and Bragg scattering data. *Phys. Rev. B* 63: 235115.

5 Gillet, J.-M. (2007). Determination of a one-electron reduced density matrix using a coupled pseudo-atom model and a set of complementary scattering data. *Acta Cryst. A* 63: 234–238.

6 Gueddida, S., Yan, Z., and Gillet, J.-M. (2018). Development of a joint refinement model for the spin-resolved one-electron reduced density matrix using different data sets. *Acta Cryst. A* 74: 131–142.

7

Electron Wavefunction Models

7.1 Introduction

According to Born's interpretation of quantum mechanics, the wavefunction is a quantity that contains all of the information of the system, and all physical quantities can be regarded as the expectation of the Hermitian operator.

The purpose of quantum chemistry is to determine the physical properties of molecules or condensed matter using fundamental quantum mechanics rules, which can be achieved by computational approaches such as ab initio calculations of the wavefunction [1] and density functional theory (DFT) [2]. It is difficult to perform calculations on large systems within the framework of Pople's ab initio calculations, where a wavefunction of more than 1000 electrons is of little significance. The DFT theory by Kohn and Hohenberg proves that electron density alone is sufficient to predict all of the ground-state chemical properties of a system. Compared to ab initio calculations, some of the fundamental electron structure parameters in DFT methods have been simplified to the point where they can be fitted with experimental observations; however, experimental electron density properties appear to be rarely used in DFT.

The X-ray constrained wavefunction (XCW) model proposed by Jayatilaka [3–10] combines the characteristics of the wavefunction and DFT method, providing a new method for studying the electron structures of atoms and molecules. In the XCW method, a statistically reasonable wavefunction of the molecular crystal can be obtained on the basis of the pure Hartree–Fock (HF) solution by adding the influence of the X-ray structure factor data, resulting in the influence of the experimental structure factor with a weighting factor. Although the wavefunction obtained by this method will be single-Slater determinant, it still captures the main properties of the electron structure to a large extent, including some electron correlation effects.

In this method, traditional ab initio quantitative calculations will yield a wavefunction, by which the structure factor amplitude can be calculated. Then, the constraints of the experimental structure factor can be gradually added to the calculation of the wavefunction by comparing the theoretical and experimental structure factors so that the theoretical and experimental structure factors are consistent, giving a statistically reasonable wavefunction by iterative calculations.

Electronic Structure Crystallography and Functional Motifs of Materials, First Edition.
Guo-Cong Guo and Xiao-Ming Jiang.

The wavefunction itself can be derived from any variational method in quantitative computation. An advantage of the XCW approach is the flexibility to adjust the degree of participation of the experimental data. Initially, a wavefunction obtained by traditional ab initio calculations can be considered non-constrained by the experimental data. The wavefunction will then be constrained by using the experimental data and the Lagrange multiplier method to improve the agreement between the experimental and theoretical structure factors. It should be particularly noted that this is not a least squares fitting procedure.

One of the advantages of the XCW approach is that once the experimental wavefunction is obtained, many properties can be directly calculated by the fundamental rules of quantum mechanics, as physical properties can be written in the form of the expectation of the wavefunction. For example, both conventional properties, such as the atomic charge and electrostatic potential, which are related to electron density, and unconventional properties such as the Fermi hole mobility function, kinetic energy density, electron localized density function, refractive index, and nonlinear optical properties can also be calculated based on the wavefunction.

The multipole model is one of the most popular methods in electron structure crystallography, where the total electron density distribution of a molecule can be expressed as the sum of non-spherical electron density functions with the atom as the center. The atomic electron density distribution will be distorted by the formation of chemical bonds. The chemical interpretation of the electron density of the multipole model is very clear; however, it also highlights a non-negligible disadvantage, namely, the types of properties obtained from the electron density of the model will be very limited because the strictly functional relationship between the ground-state electron density and the ground-state wavefunction is unknown. In addition, there is a strong correlation between the model parameters of the multipole model and atomic displacement parameter (ADP), and some physically unreasonable regions with negative electron density will appear in the electron density map.

The XCW method can obtain the XCW, which is consistent with the experimental data within a certain accuracy range. The experimental wavefunction, which contains all of the quantum mechanical properties of the wavefunction, can also be used to calculate the properties that are not directly related to the experimental structure factor. Compared to the multipole model, the XCW method can obtain the quantum-mechanically rigorous electron density, but it lacks certain chemical intuition because the canonical molecular orbitals will often be delocalized throughout the whole system. This is far from our traditional and intuitive interpretation of a molecule, where a molecule consists of atoms, chemical bonds, and functional groups, which is represented very well in the pseudo-atom model.

The X-ray-constrained extremely localized molecular orbital (XC-ELMO) method is an effective method for determining the experimental electron density [11–13]. Molecular orbitals are strictly localized on atoms, chemical bonds, and functional groups, combining the quantum mechanical rigor of the XCW method with the chemical intuition of the traditional multipole model method.

7.2 X-ray Constrained Wavefunction (XCW) Model

This chapter introduces a new approach called the XCW model, also known as the experimental wavefunction method, to obtain the electron structure from the experimental diffraction data. In this semiempirical approach, a reasonable electron wavefunction, which preserves the experimental electron density information, can be fitted from the experimental diffraction data. Many parameters need to be optimized in the experimental wavefunction refinement process, including the wavefunction parameters and nucleus coordinate parameters. Although in practical diffraction experiments, we can obtain as many diffraction points as possible by elevating the intensity of the X-ray source and increasing the angle of data collection, it is still not enough to experimentally determine all the parameters of the wavefunction, even to determine the simplest HF type electron wavefunction. The constrained wavefunction refinement approach proposed by Jayatilaka can effectively solve this problem.

7.2.1 Mathematical Framework

In the quantum mechanical ab initio calculation, the parameter $\boldsymbol{p}$ of a wavefunction $\psi(\boldsymbol{r}; \boldsymbol{p})$ can be determined using the variational method, i.e. minimizing the energy:

$$E(\boldsymbol{p}) = \langle \psi | \hat{H} | \psi \rangle. \tag{7.1}$$

Parameters $\boldsymbol{p} = (\boldsymbol{p}_e, \boldsymbol{p}_n)$ can be categorized into two types, namely $\boldsymbol{p}_e$, which describe the degrees of freedom of electrons (such as molecular orbital coefficients), and $\boldsymbol{p}_n$, which describe the degrees of freedom of the nucleus (such as nucleus coordinates).

However, in traditional least squares structure refinement, it is usually necessary to minimize the difference between the experimental and theoretical values of the structure factor:

$$\chi^2(\boldsymbol{p}, \boldsymbol{q}; \boldsymbol{F}^{\text{exp}}) = \frac{1}{N_r - N_p} \sum_k^{N_r} \frac{\left(F_k(\boldsymbol{p}, \boldsymbol{q}) - F_k^{\text{exp}}\right)^2}{\sigma_k^2}, \tag{7.2}$$

where N_r is the number of experimental data, N_p is the number of parameters in the model, σ_k is the error of the experimental data F_k^{exp}, and $F_k(\boldsymbol{p}, \boldsymbol{q})$ is the structure factor calculated from the wavefunction $\psi(\boldsymbol{r}; \boldsymbol{p})$. To better fit the experimental data, it is necessary to introduce additional phenomenological parameters $\boldsymbol{q}$, such as the temperature factor and extinction coefficient.

The constrained wavefunction method combines the least squares method and variational method. The solution must satisfy two conditions: having the minimum error χ_0^2 and simultaneously minimizing the energy of the wavefunction $E(\boldsymbol{p})$, which is equivalent to the Lagrange multiplier method, i.e. minimizing the Lagrangian $\boldsymbol{L}$:

$$L(\boldsymbol{p}, \boldsymbol{q}) = E(\boldsymbol{p}) + \lambda \chi^2(\boldsymbol{p}, \boldsymbol{q}; \boldsymbol{F}^{\text{exp}}), \tag{7.3}$$

or

$$\tilde{L}(\mathbf{p}, \boldsymbol{q}) = \mu E(\mathbf{p}) + \chi^2(\mathbf{p}, \boldsymbol{q}; \boldsymbol{F}^{\text{exp}}), \tag{7.4}$$

where λ or $\mu = 1/\lambda$ is the Lagrange multiplier.

7.2.2 Hirshfeld Atom Refinement

Obtaining accurate nucleus coordinates and the associated information is the premise of obtaining reliable electron structure information. To obtain precise nucleus coordinates, Jayatilaka and Dittrich proposed a solution in which aspherical pseudo-atoms were obtained by the atomic partitioning of molecular density. In their approach, Hirshfeld atomic partitioning can be applied to the molecular electron density obtained by ab initio calculations. Here, the molecules will be in a localized crystal field, and the resulting structure factors for aspherical Hirshfeld atoms can be subsequently used to refine the nucleus parameters. Dittrich's invariom method [14] can be considered an approximation to Hirshfeld atomic refinement, as it uses multipole-type pseudo-atoms to fit quantitative computational data. Despite using aspherical atomic densities, Hirshfeld atomic refinement cannot determine the electron density from diffraction data because, unlike pseudo-atomic methods, Hirshfeld atoms are defined by wavefunctions. Hirshfeld atomic refinement also does not use fixed, portable pseudo-atoms like other methods, in which the Hirshfeld atoms only change when the wavefunction changes after the atom changes position. The Hirshfeld atomic refinement procedure is more suitable for obtaining nucleus parameters when using constrained electron wavefunction analysis, as the same electron wavefunction model can be used in the refinement of the nucleus parameters and wavefunction parameters. The only difference is that the nucleus parameters can be obtained by the least-squares refinement, while the wavefunction parameters are obtained by the (energy) constrained least-squares refinement. Using the same wavefunction model ensures that the electron density obtained by this method reflects the properties of the true wavefunction.

7.2.2.1 Selection of Wavefunction

Because diffraction comes from periodic crystals, we should choose a wavefunction suitable for the periodic system. Traditionally, promolecule approximation has been used in crystallography. The interactions between pseudo-atoms are not considered in the promolecule; therefore, the wavefunction of the promolecule can be expressed as the Hartree product of (infinite and periodic) antisymmetric atomic wavefunctions, i.e.

$$\psi = \prod_A \psi_A(\boldsymbol{x}; \boldsymbol{R}_A). \tag{7.5}$$

The center of each atomic wavefunction ψ_A is located at nucleus coordinates $\boldsymbol{R}_A$, and all atomic coordinates can be associated with the atomic parameters in a unique set of asymmetric units by symmetry operations. Thus, the atomic wavefunction is a function of electron space and spin coordinates x.

For molecular systems, we can consider the non-correlated wavefunctions ψ of so-called procrystal fragments, which can be written in the form of the Hartree product of the wavefunctions of interaction-free molecules, i.e.

$$\psi = \prod_{M} \psi_M(\boldsymbol{x}; \boldsymbol{R}_M), \tag{7.6}$$

where the noninteracting molecules M and corresponding set of nucleus coordinates R_M may be correlated to an independent set by symmetry. However, unlike promolecular wavefunctions, the atoms in these noninteracting molecules contain at least one asymmetric unit, especially when the molecule has nontrivial symmetry. In this case, the entire molecule is used instead of the symmetry-independent parts alone. Therefore, our view is not appropriate for covalent networks or metal crystals; however, for some reason, it remains, as there is no concept of an independent molecule in these crystals.

Here, we can choose the wavefunction to be in HF or DFT form so the single-particle density matrix is idempotent, with an orbital occupancy number of 1 or 0. The real single-particle density matrix does not have this limitation, which can be eliminated by choosing different wavefunctions. For example, if we consider the exchange correlations and Coulomb interactions between the electrons of molecule M and its neighbors, ψ_M will not just be a fragment wavefunction. This wavefunction itself is sufficient to reflect all interactions in the crystal by symmetry operation. A simple alternative involves ignoring the exchange-correlation effect and approximating the periodic Coulomb interactions by placing the molecular wavefunction in a single-electron self-consistent field formed by the self-determined multipolar distribution. In this case, the wavefunction in the Hartree product can exhibit correlation. It is worth mentioning that this fragment model can also be extended to the full-period HF wavefunction, but it needs to include periodic Coulomb potential (i.e. some Ewald sum) and short-range exchange interactions.

7.2.2.2 Electron Density

All single-electron properties can be calculated from the expectation of the wavefunction, and they can also be expressed in the form of the trace of the single-particle density matrix P. In quantum mechanical calculations, the elements of the density matrix will be the functions of wavefunction parameters p such as molecular orbital coefficients and nucleus coordinates. If the atomic basis function is used to expand the molecular wavefunction, the electron density of a single molecule M in a unit cell is:

$$\rho_M(\boldsymbol{r}) = \sum_{\alpha,\beta} P_{\alpha\beta}\chi_\alpha(\boldsymbol{r} - \boldsymbol{R}_{A\alpha})\chi_\beta(\boldsymbol{r} - \boldsymbol{R}_{A\beta}). \tag{7.7}$$

Of note, the expansion of electron density contains both monocentric terms ($A_\alpha = A_\beta$) and bicentric terms ($A_\alpha \neq A_\beta$). In principle, this expanded expression can be made stricter by setting the index β to contain enough atoms in an infinite number of crystals; however, we can constrain it only in the molecules that do not interact with each other. Although plane-wave basis sets can also be used to expand molecular wavefunctions, they cannot accurately model nucleus charges whose contribution largely results in X-ray scattering. Plane-wave basis sets have often been used in the computation of pseudopotentials, which usually do not require the computation of nucleus electrons.

Because the structure factor depends on the electron density of the unit cell, we need to derive this quantity from the non-interacting molecular wavefunction. If there are N_M molecules, which are all equivalent to the molecule M (wavefunction ψ_M) in the unit cell, their electron density can be correlated by $\overline{\rho}_M(\boldsymbol{r}) = \overline{\rho}_m(\boldsymbol{S}_m\boldsymbol{r} + \boldsymbol{s}_m)$, of which $(\boldsymbol{S}_m, \boldsymbol{s}_m)$ is a crystallographic rotation-translation symmetry operation, then the electron density of the unit cell will be:

$$\rho_{\text{cell}}(\boldsymbol{r}) = \sum_m^{N_M} \rho_m(\boldsymbol{r}) = \sum_m^{N_M} \rho_M\left(\boldsymbol{S}_m^{-1}(\boldsymbol{r} - \boldsymbol{s}_m)\right). \tag{7.8}$$

Note that this equation does not consider thermal averaging effects.

7.2.2.3 Hirshfeld Atomic Partitioning Method

A basis for full space partitioning involves defining an atomic weight function w_A, which satisfies the constraints:

$$\sum_A w_A(\boldsymbol{r}) = 1. \tag{7.9}$$

The weight function defined by Hirshfeld is given by:

$$w_A(\boldsymbol{r}) = \rho_A^0(\boldsymbol{r} - \boldsymbol{R}_A) / \sum_{B \in M} \rho_B^0(\boldsymbol{r} - \boldsymbol{R}_B), \tag{7.10}$$

where ρ_A^0 is the reference electron density, usually set as the spherical average electron density of the atom. Other options are certainly possible, as long as Eq. (7.9) is satisfied. We can then obtain:

$$\rho_M(\boldsymbol{r}) = \sum_A \rho_A(\boldsymbol{r}), \tag{7.11}$$

where

$$\rho_A(\boldsymbol{r}) = w_A(\boldsymbol{r})\rho_M(\boldsymbol{r}). \tag{7.12}$$

In this way, the aspherical Hirshfeld atom can be separated from the molecular density accordingly.

7.2.2.4 Calculation of the Structure Factor

The experimentally measured structure factor can be calculated by:

$$F_{\boldsymbol{k}}^{\text{calc}} = \xi_{\boldsymbol{k}} |F_{\boldsymbol{k}}|, \tag{7.13}$$

where $|F_{\boldsymbol{k}}|$ is the scattering factor magnitude $|F_{\boldsymbol{k}}| = \left(F_{\boldsymbol{k}}^* F_{\boldsymbol{k}}\right)^{1/2}$, and $\xi_{\boldsymbol{k}}$ is the correction factor that takes into account the effects of scale factors and other systematic biases, such as X-ray absorption, extinction, multiple scattering, and thermal diffuse scattering. The structure factor is the Fourier transform of the thermally averaged unit cell electron density $\overline{\rho}_{\text{cell}}$:

$$F_{\boldsymbol{k}} = \int \overline{\rho}_{\text{cell}}(\boldsymbol{r}) e^{i\boldsymbol{k}\cdot\boldsymbol{r}} d\boldsymbol{r}. \tag{7.14}$$

Substituting Eq. (7.8) into the equation above and using variable substitution $\boldsymbol{r}' = \boldsymbol{S}_m\boldsymbol{r} + \boldsymbol{S}_m$, the structure factor can be written as a functional form of molecular density:

$$F_k = \sum_m^{N_M} e^{i\boldsymbol{k}\cdot\boldsymbol{s}_m} \int \overline{\rho}_M(\boldsymbol{r}')e^{i(S^t_m\boldsymbol{k})\cdot\boldsymbol{r}'}\,d\boldsymbol{r}', \tag{7.15}$$

following the assumption of $\boldsymbol{k}\cdot\cdot\cdot(\boldsymbol{S}\boldsymbol{r}) = \boldsymbol{S}^t\boldsymbol{k}\cdot\boldsymbol{r}$, where superscript t denotes transpose, and S_m represents the rotary axis operation, of which the determinant is ± 1, resulting in $d\boldsymbol{r} = d\boldsymbol{r}'$.

To obtain the structure factor, we need to use the thermally averaged molecular density $\overline{\rho}_M$. Under the rigid Hirshfeld atomic approximation, it can be assumed that the electron density $\rho_A(\boldsymbol{r})$ of the pseudo-atom moves along with the atomic coordinates $\boldsymbol{R}_A$, while the overall density distribution does not change and is independent of other atomic coordinates. However, this is actually not the case, because w_A depends on all the coordinates in molecule M. In addition, molecular density ρ_M also changes with changes in the atomic coordinates, and the motion of one atom can also be correlated to other atoms by phonons. However, based on this assumption, the thermally averaged electron density $\overline{\rho}_A$ of the atom A can be obtained by convolving the electron density ρ_A of atom A with the probability distribution function of the atom's position. According to the convolution theorem, the Fourier transform of the thermally averaged electron density (the form factor of the Hirshfeld atom), is the product of the Fourier transform of the electron density and the Fourier transform of the position probability density distribution:

$$\overline{\rho}_A(\boldsymbol{k}) = T_A(\boldsymbol{k}) \times e^{i\boldsymbol{k}\cdot\boldsymbol{R}_A} \int \overline{\rho}_A(\boldsymbol{r} + \boldsymbol{R}_A)e^{i\boldsymbol{k}\cdot\boldsymbol{r}}d\boldsymbol{r}, \tag{7.16}$$

where the translation factor $e^{i\boldsymbol{k}\cdot\boldsymbol{R}_A}$ describes the dependence of the thermally averaged electron density on atomic coordinates $\boldsymbol{R}_A$. In addition, T_A is the Fourier transform of the atomic position probability distribution function, which can be expressed in the form of a 3×3 temperature factor matrix $\boldsymbol{U}^{A_\alpha}$:

$$T_A(\boldsymbol{k}) = e^{-\frac{1}{2}\boldsymbol{k}^T U^A \boldsymbol{k}}. \tag{7.17}$$

Substituting $\overline{\rho}_M = \Sigma_A\overline{\rho}_A$ into Eq. (7.15) and the subsequent result into Eq. (7.13), we can obtain the final expression for the structure factor:

$$F^{\mathrm{calc}}_{\boldsymbol{k}} = \xi_{\boldsymbol{k}}(\boldsymbol{q}) \left| \sum_m^{N_M} e^{i\boldsymbol{k}\cdot\boldsymbol{s}_m} \sum_A e^{i\boldsymbol{k}\cdot\boldsymbol{R}_A} e^{-\frac{1}{2}\boldsymbol{k}^T U^A \boldsymbol{k}} \cdot \int \overline{\rho}_A(\boldsymbol{r} + \boldsymbol{R}_A)e^{i(S^t_m\boldsymbol{k})\cdot\boldsymbol{r}}d\boldsymbol{r} \right|. \tag{7.18}$$

The parameters that need to be refined include $\boldsymbol{R}_A$, $\boldsymbol{U}^A$ and other phenomenological parameters $\boldsymbol{q}$, such as scale factors. It should be noted that if the molecule has symmetry, not all positions will be independent. However, calculations must be performed on the entire molecule or larger molecular cluster. It is very useful to use the wavefunction of molecular clusters for studying intermolecular interactions, as Hirshfeld atoms happen to include the effects of intermolecular interactions.

This expression has already assumed that the Hirshefeld atom is rigid. To reduce the error caused by this assumption, the parameters obtained by the least squares method must be used to recalculate the wavefunction in the new coordinates and the parameters must be refined again until no changes are visible. The geometric optimization in this method requires information regarding the partial derivatives of the density matrix to the nucleus coordinates, which is very challenging to implement.

7.2.3 X-ray Constrained Wavefunction Refinement

The theory of XCW refinement is similar to Hirshfeld atomic refinement. For example, they both use the same wavefunction, eventually resulting in the same structure factor. However, the constrained wavefunction refinement process has two major characteristics. First, the thermal averaging process differs from Hirshfeld atom refinement, and second, the refinement process includes the partial derivative of the density matrix to the wavefunction parameters.

7.2.3.1 Special Treatment for Thermal Vibrations

As previously discussed, the processing of thermal motion requires information on the phonon density of states, including the phonon form and quantity, as well as the nucleus probability distribution function. The thermally averaged electron density can be subsequently obtained by the convolution of the electron density with the probability distribution of the nucleus coordinates. Because the electron density of a molecule (Eq. (7.7)) contains the product of the monocentric and bicentric basis functions, the processing of the thermally averaged electron density will require the use of parameters that describe the monocentric and bicentric probability density functions. The Fourier transform of the probability density distribution of a monocentric atom A can be written as (Eq. (7.17)):

$$T_\alpha(\boldsymbol{k}) \equiv T_{A_\alpha}(\boldsymbol{k}). \tag{7.19}$$

For the bicentric probability density distribution, no readily available processing method is available; therefore, some approximations are needed. Two solutions are available.

1. The bicentric probability density distribution can be ignored, similar to Hirshfeld refinement, directly projecting ρ_M on the monocentric expansion and then obtaining the structure factor. However, the multipole function of the pseudo-atom must be used as the fitting function, as this method is more in line with the characteristics of density fitting. It is also recommended to use the real-space atomic partitioning technique so that the resulting atomic properties are finer.
2. By some approximations, monocentric parameters can be used to represent a bicentric distribution, such as the approximations proposed by Stewart [15], Coppens [16], and Tanaka [17], respectively:

$$T_{\alpha\beta}(\boldsymbol{k}) = \left(e^{-\frac{1}{2}\boldsymbol{k}^T U^{A_\alpha}\boldsymbol{k}} + e^{-\frac{1}{2}\boldsymbol{k}^T U^{A_\beta}\boldsymbol{k}}\right)/2 \qquad (Coppens)$$

$$T_{\alpha\beta}(\boldsymbol{k}) = \left(e^{-\frac{1}{2}g\boldsymbol{k}^T(U^{A\alpha}+U^{B\beta})\boldsymbol{k}}\right)/2 \qquad (Stewart)$$

$$T_{\alpha\beta}(\boldsymbol{k}) = \left(e^{-\frac{1}{2}g\boldsymbol{k}^T(g_\alpha U^{A\alpha}+g_\beta U^{A\beta})\boldsymbol{k}}\right)/2 \qquad (Tanaka). \tag{7.20}$$

All of these equations satisfy the condition: $T_{\alpha\beta} = T_\alpha$ when $A_\alpha = A_\beta$.

It is worth mentioning that in these two methods, only the dependence of the basis functions on the nucleus position will be considered, completely ignoring the dependence of the density matrix on the position. Therefore, in the above projection pseudo-atom method, the influence of atomic position on the weight function must also be considered.

7.2.3.2 Density Matrix Representation of Structure Factor

Using the above results, the structure factor containing the temperature dependence can be written in the form of a density matrix:

$$F_{\boldsymbol{k}} = \mathrm{Tr}(\boldsymbol{P}\bar{\boldsymbol{I}}_{\boldsymbol{k}}) = \sum_{\alpha\beta} P_{\alpha\beta}\bar{I}_{\boldsymbol{k}\alpha\beta}, \tag{7.21}$$

where the double nucleus distribution is approximated by single nucleus distribution parameter $\bar{I}_{\boldsymbol{k}}$, which is the Fourier transform of the integral of the symmetry and thermally averaged basis function:

$$\bar{I}_{\boldsymbol{k}\alpha\beta} = \sum_m e^{i\boldsymbol{k}\cdot\boldsymbol{s}_m} T_{\alpha\beta}(\boldsymbol{k}) \int \chi_\alpha(\boldsymbol{r}-\boldsymbol{R}_{A_\alpha})\chi_\beta(\boldsymbol{r}-\boldsymbol{R}_{A_\beta})e^{i(\boldsymbol{S}_m^t\boldsymbol{k})\cdot\boldsymbol{r}}d\boldsymbol{r}. \tag{7.22}$$

The electron density ρ_M is projected on the monocentric expanded basis, using the definition of the weight function w_A:

$$\bar{I}_{\boldsymbol{k}\alpha\beta} = \sum_m e^{i\boldsymbol{k}\cdot\boldsymbol{s}_m} \sum_A T_A(\boldsymbol{k}) \int w_A(\boldsymbol{r})\chi_\alpha(\boldsymbol{r}-\boldsymbol{R}_{A_\alpha})\chi_\beta(\boldsymbol{r}-\boldsymbol{R}_{A_\beta})e^{i(\boldsymbol{S}_m^t\boldsymbol{k})\cdot\boldsymbol{r}}d\boldsymbol{r}. \tag{7.23}$$

7.2.3.3 Experimental Constrained Wavefunction Refinement

In the constrained HF case, the molecular wavefunction ψ_M is the determinant of the molecular spin orbitals, the spatial part of which ϕ_i is expanded by the basis functions:

$$\phi_i = \sum_i c_{\alpha i}\chi_\alpha, \tag{7.24}$$

where the occupied orbital coefficient $\boldsymbol{c}$ is the wavefunction parameter $\boldsymbol{p}$ ($\boldsymbol{p} \equiv \boldsymbol{c}$). To ensure the normalization of the ψ_M determinant, the molecular orbitals $\{\phi_i\}$ must be orthogonal.

Without taking any experimental data into account to obtain the orbital information, Eq. (7.1) can be minimized under the constraint of the orbitals remaining orthogonal. Thus, the differential of the following Lagrangian function is set to 0:

$$L(\boldsymbol{c}) = E(\boldsymbol{c}) - \sum_{i,j}\varepsilon_{ij}\left(\int \phi_i\phi_j - \delta_{ij}\right). \tag{7.25}$$

Therefore, we can obtain the famous HF equation:

$$\boldsymbol{F}\boldsymbol{c} = \boldsymbol{S}\boldsymbol{c}\varepsilon, \tag{7.26}$$

where $\boldsymbol{F}$ is Fock matrix, $\boldsymbol{S}$ is the overlapping matrix of the basis functions, and $\boldsymbol{\varepsilon}$ is the symmetric matrix of the correlation orthogonality of the Lagrange multiplier.

In the experimentally constrained wavefunction refinement, the addition of X-ray diffraction data must be considered. At this point, besides the same terms as HF, there is an additional term in the Lagrangian $\lambda[\chi^2(\mathbf{c}) - \Delta]$, i.e.

$$L(\mathbf{c}) = E(\mathbf{c}) - \sum_{i,j} \varepsilon_{ij} \left(\int \phi_i \phi_i - \delta_{ij} \right) - \lambda[\chi^2(\mathbf{c}) - \Delta], \tag{7.27}$$

where Δ is the error between the theoretical and experimental values, usually fixed as 1.

The statistical indicator χ^2 is defined as:

$$\chi^2 = \frac{1}{N_r - p} \sum_{\boldsymbol{h}} \left(\frac{\left|F_{\boldsymbol{h}}^c\right| - |F_{\boldsymbol{h}}|}{\sigma_{\boldsymbol{h}}} \right)^2, \tag{7.28}$$

where N_r is the number of diffraction points, and p is the number of parameters to be refined in the wavefunction model, and $\chi^2 = 0$ indicates a perfect fit, while $\chi^2 = 1$ means that the model agrees with the experimental data within the standard deviation of the experimental diffraction data $F(\boldsymbol{h})$.

Derivatives of χ^2 to the wavefunction parameters can be used in the process of minimizing the Lagrangian. Because the structure factor is a complex number, this calculation requires special attention. The experimentally constrained HF equation can be obtained as: [4, 8]

$$(\boldsymbol{F} + \lambda \boldsymbol{v})\boldsymbol{c} = \boldsymbol{S}\boldsymbol{c}\boldsymbol{\varepsilon}, \tag{7.29}$$

$$\boldsymbol{F}'\boldsymbol{C} = \boldsymbol{S}\boldsymbol{C}\boldsymbol{\varepsilon}, \tag{7.30}$$

$$\boldsymbol{F}' = \boldsymbol{F} + \lambda \frac{2}{N_r - p} \sum_{\boldsymbol{h}}^{Nr} \frac{\left|F_{\boldsymbol{h}}^c\right| - |F_{\boldsymbol{h}}|}{\sigma_{\boldsymbol{h}}^2} \frac{\text{Real}\left(\left(F_{\boldsymbol{h}}^c\right) * \bar{\boldsymbol{I}}_{\boldsymbol{h}} \right)}{\left|F_{\boldsymbol{h}}^c\right|}. \tag{7.31}$$

The experimentally constrained HF equation requires the use of conventional numerical iterative solutions to obtain the molecular orbital coefficients $\mathbf{c}$. For a given λ, as λ increases, the participation of the experimental data will also increase; thus, resembling a least squares refinement, where a large λ will reduce the fitting statistic χ^2. It should be noted that the energy of the self-consistent calculation will also increase, because an elevated λ will shift the solution from the optimal solution of energy. The best solution to the experimentally constrained HF equation is the one with χ^2 close to 1 (i.e. the model structure factor agrees with the experimental structure factor within one experimental standard deviation) and λ is at a maximum.

7.2.4 Open Shell System Method

For unconstrained HF (UHF) wavefunctions, the constraint can be expanded, by minimizing the Lagrangian: [9, 18]

$$L(\mathbf{c}^\alpha, \mathbf{c}^\beta, \boldsymbol{\varepsilon}^\alpha, \boldsymbol{\varepsilon}^\beta, \lambda) = E(\mathbf{c}^\alpha, \mathbf{c}^\beta) - Tr[(\mathbf{c}^\alpha)^+ S\mathbf{c}^\alpha - 1]\boldsymbol{\varepsilon}^\alpha - Tr[(\mathbf{c}^\beta)^+ S\mathbf{c}^\beta - 1]\boldsymbol{\varepsilon}^\beta + \lambda[\chi^2(\mathbf{c}^\alpha, \mathbf{c}^\beta) - \Delta], \tag{7.32}$$

where $\mathbf{c}^\alpha$ and $\mathbf{c}^\beta$ are the molecular orbital coefficients of α and β, respectively, E is the UHF energy, S is the overlapping integral of the selected orbitals, and $\boldsymbol{\varepsilon}^\alpha$, $\boldsymbol{\varepsilon}^\beta$are the conventional Lagrange multipliers, whose role is to introduce the wavefunction orthogonalization condition into the UHF equation. In addition, Δ is the error of χ^2, which is the statistical index of consistency between the theoretical and experimental structure factors.

Setting the derivative of Eq. (7.32) to $\mathbf{c}^\alpha$ and $\mathbf{c}^\beta$ to 0, the experimentally constrained UHF equation can be obtained as:

$$\begin{aligned} \tilde{\boldsymbol{F}}^\alpha \mathbf{c}^\alpha &= S\mathbf{c}^\alpha \boldsymbol{\varepsilon}^\alpha \\ \tilde{\boldsymbol{F}}^\beta \mathbf{c}^\beta &= S\mathbf{c}^\beta \boldsymbol{\varepsilon}^\beta, \end{aligned} \tag{7.33}$$

where $\tilde{\boldsymbol{F}}^\alpha$ and $\tilde{\boldsymbol{F}}^\beta$ are the modified UHF matrices:

$$\begin{aligned} \tilde{\boldsymbol{F}}^\alpha &= \boldsymbol{F}^\alpha - \lambda \boldsymbol{C} \\ \tilde{\boldsymbol{F}}^\beta &= \boldsymbol{F}^\beta - \lambda \boldsymbol{C}. \end{aligned} \tag{7.34}$$

$\boldsymbol{F}^\alpha$ and $\boldsymbol{F}^\beta$ are the conventional UHF matrices, where the constraint matrix $\boldsymbol{C}$ is:

$$\boldsymbol{C} = \frac{2}{N_r - N_p} \sum_j^{N_r} \left(\frac{F_j - F_j^X}{\sigma_j} \right) \left[\frac{\left(F_j^X\right)^* \boldsymbol{I}_j - F_j^X \boldsymbol{I}_j^\dagger}{2\left|F_j^X\right|} \right], \tag{7.35}$$

where the matrix $\boldsymbol{I}_j$, which contains the Fourier transform of the basis function pairs $g_\mu(\boldsymbol{r})g_\nu(\boldsymbol{r})$ of all symmetric molecules in the unit cell, and is multiplied by an appropriate Debye–Waller factor, namely:

$$I_{\mu\nu j} = \sum_{m=1}^{N_m} \exp(2\pi i \boldsymbol{q}_j \cdot \boldsymbol{r}_m) T_{\mu\nu} \int g_\nu^*(\boldsymbol{r}) g_\mu(\boldsymbol{r}) \exp\left(i\boldsymbol{R}_m^T \boldsymbol{q}_j \cdot \boldsymbol{r}\right) \mathrm{d}\boldsymbol{r}, \tag{7.36}$$

where the matrix $\boldsymbol{I}_j$ is independent of molecular orbital coefficients, $\boldsymbol{R}_m^T$ is the rotation matrix of the symmetry operation of the mth molecule, $\boldsymbol{r}_m$ is the reference origin coordinates of the mth symmetric molecule in the unit cell, $T_{\mu\nu}$ is the Debye–Waller factor of the basis function pair $(\mu\nu)$, and $\boldsymbol{q}_j$ is the scattering vector of the jth diffraction point.

For open-shell systems, the unpaired electron density (spin density) can be calculated directly based upon the refined experimentally constrained wavefunction:

$$\rho_s = \rho_\alpha - \rho_\beta, \tag{7.37}$$

where the electron density of spin α, ρ_α and the electron density of spin β, ρ_β can be calculated by the following equations:

$$\rho_\alpha = \sum_{\mu,\nu} D_{\mu\nu}^\alpha g_\mu(r) g_\nu(r)$$

$$\rho_\beta = \sum_{\mu,\nu} D^\beta_{\mu\nu} g_\mu(r) g_\nu(r), \tag{7.38}$$

where $g_\mu(r)$ and $g_\nu(r)$ are the atomic orbital basis functions, and D^α and D^β are the density matrices of electrons α and β, respectively, according to:

$$D^\alpha_{\mu\nu} = \sum_i^{n_\alpha} c^{\alpha *}_{i\mu} c^\alpha_{i\nu}$$

$$D^\beta_{\mu\nu} = \sum_i^{n_\beta} c^{\beta *}_{i\mu} c^\beta_{i\nu}, \tag{7.39}$$

where $c^\alpha_{i\mu}$, $c^\alpha_{i\nu}$, $c^\beta_{i\mu}$, and $c^\beta_{i\nu}$ are the molecular orbital coefficients. The integral of the spin density gives the number of unpaired electrons:

$$n_\alpha - n_\beta = \int \rho_s(\mathbf{r}) d\mathbf{r}. \tag{7.40}$$

7.2.5 Treatment of Relativistic Effects

We usually do not consider core electrons in structural chemistry studies because these core electrons are not involved in bond formation; however, in X-ray diffraction experiments, core electrons make a major contribution to the diffraction density, playing an important role in determining the experimental electron structure based on the diffraction data. Therefore, they must be treated correctly. As the atomic weight increases, the relativistic effect, which becomes increasingly prominent [9, 18], also affects the electron density; therefore, the model describing the core electrons needs to be improved accordingly for heavy atoms.

The scalar second-order Douglas–Kroll–Hess (DKH) Hamiltonian is an approximation of the widely used Dirac-Coulomb Hamiltonian (DCH). One-electron DCH can be written as:

$$\hat{\boldsymbol{h}}_{\mathrm{DCH}} = \begin{pmatrix} V & c\boldsymbol{\sigma} \cdot \hat{\mathbf{p}} \\ c\boldsymbol{\sigma} \cdot \hat{\mathbf{p}} & V - 2mc^2 \end{pmatrix}, \tag{7.41}$$

where c is the speed of light, $\hat{\mathbf{p}}$ is the momentum operator, m is the electron rest mass, V is the potential energy of electron-nucleus Coulomb interactions, and $\boldsymbol{\sigma}$ is the Pauli spin matrix:

$$\sigma_x = \begin{pmatrix} 0 & 1 \\ 1 & 0 \end{pmatrix}, \quad \sigma_y = \begin{pmatrix} 0 & -i \\ i & 0 \end{pmatrix}, \quad \sigma_z = \begin{pmatrix} 1 & 0 \\ 0 & -1 \end{pmatrix}. \tag{7.42}$$

Unlike nonrelativistic Hamiltonian, DCH is a 4×4 matrix that strictly treats spins. However, DCH introduces negative energy states while treating positive energy states, leading to a significant amount of computation.

A key approach involves transforming the DCH into a diagonal block matrix by a unitary matrix that separates the positive and negative energy states:

$$\hat{\boldsymbol{H}}_{\mathrm{decoupled}} = \boldsymbol{U}\hat{\boldsymbol{h}}_{\mathrm{DCH}}\boldsymbol{U}^{+} = \begin{pmatrix} h_+ & 0 \\ 0 & h_- \end{pmatrix}, \tag{7.43}$$

where the h_+ block corresponds to a two-component Hamiltonian containing only positive-energy eigenstate, which is a major consideration in chemical research, and the h_- block that corresponds to a negative energy state can be ignored. The two-component Hamiltonian h_+ can be further divided into spin-independent and spin-dependent terms, of which only the spin-independent terms in h_+ can be retained to obtain the single-component scalar DKH Hamiltonian.

The conversion from DCH to DKH Hamiltonian will be accompanied by a change in the wavefunction of the system:

$$\begin{aligned}\left\langle \psi \middle| \hat{h}_{\text{DCH}} \middle| \psi \right\rangle &= \left\langle \psi \middle| \boldsymbol{U}^{+}\boldsymbol{U}\hat{\boldsymbol{h}}_{\text{DCH}}\boldsymbol{U}^{+}\boldsymbol{U} \middle| \psi \right\rangle \\ &= \left\langle \boldsymbol{U}\psi \middle| \boldsymbol{U}\hat{\boldsymbol{h}}_{\text{DCH}}\boldsymbol{U}^{+} \middle| \boldsymbol{U}\psi \right\rangle \\ &= \left\langle \tilde{\psi} \middle| \hat{\boldsymbol{H}}_{\text{decoupled}} \middle| \tilde{\psi} \right\rangle, \end{aligned} \tag{7.44}$$

where $\tilde{\psi} = \boldsymbol{U}\psi$, and ψ corresponds to $\hat{\boldsymbol{h}}_{\text{DCH}}$, $\tilde{\psi}$ to the transformed Hamiltonian $\hat{\boldsymbol{H}}_{\text{decoupled}} = \boldsymbol{U}\hat{h}_{\text{DCH}}\boldsymbol{U}^{+}$. If ψ and operator $\hat{\boldsymbol{X}}$ are used simultaneously, $\tilde{\psi}$ needs to be used at the same time as $\boldsymbol{U}\hat{\boldsymbol{X}}\boldsymbol{U}^{+}$. However, in practice, $\tilde{\psi}$ is often used simultaneously with $\hat{\boldsymbol{X}}$, resulting in a calculation error in the expectation of $\hat{\boldsymbol{X}}$, which is called the picture-change effect (PCE), and can be written as:

$$PCE(\boldsymbol{X}) = \left\langle \tilde{\psi} | \boldsymbol{U}\hat{\boldsymbol{X}}\boldsymbol{U}^{+} | \tilde{\psi} \right\rangle - \left\langle \tilde{\psi} | \hat{\boldsymbol{X}} | \tilde{\psi} \right\rangle . \tag{7.45}$$

PCE has a prominent effect on the properties related to electrons localized in the nucleus region (the high-energy part of the atom), such as the electric field gradient vector or hyperfine coupling coefficient. These properties must be PCE-corrected, while those related to the valence electron region, such as the electron density and related properties (dipole moment and hyperpolarizability), will be less affected by PCE.

7.3 The X-ray-Constrained Extremely Localized Molecular Orbital Method

The X-ray-constrained extremely localized molecular orbital (XC-ELMO) method is an effective approach for determining the experimental electron density. Molecular orbitals are strictly localized on atoms, chemical bonds, and functional groups, combining the quantum mechanical rigor of the XCW method with the chemical intuition of the traditional multipole model method.

7.3.1 Theoretical Extremely Localized Molecular Orbitals

We can consider a closed shell system containing $2N$ electrons and divide it into f fragments, which can be atoms, chemical bonds, or functional groups. Each fragment i can use a set of localized basis functions $\{|\chi_{i\mu}\rangle\}_{\mu=1}^{M_i}$ defined on the atoms belonging to the fragment. The αth localized molecular orbital of the ith fragment can be defined as:

$$| \varphi_{i\alpha} \rangle = \sum_{\mu=1}^{M_i} C_{i\mu,i\alpha} \mid \chi_{i\mu} \rangle . \tag{7.46}$$

Therefore, the localized molecular orbitals here will be strictly localized to the fragment to which they belong. It should be noted that localized molecular orbitals belonging to different fragments may be non-orthogonal due to some shared portion of the localized basis functions of the overlapping fragments, causing convergence problems in the solution of the localized molecular orbitals.

The coefficients in Eq. (7.46) can be determined by minimizing the system energy, and the wavefunction of the system can be written as:

$$| \Psi_{\text{ELMO}} \rangle = \hat{A} \left[\prod_{i=1}^{f} \prod_{\alpha=1}^{n_i} \varphi_{i\alpha} \overline{\varphi}_{i\alpha} \right] , \tag{7.47}$$

where $\hat{A}$ is the antisymmetrizer, n_i is the number of orbitals occupied in the fragment i, and $\phi_{i\alpha}$ is the spin-orbital containing the space part of $\varphi_{i\alpha}$ and β spin part. Localized molecular orbitals can be determined by minimizing the following energy functional:

$$E[\boldsymbol{\varphi}] = \langle \psi_{\text{ELMO}} \mid \hat{H} \mid \psi_{\text{ELMO}} \rangle , \tag{7.48}$$

where $\hat{H}$ is the nonrelativistic Hamiltonian of the system.

We can define dual orbitals $| \tilde{\varphi}_{j\beta} \rangle$ as:

$$| \tilde{\varphi}_{j\beta} \rangle = \sum_{k=1}^{f} \sum_{\gamma=1}^{n_k} [\boldsymbol{S}^{-1}]_{k\gamma,j\beta} \mid \varphi_{k\gamma} \rangle, \tag{7.49}$$

where $\boldsymbol{S}$ is the overlapping matrix occupying the localized molecular orbitals. Stoll [19] indicated that by the variation of $|\varphi_{j\beta}\rangle$, the differential of the energy functional $E[\boldsymbol{\varphi}]$ could be expressed as:

$$\delta_{(j\beta)} E = 4 \langle \delta\varphi_{j\beta} \mid (1 - \hat{\rho})\hat{F} \mid \tilde{\varphi}_{j\beta} \rangle , \tag{7.50}$$

where $\hat{F}$ is the Fock operator and $\hat{\rho}$ is the system density operator:

$$\hat{\rho} = \sum_{j=1}^{f} \sum_{\beta=1}^{n_j} |\tilde{\varphi}_{j\beta}\rangle\langle\varphi_{j\beta}| = \sum_{j=1}^{f} \sum_{\beta=1}^{n_j} |\varphi_{j\beta}\rangle\langle\tilde{\varphi}_{j\beta}|. \tag{7.51}$$

When $\delta_{(j\beta)}E = 0$, the energy functional can obtain a minimum value; therefore, the localized molecular orbitals will satisfy the following equation:

$$(1 - \hat{\rho})\hat{F}|\tilde{\varphi}_{j\beta}\rangle = 0. \tag{7.52}$$

This equation can be written in the form of the Eigen equation:

$$\hat{F}^j |\varphi_{j\beta}\rangle = \varepsilon_{j\beta}|\varphi_{j\beta}\rangle, \tag{7.53}$$

where $\hat{F}^j$ is the Fock operator of the modified fragment j:

$$\hat{F}^j = \left(1 - \hat{\rho} + \hat{\rho}_j^{\dagger}\right) \hat{F}(1 - \hat{\rho} + \hat{\rho}_j). \tag{7.54}$$

Obviously, this relies on the localized density operator $\hat{\rho}_j$:

$$\hat{\rho}_j = \sum_{\beta=1}^{n_j} |\tilde{\varphi}_{j\beta}\rangle\langle\varphi_{j\beta}|. \tag{7.55}$$

To determine the extremely localized molecular orbitals of the system, the Eigen Eq. (7.53) needs to perform the self-consistent calculation for each fragment. It should be noted that the equations of the different fragments must be coupled together because the Fock operator $\hat{F}^j$ of each fragment is related to the overall density operator of the system.

7.3.2 Refinement of the Experimentally Constrained Extremely Localized Molecular Orbitals

The process of the experimentally constrained extremely localized molecular orbital refinement [13] is similar to the experimentally constrained wavefunction refinement process. Hence, the extremely localized molecular orbitals can be obtained by minimizing the Lagrangian $J[\boldsymbol{\varphi}]$. Because this method has no orbital orthogonality condition, $J[\boldsymbol{\varphi}]$ can be written as:

$$\begin{aligned} J[\boldsymbol{\varphi}] &= \langle\psi_0|\hat{H}_0|\psi_0\rangle + \lambda(\chi^2 - \Delta) \\ &= E_0[\boldsymbol{\varphi}] + \lambda(\chi^2[\boldsymbol{\varphi}] - \Delta), \end{aligned} \tag{7.56}$$

where ψ_0 and $\hat{H}_0$ are the wavefunction and nonrelativistic Hamiltonian, respectively, λ is the Lagrange multiplier, and Δ is the error between the theoretical and experimental values. In addition, χ^2 is the fit statistics of both the magnitude of the calculated value $\left|F_{\boldsymbol{h}}^{\text{calc}}\right|$ and that of the experimental value $\left|F_{\boldsymbol{h}}^{\text{exp}}\right|$ for the structure factor, according to:

$$\chi^2 = \frac{1}{N_r - N_p} \sum_{\boldsymbol{h}} \frac{\left(\eta\left|F_{\boldsymbol{h}}^{\text{calc}}\right| - \left|F_{\boldsymbol{h}}^{\text{exp}}\right|\right)^2}{\sigma_{\boldsymbol{h}}^2}, \tag{7.57}$$

where N_r and N_p are the number of diffraction data and refinement parameters, respectively, and $\sigma_{\boldsymbol{h}}$ is the experimental error and η is the scale factor.

The differential of $J[\boldsymbol{\varphi}]$ to the orbital $|\tilde{\varphi}_{j\beta}\rangle$ can be written as:

$$\delta_{(\varphi_{j\beta})} J = \delta_{(\varphi_{j\beta})} E + \lambda\delta_{(\varphi_{j\beta})}\chi^2, \tag{7.58}$$

where $\delta_{(j\beta)}$ can be given by Eq. (7.50). Considering $\left|F_{\text{h}}^{\text{calc}}\right| = \left[F_{\text{h}}^{\text{calc}}\left(F_{\text{h}}^{\text{calc}}\right)^*\right]^{1/2}$,

$$\delta_{(\varphi_{j\beta})}\chi^2 = \frac{\eta}{N_r - N_p} \sum_{\boldsymbol{h}} \frac{\eta\left|F_{\boldsymbol{h}}^{\text{calc}}\right| - \left|F_{\boldsymbol{h}}^{\text{exp}}\right|}{\sigma_{\boldsymbol{h}}^2\left|F_{\boldsymbol{h}}^{\text{calc}}\right|} \left\{\left(F_{\boldsymbol{h}}^{\text{calc}}\right)^* \delta_{(\varphi_{j\beta})} F_{\boldsymbol{h}}^{\text{calc}} + F_{\boldsymbol{h}}^{\text{calc}} \delta_{(\varphi_{j\beta})} \left(F_{\boldsymbol{h}}^{\text{calc}}\right)^*\right\}. \tag{7.59}$$

We can define this structure factor operator by:

$$\hat{I}_{\boldsymbol{h}} = \sum_{j=1}^{N_m} \mathrm{e}^{i2\pi(\boldsymbol{R}_j\boldsymbol{r}+\boldsymbol{r}_j)\cdot(\boldsymbol{Bh})} = \hat{I}_{\boldsymbol{h},R} + i\hat{I}_{\boldsymbol{h},C}, \tag{7.60}$$

where $\hat{I}_{\boldsymbol{h},R}$ and $\hat{I}_{\boldsymbol{h},C}$ are Hermitian; thus,

$$F_{\boldsymbol{h}}^{\text{calc}} = 2\sum_{i=1}^{f}\sum_{\alpha=1}^{n_i} \langle \varphi_{i\alpha} | \hat{I}_h | \tilde{\varphi}_{i\alpha} \rangle, \tag{7.61}$$

$$\left(F_{\boldsymbol{h}}^{\text{calc}}\right)^* = 2\sum_{i=1}^{f}\sum_{\alpha=1}^{n_i} \langle \tilde{\varphi}_{i\alpha} | \hat{I}_{\mathrm{h}}^{\dagger} | \varphi_{i\alpha} \rangle. \tag{7.62}$$

Stoll indicated the relationship between $|\delta_{(j\beta)}\tilde{\varphi}_{i\alpha}\rangle$ and $|\delta\varphi_{j\beta}\rangle$: [19]

$$| \delta_{(j\beta)}\tilde{\varphi}_{i\alpha} \rangle = (1 - \hat{\rho}) \mid \delta\varphi_{j\beta} \rangle [\boldsymbol{S}^{-1}]_{j\beta,i\alpha} - \mid \tilde{\varphi}_{j\beta} \rangle \langle \delta\varphi_{j\beta} \mid \tilde{\varphi}_{i\alpha} \rangle. \tag{7.63}$$

Therefore,

$$\delta_{(j\beta)} F_{\mathrm{h}}^{\text{calc}} = 2\{ \langle \delta\varphi_{j\beta} | (1 - \hat{\rho}) \hat{I}_{\mathrm{h}} | \tilde{\varphi}_{j\beta} \rangle + \langle \tilde{\varphi}_{j\beta} | \hat{I}_{\mathrm{h}} (1 - \hat{\rho}) | \delta\varphi_{j\beta} \rangle \}, \tag{7.64}$$

$$\delta_{(j\beta)}\left(F_{\boldsymbol{h}}^{\text{calc}}\right)^* = \left(\delta_{(j\beta)} F_{\boldsymbol{h}}^{\text{calc}}\right)^*. \tag{7.65}$$

Substituting Eq. (7.65) into Eq. (7.59) gives:

$$\delta_{(j\beta)}\chi^2 = \frac{2\eta}{N_r - N_p} \sum_{\boldsymbol{h}} \frac{\eta\left|F_{\boldsymbol{h}}^{\text{calc}}\right| - \left|F_{\boldsymbol{h}}^{\text{exp}}\right|}{\sigma_{\boldsymbol{h}}^2 \left|F_{\boldsymbol{h}}^{\text{calc}}\right|} \times \mathrm{Re}\left\{\left(F_{\boldsymbol{h}}^{\text{calc}}\right) * \delta_{(j\beta)} F_{\boldsymbol{h}}^{\text{calc}}\right\}, \tag{7.66}$$

where $\delta_{(j\beta)} F_{\boldsymbol{h}}^{\text{calc}}$ can be written as:

$$\delta_{(j\beta)} F_{\boldsymbol{h}}^{\text{calc}} = 4\left\{\left\langle \delta\varphi_{j\beta} \right| (1 - \hat{\rho}) \hat{I}_{\boldsymbol{h},R} \left| \tilde{\varphi}_{j\beta} \right\rangle + i \left\langle \delta\varphi_{j\beta} \right| (1 - \hat{\rho}) \hat{I}_{\boldsymbol{h},C} \left| \tilde{\varphi}_{j\beta} \right\rangle\right\}. \tag{7.67}$$

Substituting Eqs. (7.62) and (7.67) into Eq. (7.66) gives:

$$\begin{aligned} \delta_{(j\beta)}\chi^2 = 4 \Bigg\{ & \sum_{\boldsymbol{h}} K_{\boldsymbol{h}} \mathrm{Re}\left\{F_{\boldsymbol{h}}^{\text{calc}}\right\} \left\langle \delta\varphi_{j\beta} \right| (1 - \hat{\rho}) \hat{I}_{\boldsymbol{h},R} \left| \tilde{\varphi}_{j\beta} \right\rangle \\ & + \sum_{\boldsymbol{h}} K_{\boldsymbol{h}} \mathrm{Im}\left\{F_{\boldsymbol{h}}^{\text{calc}}\right\} \langle \delta\varphi_{j\beta} | (1 - \hat{\rho}) \hat{I}_{\boldsymbol{h},\mathrm{C}} | \tilde{\varphi}_{j\beta} \rangle \Bigg\}, \end{aligned} \tag{7.68}$$

where

$$K_{\boldsymbol{h}} = \frac{2\eta}{N_r - N_p} \frac{\eta\left|F_{\boldsymbol{h}}^{\text{calc}}\right| - \left|F_{\boldsymbol{h}}^{\text{exp}}\right|}{\sigma_{\boldsymbol{h}}^2 \left|F_{\boldsymbol{h}}^{\text{calc}}\right|}. \tag{7.69}$$

Therefore, the differential of $J[\boldsymbol{\varphi}]$ can be written as:

$$\begin{aligned} \delta_{(j\beta)} J = 4 \Big\{ & \left\langle \delta\varphi_{j\beta} \mid (1 - \hat{\rho}) \hat{F} \mid \tilde{\varphi}_{j\beta} \right\rangle \\ & + \lambda \sum_{\boldsymbol{h}} K_{\boldsymbol{h}} Re\left\{F_{\boldsymbol{h}}^{\text{calc}}\right\} \left\langle \delta\varphi_{j\beta} \right| (1 - \hat{\rho}) \hat{I}_{\boldsymbol{h},R} \left| \tilde{\varphi}_{j\beta} \right\rangle \\ & + \lambda \sum_{\boldsymbol{h}} K_{\boldsymbol{h}} Im\left\{F_{\boldsymbol{h}}^{\text{calc}}\right\} \left\langle \delta\varphi_{j\beta} \right| (1 - \hat{\rho}) \hat{I}_{\boldsymbol{h},C} \left| \tilde{\varphi}_{j\beta} \right\rangle \Big\}. \end{aligned} \tag{7.70}$$

We can set $\delta_{(j\beta)}J = 0$, where the experimentally constrained extremely localized molecular orbitals satisfy the equation:

$$\Big\{(1-\hat{\rho})\hat{F} + \lambda\sum_{h} K_{h}\mathrm{Re}\left\{F_{h}^{\mathrm{calc}}\right\}(1-\hat{\rho})\hat{I}_{h,R} + \lambda\sum_{h} K_{h}\mathrm{Im}\left\{F_{h}^{\mathrm{calc}}\right\}(1-\hat{\rho})\hat{I}_{h,C}\Big\}\,|\tilde{\varphi}_{j\beta}\rangle = 0, \tag{7.71}$$

which is equivalent to:

$$\hat{F}^{j,\mathrm{exp}}\mid\varphi_{j\beta}\rangle = \varepsilon_{j\beta}^{\mathrm{exp}}\mid\varphi_{j\beta}\rangle, \tag{7.72}$$

where $\hat{F}^{j,\mathrm{exp}}$ is the modified Fock operator of the fragment j:

$$\begin{aligned}\hat{F}^{j,\mathrm{exp}} &= \left(1-\hat{\rho}+\hat{\rho}_{j}^{\dagger}\right)\hat{F}(1-\hat{\rho}+\hat{\rho}_{j}) \\ &+ \lambda\sum_{h} K_{h}\mathrm{Re}\left\{F_{h}^{\mathrm{calc}}\right\}\left(1-\hat{\rho}+\hat{\rho}_{j}^{\dagger}\right)\hat{I}_{h,R}(1-\hat{\rho}+\hat{\rho}_{j}) \\ &+ \lambda\sum_{h} K_{h}\mathrm{Im}\left\{F_{h}^{\mathrm{calc}}\right\}\left(1-\hat{\rho}+\hat{\rho}_{j}^{\dagger}\right)\hat{I}_{h,C}(1-\hat{\rho}+\hat{\rho}_{j}).\end{aligned} \tag{7.73}$$

To experimentally obtain the constrained extremely localized molecular orbitals, Eq. (7.72) must perform self-consistent calculations on each fragment, similar to the process for obtaining the theoretical extremely localized molecular orbitals. In addition, the equations for the different fragments will be coupled together since the Fock operator for each fragment $\hat{F}^{j,\mathrm{exp}}$ is related to the overall density operator of the system.

References

1 Helgaker, T., Jorgensen, P., and Olsen, J. (2000). *Molecular Electronic Structure Theory*. Chichester: Wiley & Sons, LTD.

2 Parr, R.G. and Yang, W. (1989). *Density Functional Theory of Atoms and Molecules*. New York: Oxford University Press.

3 Jayatilaka, D. (1998). Wave function for beryllium from x-ray diffraction data. *Phys. Rev. Lett.* 80: 798–801.

4 Jayatilaka, D. and Grimwood, D.J. (2001). Wavefunctions derived from experiment. i. motivation and theory. *Acta Cryst. A* 57: 76–86.

5 Grimwood, D.J. and Jayatilaka, D. (2001). Wavefunctions derived from experiment. ii. a wavefunction for oxalic acid dihydrate. *Acta Cryst. A* 57: 87–100.

6 Bytheway, I., Grimwood, D., and Jayatilaka, D. (2002). Wavefunctions derived from experiment. iii. topological analysis of crystal fragments. *Acta Cryst. A* 58: 232–243.

7 Bytheway, I., Grimwood, D.J., Figgis, B.N. et al. (2002). Wavefunctions derived from experiment. iv. investigation of the crystal environment of ammonia. *Acta Cryst. A* 58: 244–251.

8 Grimwood, D.J., Bytheway, I., and Jayatilaka, D. (2003). Wavefunctions derived from experiment. v. investigation of electron densities, electrostatic potentials, and electron localization functions for noncentrosymmetric crystals. *J. Comput. Chem.* 24: 470–483.

9 Hudák, M., Jayatilaka, D., Perašínová, L. et al. (2010). X-ray constrained unrestricted hartree-fock and douglas-kroll-hess wavefunctions. *Acta Cryst. A* 66: 78–92.

10 Jayatilaka, D. (2012). Using wavefunctions to get more information out of diffraction experiments. In: *Modern Charge-Density Analysis* (ed. C. Gatti and P. Macchi), 213–257. Berlin: Springer.

11 Dos Santos, L.H.R., Genoni, A., and Macchi, P. (2014). Unconstrained and X-ray constrained extremely localized molecular orbitals: analysis of the reconstructed electron density. *Acta Cryst. A* 70: 532–551.

12 Genoni, A. (2013). X-ray constrained extremely localized molecular orbitals: theory and critical assessment of the new technique. *J. Chem. Theory Comput.* 9: 3004–3019.

13 Genoni, A. and Meyer, B. (2016). X-ray constrained wave functions: fundamentals and effects of the molecular orbitals localization. *Adv. in Quantum Chem.* 73: 333–362.

14 Dittrich, B., Koritsanszky, T., and Luger, P. (2004). A simple approach to nonspherical electron densities by using invarioms. *Angew. Chem. Int. Ed.* 43: 2718–2721.

15 Stewart, R.F. (1969). Generalized X-ray scattering factors. *J. Chem. Phys.* 51: 4569–4577.

16 Coppens, P., Willoughby, T.V., and Csonka, L.N. (1971). Electron population analysis of accurate diffraction data. I. Formalisms and restrictions. *Acta Cryst. A* 27: 248–256.

17 Tanaka, K. (1988). X-ray analysis of wavefunctions by the least-squares method incorporating orthonormality. I. General formalism. *Acta Cryst. A* 44: 1002–1008.

18 Bucinsky, L., Jayatilaka, D., and Grabowsky, S. (2016). Importance of relativistic effects and electron correlation in structure factors and electron density of diphenyl mercury and triphenyl bismuth. *J. Phys. Chem. A* 120: 6650–6669.

19 Stoll, H., Wagenblast, G., and Preuss, H. (1980). On the use of local basis sets for localized molecular orbitals. *Theor. Chim. Acta* 57: 169–178.

8

Functional Electronic Structures and Functional Motif of Materials

8.1 Introduction

The invention and application of new materials serve as milestones in human history. However, the traditional manner of cooking to discover new materials lacks theoretical guidelines, resulting in a long research cycle, and this approach no longer meets growing needs. Therefore, researchers are eager to develop new theories and methods to guide new materials research with specific functions and pursue changes in the research of materials science.

Functional motif refers to the microstructural unit that plays a key role in generating material functions. The performance of a material will depend on the functional motif and its arrangement in the material structure, and understanding the functional motif and its arrangement is crucial for understanding material properties and rationally designing new materials with specific functions [1]. Based on the functional motif philosophy, we proposed a new materials science research paradigm. This new paradigm studies the crystal and electronic structures of a material, along with its macroscopic physiochemical properties, and then uses this information to derive and understand the dominating functional motif of a certain property. A quantitative structure-property relationship can then be established, and an in-depth study and summary of the functional motif and its quantitative structure–property relationship will eventually form the functional motif theory. This theory can be used to efficiently and systematically design new materials with certain functions, or predict new functions of known materials. Eventually, high-performance materials can be obtained through the modular assembly and chemical synthesis of functional motifs. Therefore, the development cycle of new materials can be greatly shortened.

This chapter first introduces the concept of material functional motif, where six types of microstructures (crystal structure, electronic structure, magnetic structure, and defect structures with modulated statistical and local defects) that determine material performances are systematically elaborated. Typical functional materials with corresponding structural characteristics are used as examples and explained, setting the foundation of materials design. Subsequently, we introduce the functional electronic structure concept, defined as the electronic structures that

Electronic Structure Crystallography and Functional Motifs of Materials, First Edition.
Guo-Cong Guo and Xiao-Ming Jiang.

dominate specific material functions, as studying the functional electronic structure is the key to revealing the structure–property relationship.

8.2 Material Functional Motif

The structure of a substance or material will determine its properties, and materials with a certain structure often share similar properties. Similar to the concept of functional groups in organic compounds and genes in biology, a structural unit with a specific structure can be defined as a functional motif. Functional motif refers to the microscopic structural units, such as components and structural modules, that play a key role in creating material functions and cannot be replaced by other structural units without loss or significantly suppressing the corresponding functions. Functional motifs are usually structurally independent and chemically stable and can be obtained relatively easily through chemical reactions.

The intrinsic properties and functions of materials can be mostly determined by the microscopic structure. In terms of the length scale of structural features, material structures can be classified into macroscopic (>~1 μm), mesoscopic (~10 nm–1 μm), and microscopic (<~10 nm), with the microscopic structure determining the intrinsic material properties, and the macroscopic and mesoscopic structures (i.e. superlattice and domain structure) playing an important role in generating new extrinsic functions that are not inherent to the materials but enhance the intrinsic properties of materials. The microscopic structure is usually the main focus and can be classified into different types based on the functional motif (featured electron density, spin, molecule, atom, and functional group) and the structure order (the arrangement of the functional motif), such as the electronic structure (i.e. electron density distribution, electron wave function, electron cloud, and orbital), magnetic structure (spin order), crystal structure (the three-dimensional (3D) periodic order of molecules and atoms), and the defect structure (the arranged structures of certain molecules or atoms without strict 3D periodicity). The defect structure, based on the degree of defects, can be further classified into modulated defects, statistical defects, and local defects. Modulated defects are more commonly referred to as modulated or aperiodic structures, as modulated structures are essentially lattice defects (relative to a perfect crystal). Modulated defects refer to the regular distortion of structural parameters, such as the position, occupancy number, or temperature factor of some atoms in the crystal structure. This distortion can be described by using a periodic function dissimilar to the basic lattice period, where the overall structure has no 3D translation periodicity, but still has long-range order. Statistical defects refer to the fact that the degree of distortion of structural parameters, such as the position, occupancy number, or temperature factor of some atoms, cannot be described by periodic functions, but will still retain a certain long-range order statistically. Local defects signify that the degree of disorder of the atoms or molecules in the structure is so large that no long-range statistical order can be observed. Therefore, the structure can only be described within a few coordination atomic layers. The typical characteristics of

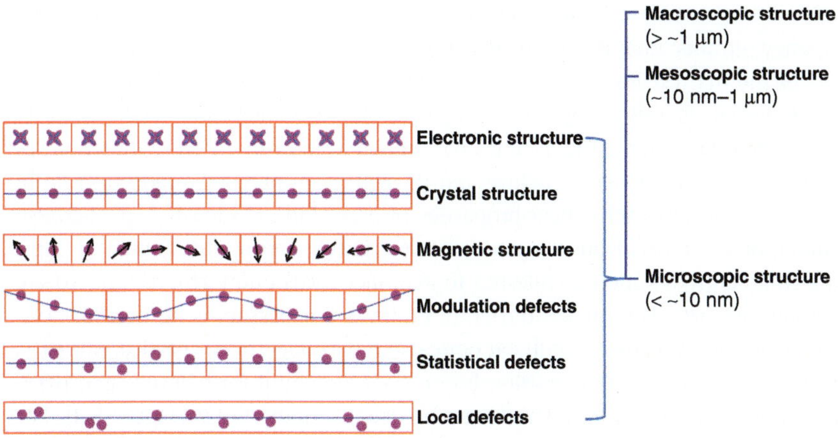

Figure 8.1 Schematic diagram showing the types and main features of material microstructures. Squares represent unit cells, circles represent atoms, arrows represent spins, and butterfly like shapes represent electron orbitals [1].

these microstructures are shown in Figure 8.1. Among these microstructures, the electronic structure will be on the subatomic or electronic orbital scale, with others arranged structures on the molecular or atomic scale.

Modulated defects, statistical defects, and local defects have a common structural and chemical origin; namely, they are all a result of the microscopic disorder of chemical bonds, atoms, or molecules in the material. These three defect structure types are simply different defect forms of the same microscopic disorder at different scales, and in most cases, they coexist in the same material. Additionally, different analysis theories and methods need to be applied to these three defect structures. The following sections briefly introduce the structural characteristics of these microstructure types and the material properties that they affect so that readers can obtain a comprehensive understanding of material microstructures, which facilitate our understanding of the concept of material functional motif.

8.2.1 Crystal Structure

In crystalline materials, the structure units (e.g. atoms, ions, or molecules) are arranged in a highly ordered manner, forming a periodic repeated lattice that extends in three dimensions. The crystal structure is the most common way to describe the microstructure of a crystalline material, and it contains structural information such as atom positions, their occupancy, and the thermal factor. The symmetry of an isolated structural unit can be described by 32-point groups. A typical feature of crystal structure is the arrangement of structural units with translational symmetry, which can be described by space groups (a total of 230 space groups). The smallest structural unit repeated in 3D space is known as the unit cell, with the length (a, b, c) of the principal axis or edge of the unit cell and the angle (α, β, γ) between them known as the unit cell parameters (or lattice constants).

Atoms within a unit cell are typically described by the fractional coordinate, occupancy number, and temperature factor.

One of the main features that distinguishes crystalline materials from noncrystalline materials is anisotropy. For example, parameters such as electrical conductivity, dielectric constant, and Young's modulus may vary in different directions of a crystal. Thus, the crystal structure and its symmetry play an important role in determining the physiochemical properties of crystalline materials (e.g. cleavage, nonlinear optics, piezoelectricity, and ferroelectricity). Among the 32-point groups, 21 are non-centrosymmetric (referred to as noncentral) point groups, and 10 are polar point groups. The noncentral point group is a necessary condition for crystals to exhibit second-order nonlinear optical and piezoelectric properties, while the point group is a necessary condition for crystals to exhibit ferroelectric and pyroelectric properties. Crystal materials can be used in many fields (Figure 8.2). For example, the NaCl crystal is the most commonly used conventional spectral window material. Its optical transparency is mainly due to the electron configuration of Na^+ and Cl^- in its crystal structure, similar to that of inert gas. The β-BaB_2O_4 crystal exhibits strong ultraviolet-visible-near-infrared second-order nonlinear optical properties, which are mainly due to the non-centrosymmetric arrangement of the nonlinear optical active group $(B_3O_6)^{3-}$. The quartz crystal (SiO_2) is a well-known piezoelectric material, with its piezoelectric properties mainly determined by the electric dipole moments (Si—O bonds) arranged non-centrosymmetric in its structure. $BaTiO_3$ shows prominent ferroelectricity at room temperature, which is mainly derived from the distorted TiO_6 unit in its crystal structure. The zeolite crystal has a 3D pore structure, which enables it to absorb and separate certain molecules or

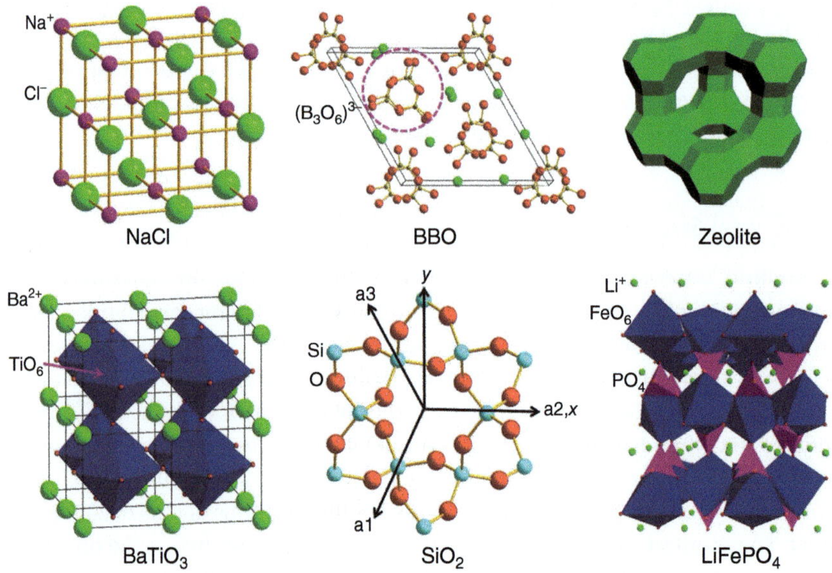

Figure 8.2 The crystalline structures of the above-mentioned crystal materials [1].

ion groups. $LiFePO_4$ is a typical anode material for lithium-ion batteries, and its charge–discharge performance is mainly determined by its porous crystal structure and multivalent iron ions.

8.2.2 Electronic Structure

Electrons in materials are essentially de Broglie (probability) waves, and as a result, they fit into the Pauli exclusion principle, with density distribution in real space (referred to as electron density) and density distribution in momentum space (referred to as momentum density). The electron density and momentum density complement each other and jointly determine the intrinsic properties of a material. The material electronic structure generally refers to electron density and momentum density, as these determine the collective behavior of electrons. The electron wave function and density matrix are fundamental physical quantities that describe electronic structure and are equivalent in the pure state. Concepts such as orbitals (e.g. s, p, and d orbitals) and chemical bonds (e.g. σ bonds and π bonds) describe the electron properties in real space, and concepts that describe the atomic energy level, energy bonds, and density state can all be used to explain the many physical and chemical phenomena and principles observed in materials.

In real (or position) space, the electronic structure information can be obtained through topological analysis of the electron density or wave function. This information includes topological atomic indicators (i.e. atomic integral charge, topological volume, atomic energy, atomic force, dipole moment, quadrupole moment, and octupole moment) and topological chemical bond indicators (i.e. charge at the critical bond point, Laplace value, ellipsoid, and bond energy). Material properties are related to some of the topological properties of the electronic structure. For example, the Laplacian of the electron density is the second derivative of the electron density of the space coordinate, with its sign determining whether the electron density at a certain position is electron-depleted or enriched relative to the adjacent electron density. When $\nabla^2\rho(\boldsymbol{r}) > 0$, the electron density at position $\boldsymbol{r}$ will be smaller than the mean electron density in its neighborhood state, and vice versa, while $\nabla^2\rho(\boldsymbol{r}) < 0$ indicates that the region at $\boldsymbol{r}$ will be an electron-enriched region. According to Lewis acid-base electron theory, all substances (molecules, ions, or atomic groups) that accept electron pairs are acids, while all substances (molecules, ions, or atomic groups) that donate electron pairs are bases. Electron-enriched regions can be classified as a Lewis base (electron donor), and electron-depleted regions can be classified as a Lewis acid (electron acceptor).

The real-space electronic structure of crystalline materials can be characterized by high-precision X-ray diffraction, and the ionic and covalent states of the chemical bonds can be observed from the deformation density (the difference between the experimental electron density and the model density). For example, LiB_3O_5 is the most widely used ultraviolet-visible-near-infrared nonlinear optical crystal. Figure 8.3 shows the experimental deformation electron density and electrostatic potential of LiB_3O_5 in real space, where the lone pair of electrons on the O atom and the bond charge of the B—O bond can be observed from the deformation density.

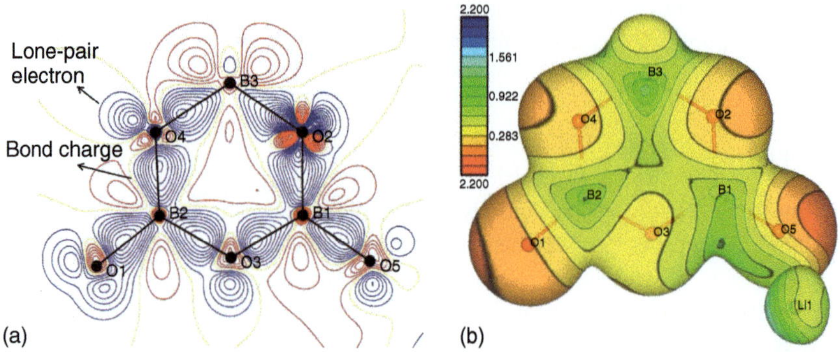

Figure 8.3 (a) Static deformation electron density and (b) electrostatic potential maps of the asymmetric unit of LBO. Source: Data were obtained by multipole refinement of X-ray diffraction data using Mo Kα radiation (λ = 0.71073 Å) at 85 K [1].

No obvious shared electrons were found between Li and O, indicating that B—O and Li—O are mainly covalent and ionic bonds, respectively. The corresponding electrostatic potential diagram shows that B and O have a net positive charge and net negative charge, respectively, consistent with the classical B—O bond characteristics. The Li atom has negligible valence electrons and a net positive charge.

The electronic structure in the *k*-space of crystalline materials includes energy level of occupied and unoccupied states, energy bands, the density of states, and so on. The intrinsic material properties (such as electrical transport, optics, magnetism, and superconductivity) will be closely related to its electronic structure. For example, according to the flat/steep band model of superconductors, the electronic band structure of potential superconductors will show a steep band passing through the Fermi level and a flat band close to the Fermi level, such as the case of Y_2O_2Bi [2]. Additionally, the energy band structure of topological insulators will exhibit topological order protected by symmetry, causing the interior of the bulk phase to behave like an insulator, while the surface will be in a metallic state, implying that electrons may move along the material surface [3].

Figure 8.4 shows a schematic diagram of the electronic band structure for some typical materials. Metals have partially occupied energy bands, with no gap between the highest occupied state and the lowest occupied state. In the band structure of an insulator, the valence and conduction bands are separated by a band gap, semiconductors have relatively narrow band gaps, and semimetals have small overlaps in their valence and conduction bands. One spin of a half-metal is a metal state, while the other spin is a semiconductor state, and when the conduction band and valence band meet at the Fermi level, the material will be a gapless semiconductor or a zero-gap material. Zero-gap materials may have quadratic linear or asymmetric *E*-*k* dispersion relations. If the valence band of one spin meets the conduction band of another spin at the Fermi level, it will be a spin-gapless semiconductor [4].

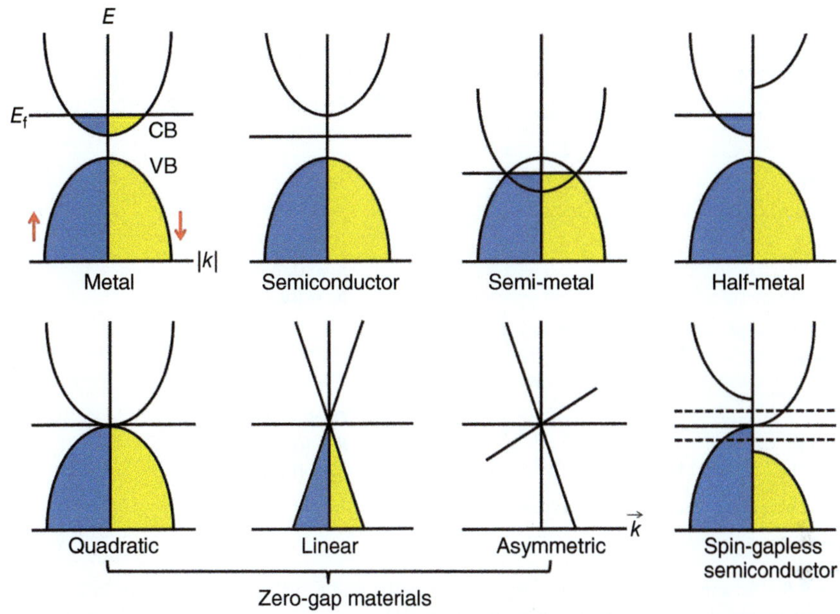

Figure 8.4 Electronic band structures of various classes of materials in terms of the band dispersions around the Fermi level. Blue and yellow areas indicate the spin-up and spin-down states, respectively. In intrinsic zero-gap materials, the conduction and valence band edges touch the Fermi level. The band structure of spin-gapless semiconductors is dependent on spin. Source: The schemes are redrawn with the Ref. [4].

8.2.3 Magnetic Structure

Magnetic structure often refers to the ordered arrangement of static magnetic moments in a crystal lattice below a particular temperature (i.e. the magnetic phase transition temperature). When the ratio of the magnetic structure period to the lattice period length is a rational number, then it will be a commensurate magnetic structure, and when the value is irrational, it will be an incommensurate magnetic structure. Incommensurate structures can be described by the propagation vector method, while commensurate structures can be described either by the propagation vector method or the magnetic space group method. When the compound contains only one magnetic ion, or the magnetic moment arrangement of the ion is colinear (i.e. ferromagnetic or antiferromagnetic coupling), the magnetic structure will be relatively simple. When two or more magnetic ions are contained, or the spin arrangement is non-colinear, the magnetic structure will be more complex and may be sinusoidal, helical, cycloid, or conical [5]. Figure 8.5 shows typical sinusoidal, helical, and cycloidal magnetic structures. In helical, cycloidal, and conical orders, the magnetic moments will be the same at all positions, but in different directions. In the helical order, the helical rotation axis will be consistent with the propagation direction, while in the cycloidal order, the helical rotation axis will be perpendicular to the propagation direction. The sinusoidal structure can also

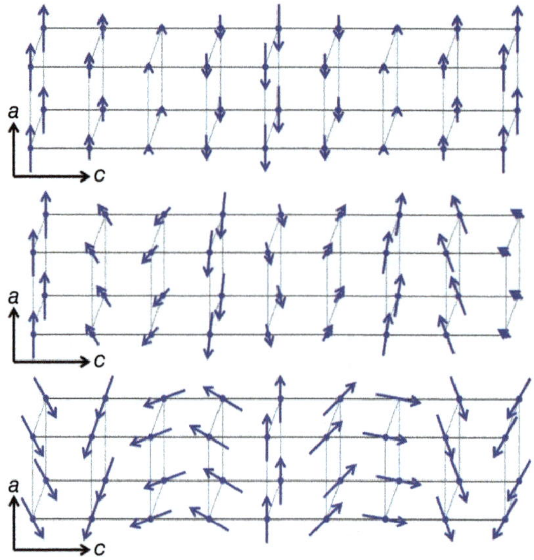

Figure 8.5 Schematic of a spin density wave, a helix, and a cycloid that all occur in a magnetic material with moderate spin frustration [1].

be defined as a spin-density wave. In sinusoidal structures, the magnetic moment direction is colinear, but with different magnitudes. These magnetic structures are often found in frustrated spin systems, where not all spin alignments are colinear. When the degree of spin frustration is moderate, the system will usually adopt a non-colinear magnetic structure, such as helical or cycloidal orders. Both helical and cycloidal orders are chiral, and the spins can rotate around the rotation axis either clockwise or counterclockwise, resulting in two magnetic structures with different chirality but the same energy level. When two chiral magnetic structures with equal probability appear below a critical temperature, the two orders will overlap and form a spin density wave. Below the critical temperature, the material magnetic structure may also adopt one of the two chiral structures. With a high degree of spin frustration, the magnetic atomic spins will compete with each other and will not be able to form long-range order. Subsequently, spin liquid or spin glass behavior may occur. In this state, the magnetic ground state will present as a large number of different spin configurations.

The properties of magnetic materials are closely related to their magnetic structures. Ferromagnetic materials, such as transition metal irons, cobalt, nickel, and rare earth metals with atomic numbers from 64 to 69, as well as some ferromagnetic alloys such as MnBi and Cu_2MnAl, can achieve permanent magnetization. Magnetoelectric compounds, such as CuO, $CuBr_2$, $BiMnO_3$, $BiFeO_3$, and $RMnO_3$ (R=Ho—Lu, Y), are type-I multiferroic materials and can exhibit concurrent magnetism and magnetic-driven ferroelectricity. Some special magnetic structures can exist in these compounds, such as cycloidal or E-type antiferromagnetic magnetic structures, leading to spatial symmetry breaking and ferroelectricity.

8.2.4 Modulated Defects

Translational symmetry is one of the main features of crystal structures and a typical manifestation of long-range structural order. However, the long-range order of a crystalline material is not only presented as translational symmetry, but also as the modulated structure, which is the 3D periodic modulation resulting from the deviation of the atomic position, occupancy number, or temperature factor from its equilibrium value, on the basis of the basic lattice [6, 7]. Modulated structures include commensurate modulation, incommensurate modulation, and composite structures. Modulated structures can be considered the result of introducing regular defects that can be described by the modulation functions in a perfect lattice structure. Therefore, in this book, they are also referred to as modulated defects.

Common modulation functions used to describe modulation structures include harmonic functions, square wave functions, and triangular wave functions. The directions and periodicity of the modulation function can be described by the modulation vector in the reciprocal space. When the modulation vector is a rational multiple of its lattice period, the modulation structure will be commensurate, whereas when the modulation vector is an irrational multiple of its lattice period, the modulation structure will be incommensurate. Commensurate modulation structures can also be described by supercells. When modulation occurs due to a shift in atomic position, it is known as displacement modulation (Figure 8.6a), and when it is due to a shift in atomic occupancy probability, it is referred to as occupancy modulation (Figure 8.6b). The modulated structure can be mathematically described as a section of the $3 + n$-dimensional superspace. The reciprocal space of modulated structure is a projection from the $3+n$-dimensional super-reciprocal space to the 3D reciprocal space, where n is the number of modulation vectors.

Modulated defects can typically be found in compounds with phase transition properties, such as $NaNO_2$, K_2SeO_4, and $LiAlSiO_4$. Those compounds undergo

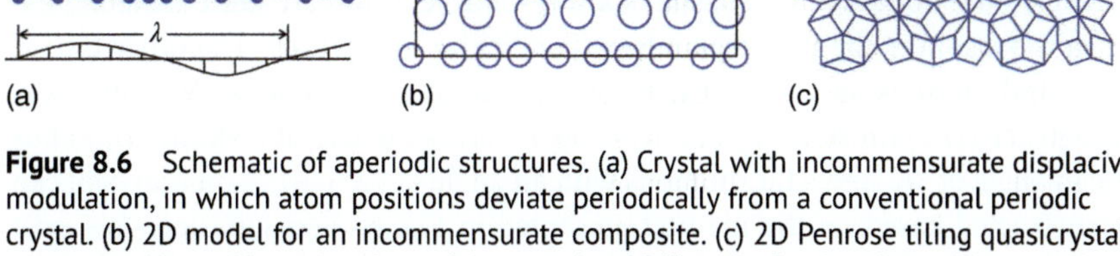

Figure 8.6 Schematic of aperiodic structures. (a) Crystal with incommensurate displacive modulation, in which atom positions deviate periodically from a conventional periodic crystal. (b) 2D model for an incommensurate composite. (c) 2D Penrose tiling quasicrystal constructed by a skinny rhombus with an angle of 36° and a fat rhombus with an angle of 72°. Source: The schemes are redrawn with the Ref. [6].

phase transition at a critical temperature, with the modulated structure emerging in an intermediate state between the low-symmetry low-temperature phase and high-symmetry high-temperature phase. Two interpenetrating periodic lattices, when arranged incommensurately with each other, will form an incommensurate composite structure. This type of structure can easily form in mismatched layered compounds, such as $[LaS]_{1.14}[NbS_2]$ [8]. Other cases where incommensurate composite structures may form include two colinear arrangements of atoms with their own periodicities, such as $[Sr]_{1+x}[TiS_3]$ ($x \approx 0.1$) [9].

Modulated structures can be observed in many crystalline materials, such as minerals, metals, organic and inorganic compounds, and some macromolecules, with the properties of some materials depending on the aperiodic modulation order in their structure. For example, when some metals are cooled, their charge density will exhibit modulation of the metallic state, leading to a charge density wave (CDW) state. Driven by electro-acoustic coupling associated with the metal-nested Fermi surface, the formation of CDW will introduce an insulating bandgap at the metallic Fermi surface, where the superconducting state is similar to CDW. Due to the formation of Cooper pairs associated with superconductivity, a bandgap will also be introduced at the Fermi surface, preventing the transition from the superconducting state to the metallic state unless the temperature is increased [10]. The superconducting state will compete with the CDW [11], and this competition will also occur in some transition metal chalcogenides, such as $NbSe_2$, $HfTe_3$, $ZrTe_{3-x}Se_x$, and $YBa_2Cu_3O_y$. Modulated structures are also found in some nonlinear optical crystals, such as $Cs_2TB_4O_9$ (T=Ge, Si) [12] and A_2SnS_5 (A=Ba, Sr) [13], and these modulated structures will have an impact on their second harmonic generation properties. Compared to α-Ba_2SnSe_5 (without structure modulation), a 44% or 25% distortion of the Sn/S unit in A_2SnS_5 (A=Ba, Sr) can significantly enhance the second harmonic signal. Modulated structures are also present in other material systems, such as relaxor ferroelectrics $Ca_{0.24}Ba_{0.76}Nb_2O_6$ and $Ca_{0.31}Ba_{0.69}Nb_2O_6$ [14], and multiferroic $CaMn_7O_{12}$ with low Cu-doping [15].

Quasicrystals are similar to incommensurate modulated structures and have long-range order, but no translational symmetry. Compared to traditional 3D translational periodic crystal structures, the quasicrystal structure can be regarded as a generalized lattice defect, but it is fundamentally different from traditional crystal structures. Quasicrystals have a unique diffraction spectrum that exhibits quintuple or other symmetries, which do not belong to the point groups of the crystal. There are two types of known quasicrystals. One is polygonal quasicrystals, which have local 8-fold, 10-fold, or 12-fold rotational symmetry, corresponding to octagonal, decagonal, and dodecagonal quasicrystals, respectively, and are periodic along the axes and quasi-periodic in the planes perpendicular to the axes. The other type is icosahedral quasicrystals, which are nonperiodic in all directions. A simple way to construct 2D Penrose tiling quasicrystals involves the use of a skinny rhombus with an angle of 36° and a fat rhombus with an angle of 72° [16]. A quasicrystal can be constructed based on the method presented in Figure 8.6c. Quasicrystals exist in many metal alloys (e.g. Al—Li—Cu, Al—Mn—Si, Al—Ni—Co, Al—Pd—Mn, Al—Cu—Fe, and Al—Cu—V) and polymers.

8.2.5 Statistical Defects

The translation periodicity of a perfect crystal structure can be described by the lattice base vector and translation vector. Although modulated defects (structures) have no translation periodicity, they can use the modulation function and vector to describe long-range order. A material with a high degree of disorder will lose the strict long-range order, but still retain the statistical long-range order, which can be defined as statistical defects. Here, we provide an example to distinguish between crystal structure, modulated defects, statistical defects, and local defects (see next section). Assuming that 1 represents the atoms in the equilibrium position, and 2 and 3 represent the atoms in different nonequilibrium positions, each unit cell containing only one atom with a perfect crystal structure can be shown as 111111111. The modulation defects, statistical defects, and local defects can be shown as 123123123, 123312213, and 113111211, respectively. The defect structure 123312213 has no strict long-range order, but the three adjacent segments, namely, 123, 312, and 213, will have the same mean value of 2 (assuming 1, 2, and 3 can be added up); thus, indicating statistical long-range order. In local defect 113111211, neither strict nor statistically long-range order will exist.

The structural characteristics of statistical defects are more complicated than other microstructures. Due to the variety of statistical defects, it is difficult to obtain a unified description method. The structural parameters used to describe defect structures significantly differ from other microstructures [17]. In principle, any model parameter that facilitates defect structure description can be used in simulation and refinement calculations. Correlation coefficients (short-range parameters) and adjacency probabilities are common parameters for describing defect structures. In crystal structure analysis, to conveniently describe statistical defects, mean defect structure parameters, such as the split atom position, mixed or partial occupancy number, and abnormal temperature factor, are often used. However, using average defect structure parameters to describe defects may exclude important information. Different defect structures with the same average defect structure parameters may have different atomic positions and occupancy distributions. For example, with the same 30% vacancy defects, different vacancies may be present depending on the correlation coefficients between adjacent atoms along different directions (Figure 8.7), which can further lead to different physical and chemical properties.

Many material properties are related to defects, such as the mechanical properties, light damage threshold of optical materials, and the carrier lifetime of semiconductors. The impedance of semiconductors and the diffusion of ionic crystals are dominated by defects. Defects strongly affect electron and phonon behaviors, and play an important role in the regulation of thermal and electrical properties of thermoelectric materials. Defects such as cracks, wrinkles, and domain boundaries in graphene prepared via the vapor transport deposition method will affect its electrical properties. Defects can also affect the electrochemical performance of electrode materials. For example, $LiTMO_2$ (TM = Ni, Mn, Co, or $Ni_xMn_yCo_z$, where $x+y+z=1$) is an important anode material in high-energy-density lithium-ion batteries. In the defect structure of the anode, the highly concentrated Ni/Li disordered

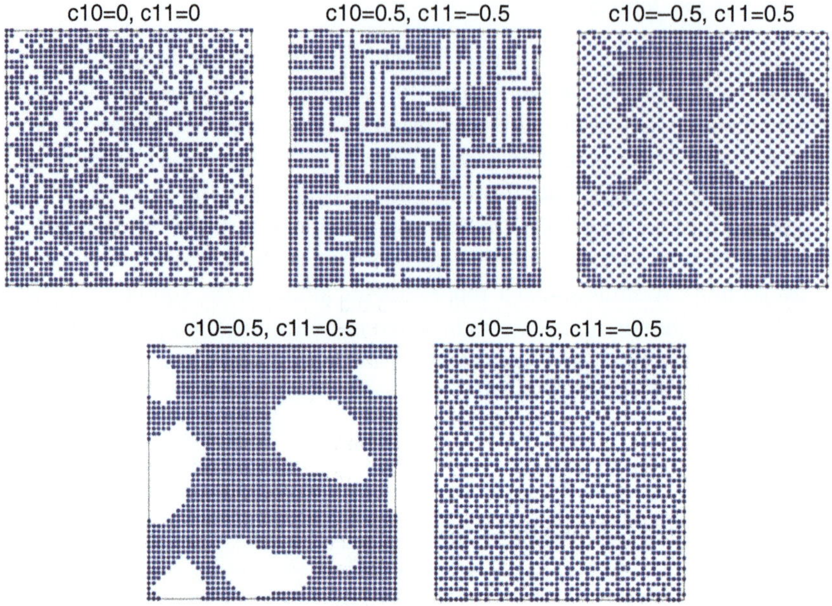

Figure 8.7 2D models using 50 × 50 cells (one atom in each cell) with different correlation coefficients (c10 and c11) between nearest neighboring atoms along cell axis [(1, 0), (0, 1), (−1, 0), and (0, −1) directions for c10] and diagonal directions [(1, 1), (−1, 1), (1, −1), and (−1, −1) directions for c11]. The zero, positive, and negative values of c10 and c11 are for random, positive, and negative correlation cases, respectively [1].

structure (Ni/Li anti-site) will have a negative impact on the Li migration ability, cycle stability, first-cycle efficiency, and overall electrode performance [18]. However, less concentrated Ni/Li disordered structures can improve the thermal and cyclic structural stability of Ni-rich $LiTMO_2$ materials.

8.2.6 Local Defects

In materials with a high degree of structural disorder, such as liquids and glass, only the structure within a few coordination layers of atoms will be considered in research, and defined as local structure. The positions of atoms or molecules on a larger scale will show an almost random distribution. In contrast to the complete coordination situation, the defects in the coordination environment can be defined as local defects, which include abnormalities or distortions of coordination numbers, bond lengths, bond angles, and coordination polyhedrons. It should be noted that local defects are not only studied in long-range structures. In many cases, local structures and defects need to be studied in other microstructures, depending on the specific problem.

Local defects widely exist in various materials, affecting the properties and functions of some specific materials, such as negative thermal expansion materials, amorphous materials, and shape memory alloys. Most existing functional materials contain impurities, as they are doped with different elements to achieve different

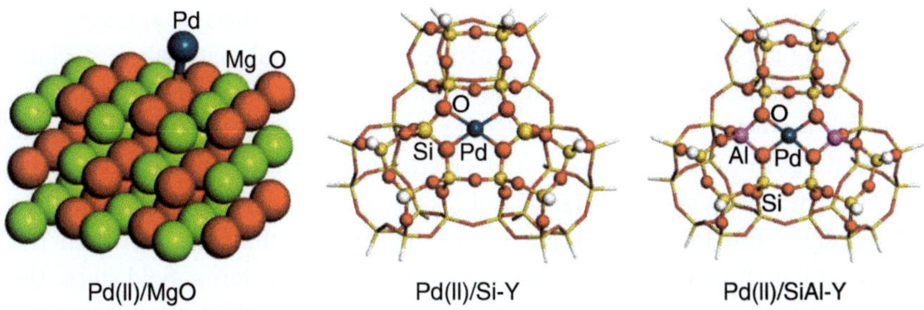

Figure 8.8 Optimized geometries of model-supported Pd(II) catalysts with local structures, Pd(II)/MgO, Pb(II)/Si-Y and Pb(II)/SiAl-Y, formed by divalent palladium and MgO, Y-type Si, and SiAl zeolites, respectively. Source: Reproduced from Ref. [19] with permission. © 2019 the American Chemical Society.

functions. In some specific cases, local defects play a dominant role in determining material performance. For example, doping silicon with boron or phosphorus at a concentration of 0.1% or less will result in p-type or n-type semiconductors. Some solid-phase catalysts used in the chemical industry contain special metal sites that are optimal active sites for catalytic reactions. Therefore, it is necessary to study the local structure and defects around active sites. For example, palladium-based catalysts need to be used in the catalytic reaction to produce dimethyl carbonate. Palladium atoms are used on different bases to form active catalytic sites, and the local structures formed by palladium atoms and different bases will differ. Figure 8.8 shows the local structures of catalytic active sites formed by divalent palladium and MgO, Y-type Si, and SiAl zeolites, respectively, with the three local structures showing different electron adsorption and HOMO-LUMO energy bandwidths, resulting in different stabilities of the formed palladium-based catalysts [19].

8.3 Functional Electronic Structures

Among the above-mentioned typical microstructures, the electronic structure will be particularly important for several reasons. First, the electronic structure is considered an essential determinant factor of material properties, especially chemical and physical properties related to quantum effects (e.g. optics, electricity, quantum phase transition, magneticity, and superconductivity). Many studies in the field of material, physics, and chemistry are closely related to electronic structures. Second, experimental measurements of electronic structure are challenging regardless of whether they are performed in real space or momentum space. Therefore, from the perspective of rational design of functional materials, we are more concerned with the electronic structure characteristics that play a dominant role in specific functions, which are known as functional electronic structures. Studying functional electronic structures is the key to revealing the principle of the structure–property relationships.

When applied, materials will become excited and produce detectable physical properties, achieving certain functions under external fields such as light, electricity, magneticity, temperature, and force. In general, the internal structure (including the crystal and electronic structure) of functional materials, especially functional crystalline materials, will undergo slight changes under an external field, and exhibit corresponding functions. Under an external field, when the material function starts to form, the crystal and electronic structures at the initial state (defined as the state when the external field is not applied or stopped) will undergo subtle changes and form a functional state (defined as the state in which the material exhibits functionality). When the action from the external field stops, the functional state of the material will reversibly return to the initial state. Therefore, the electronic structure of a functional material studied in situ can be compared to that in the initial state to better understand the topological characteristic difference. The region where a large difference is observed can be identified as the functional electronic structure, while molecular fragments or structural units with functional electronic structures can be identified as functional motifs.

An example of second-order nonlinear optical (NLO) crystal is presented here to demonstrate the potential applications of electronic structure crystallography in the research of functional electronic structure and functional motif. NLO crystals, such as BaB_2O_4 (BBO), LiB_3O_5 (LBO), and K_2HPO_4 (KDP), $KTiOPO_4$ (KTP), $AgGaS_2$, $ZnGeP_2$, and $LiInS_2$, can modulate laser frequency as their electrons undergo nonlinear polarization under the field of a strong laser beam because of second harmonic generation effect; thus, they can generate a frequency-doubled laser beam, with this feature earning them extensive applications in high-technology fields. NLO functional motifs refer to material structural units that have large microscopic polarizability and are the main contributor to the macroscopic NLO effects of crystals.

Jiang and Guo proposed a method of experimentally determining NLO functional motifs [20], i.e. by measuring the electronic structures of NLO crystals in situ at dark (initial state) and under laser irradiation (functional state), comparing their topological features, and then identifying the electronic structure regions with large topological feature changes as NLO functional motifs. Furthermore, the researchers used a UV-visible-NIR NLO crystal LiB_3O_5 (LBO) as an example to study its electron density and wavefunction in situ without laser irradiation versus under 1064 nm laser irradiation, which involved the following steps:

(1) High-resolution single-crystal diffraction data of the LBO crystals from the entire Ewald sphere were collected at ~90 K in situ under two conditions sequentially: the absence of laser irradiation and the presence of 1064-nm laser irradiation. The intensity and error of diffraction data satisfied the requirements for general electronic structure refinement. This was indicated by a number of statistical metrics (Figure 8.9), namely consistency factors R_{int} and R_{sigma} around 0.03 in the full resolution range, χ^2 around 1, $I/\text{sig}(I)\sim\log10(I)$ curves in a sinusoidal pattern, and the normalized scale factors within ±0.3 around 1.
(2) The unit cell electron density distribution was obtained sequentially through crystal structure analysis, independent-atom model refinement, kappa refinement, and multipole model refinement (Figure 8.5); Figure 8.10 presents the statistical metrics for the multipole model refinement of the LBO diffraction

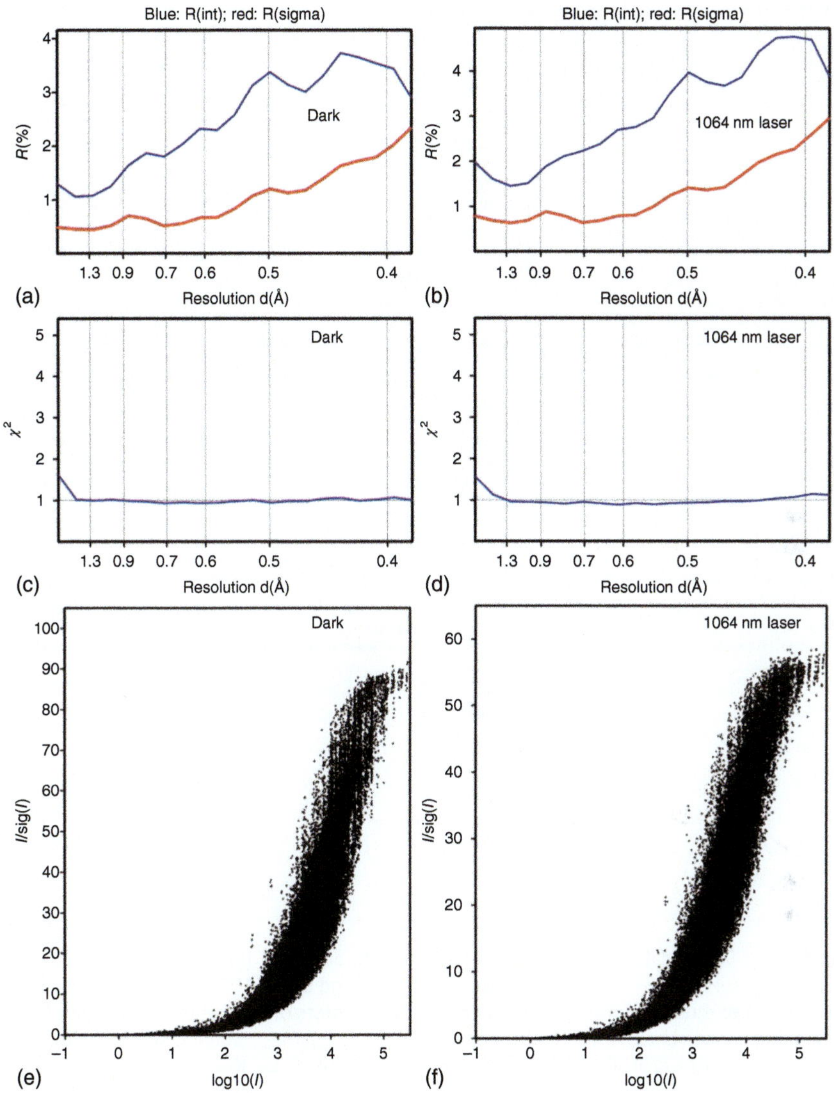

Figure 8.9 Statistical metrics for the diffraction data of LiB_3O_5 (LBO) collected from the whole Ewald sphere at ~90 K under no laser irradiation or under 1064 nm laser irradiation after scale factor correction and absorption correction: (a, b) data consistency factors R_{int} and R_{sigma}; (c, d) χ^2; (e, f) I/sig(I)~$\log_{10}$ (I) curves. Source: Reproduced with permission from Ref. [20]. © 2018 WILEY-VCH Verlag GmbH & Co. KGaA, Weinheim.

datasets. The statistical metrics had the following features: symmetric DRK plots with their central parts passing through the origin, mirror-symmetric plots of fractal dimension of residual electron density, and errors within ±0.05 for the curve of F_{obs}/F_{calc} versus the structure resolution.

(3) The electron wavefunction was obtained through Hirshfeld refinement and wavefunction refinement, and topological analysis was performed on the

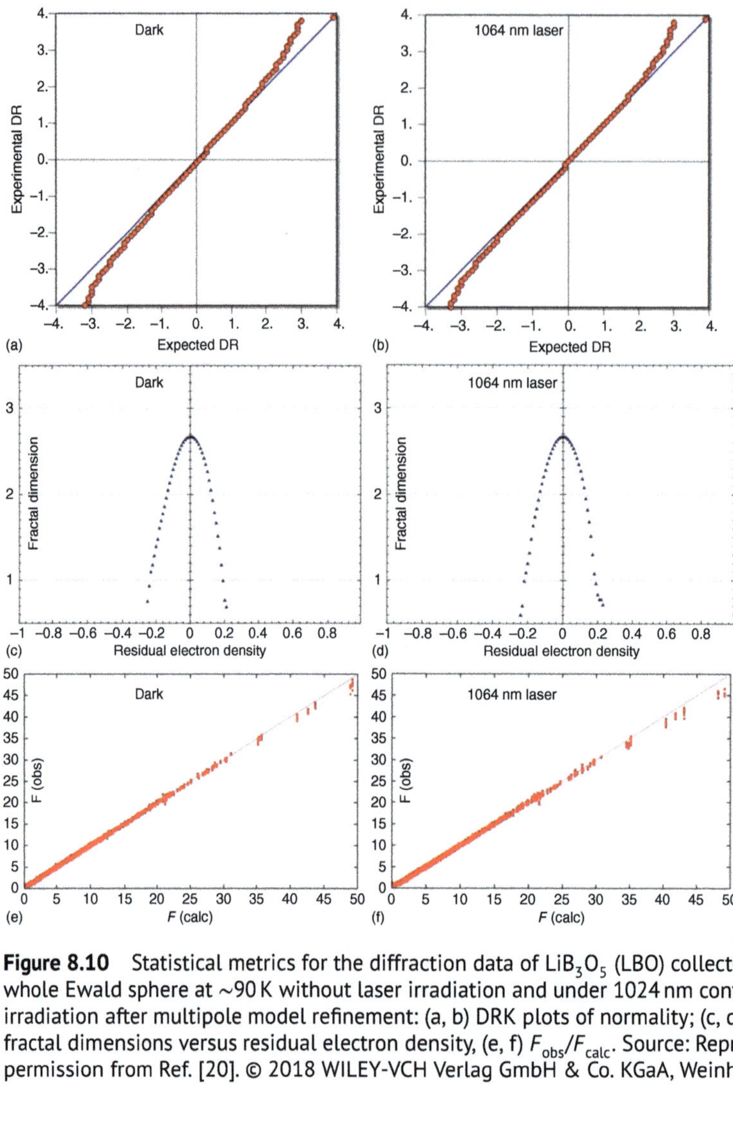

Figure 8.10 Statistical metrics for the diffraction data of LiB_3O_5 (LBO) collected from the whole Ewald sphere at ~90 K without laser irradiation and under 1024 nm continuous laser irradiation after multipole model refinement: (a, b) DRK plots of normality; (c, d) plots of fractal dimensions versus residual electron density, (e, f) F_{obs}/F_{calc}. Source: Reproduced with permission from Ref. [20]. © 2018 WILEY-VCH Verlag GmbH & Co. KGaA, Weinheim.

refined electron density and wavefunction according to the quantum theory of atoms in molecules (QTAIM) to obtain the properties of topological atoms and bonds (Figures 8.11 and 8.12). Compared with the case at dark, the topological atomic charge, atomic volume, and dipole moment of the $[B_3O_5]^-$ group changed significantly under laser irradiation. Specifically, electrons were transferred from the O atoms toward the B atoms, and the electronic structure of the triangular BO_3 unit was more responsive to the external field than the tetrahedral BO_4 unit. By contrast, the changes in the electron cloud around Li were negligible. These observations confirmed that the B–O group $(B_3O_5)^-$ was the NLO functional motif of LBO.

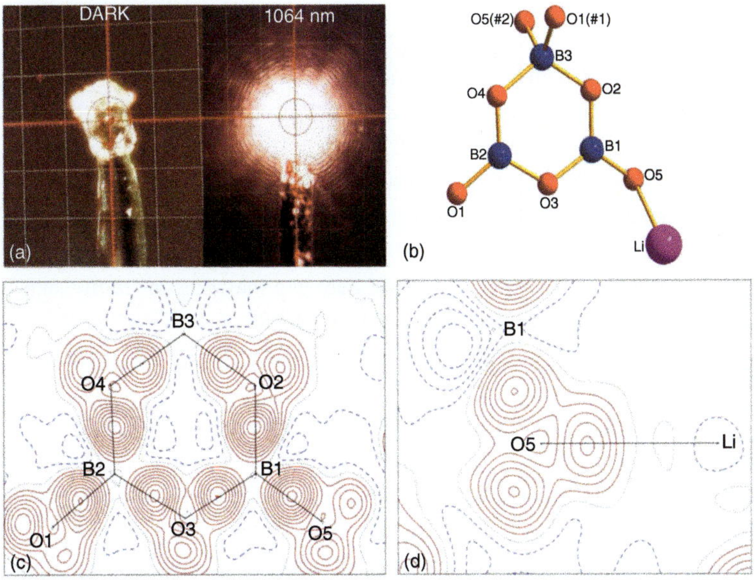

Figure 8.11 (a) LBO crystal in the initial state versus in the functional state under 1064 nm laser irradiation; (b) asymmetric unit of LBO; (c, d) deformation density of LBO. Source: Reproduced with permission from Ref. [20]. © 2018 WILEY-VCH Verlag GmbH & Co. KGaA, Weinheim.

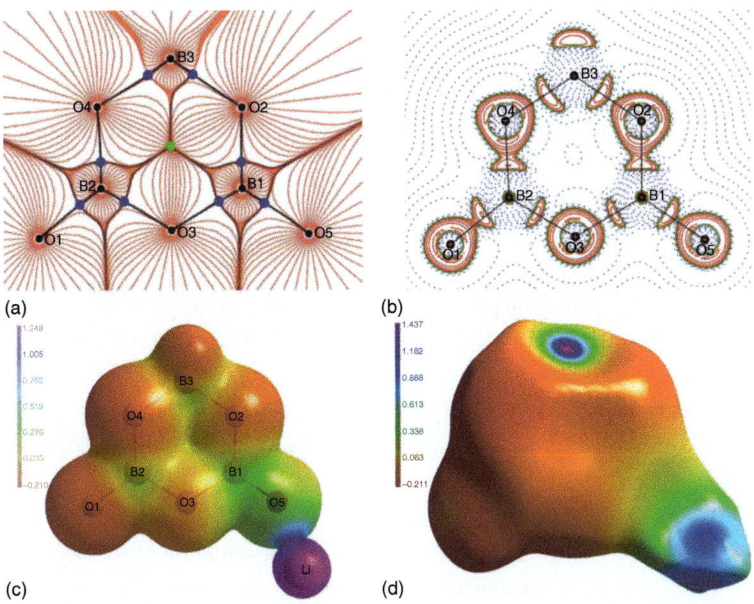

Figure 8.12 (a) electron density gradient of LBO; (b) Laplacian of LBO electron density; (c) electrostatic potential of LBO; (d) Hirshfeld surface of LBO. Source: Reproduced with permission from Ref. [20]. © 2018 WILEY-VCH Verlag GmbH & Co. KGaA, Weinheim.

References

1 Jiang, X.-M., Deng, S., Whangbo, M.-H., and Guo, G.-C. (2022). Material research from the viewpoint of functional motifs. *Natl. Sci. Rev.* 9: nwac017.

2 Cheng, X.Y., Gordon, E.E., Whangbo, M.-H. et al. (2017). Superconductivity induced by oxygen doping in Y_2O_2Bi. *Angew. Chem. Int. Ed.* 56: 10123–10126.

3 Qi, X.-L. and Zhang, S.-C. (2011). Topological insulators and superconductors. *Rev. Mod. Phys.* 83: 1057–1110.

4 Wang, X.-L., Dou, S.X., and Zhang, C. (2010). Zero-gap materials for future spintronics, electronics and optics. *NPG Asia Mater.* 2: 31–38.

5 Rodríguez-Carvajal, J. and Bourée, F. (2012). Symmetry and magnetic structures. *EPJ Web Conf.* 22: 00010.

6 van Smaalen, S. (2012). *Incommensurate Crystallography*. Oxford University Press.

7 Janssen, T., Chapuis, G., and de Boissieu, M. (2007). *Aperiodic Crystals. From Modulated Phases to Quasicrystals*. Oxford University Press.

8 Wiegers, G.A. (1996). Misfit layer compounds: structures and physical properties. *Prog. Solid State Chem.* 24: 1–139.

9 Onoda, M., Saeki, M., Yamamoto, A. et al. (1993). Structure refinement of the incommensurate composite crystal $Sr_{1.145}TiS_3$ through the Rietveld analysis process. *Acta Crystallogr., Sect. B: Struct. Sci.* 49: 929–936.

10 Whangbo, M.-H., Canadell, E., Foury, P. et al. (1991). Hidden Fermi surface nesting and charge density wave instability in low-dimensional metals. *Science* 252: 96–98.

11 Whangbo, M.-H., Deng, S.Q., Köhler, J. et al. (2018). Interband electron pairing for superconductivity from the breakdown of the Born-Oppenheimer approximation. *ChemPhysChem* 19: 3191–3195.

12 Zhou, Z., Xu, X., Fei, R. et al. (2016). Structure modulations in nonlinear optical (NLO) materials $Cs_2TB_4O_9$ (T = Ge, Si). *Acta Cryst.* B72: 194–200.

13 Li, R.-A., Zhou, Z., Lian, Y.-K. et al. (2020). A_2SnS_5: a structural incommensurate modulation exhibiting strong second-harmonic generation and a high laser-induced damage threshold (A=Ba, Sr). *Angew. Chem. Int. Ed.* 59: 11861–11865.

14 Graetsch, H.A., Pandey, C.S., Schreuer, J. et al. (2014). Incommensurate modulations of relaxor ferroelectric $Ca_{0.24}Ba_{0.76}Nb_2O_6$ (CBN24) and $Ca_{0.31}Ba_{0.69}Nb_2O_6$ (CBN31). *Acta Cryst.* B70: 743–749.

15 Sławiński, W., Przeniosło, R., Sosnowska, I. et al. (2012). Helical screw type magnetic structure of the multiferroic $CaMn_7O_{12}$ with low Cu-doping. *Acta Cryst.* B68: 240–249.

16 Penrose, R. (1974). The role of aesthetics in pure and applied mathematical research. *Bull. Inst. Math Appl.* 10: 266–271.

17 Welberry, T.R. (2004). *Diffuse X-ray Scattering and Models of Disorder. Oxford University Press.*

18 Zheng, J., Ye, Y., Liu, T. et al. (2019). Ni/Li Disordering in layered transition metal oxide: electrochemical impact, origin, and control. *Acc. Chem. Res.* 52: 2201–2209.

19 Tan, H.-Z., Chen, Z.-N., Xu, Z.-N. et al. (2019). Synthesis of high-performance and high-stability Pd(II)/nay catalyst for co direct selective conversion to dimethyl carbonate by rational design. *ACS Catal.* 9: 3595–3603.

20 Jiang, X.-M., Lin, S.-J., He, C. et al. (2021). Uncovering functional motif of non-linear optical material by in situ electron density and wavefunction studies under laser irradiation. *Angew. Chem. Int. Ed.* 60: 11799–11803.

Index

Electronic Structure Crystallography and Functional Motifs of Materials, First Edition.
Guo-Cong Guo and Xiao-Ming Jiang.

f

m

n

t

u